EXOTIC
CLUSTERING

Related Titles from AIP Conference Proceedings

To learn more about these titles, or the AIP Conference Proceedings Series, please visit the
webpage **http://proceedings.aip.org/proceedings**

EXOTIC CLUSTERING

4th Catania Relativistic Ion Studies
CRIS 2002

Catania, Italy 10–14 June 2002

EDITORS
Salvatore Costa
Antonio Insolia
Cristina Tuvè
Università di Catania and INFN
Catania, Italy

SPONSORING ORGANIZATIONS
Istituto Nazionale di Fisica Nucleare
Università di Catania

AMERICAN INSTITUTE OF PHYSICS

Melville, New York, 2002
AIP CONFERENCE PROCEEDINGS ■ VOLUME 644

Editors:

Salvatore Costa
Antonio Insolia
Cristina Tuvè

Dipartimento di Fisica e Astronomia
Università di Catania and
Istituto Nazionale di Fisica Nucleare
Via S. Sofia 64
95123 Catania
ITALY

E-mail: Salvatore.Costa@ct.infn.it
           Antonio.Insolia@ct.infn.it
           Cristina.Tuve@ct.infn.it

SEP/AE
PHY S

L.C. Catalog Card No. 2002114107
ISBN 0-7354-0099-7
ISSN 0094-243X
Printed in the United States of America

cm 1/28/03

CONTENTS

PROTON AND ALPHA CLUSTER EMISSION

ALPHA EMISSION AND CLUSTERING FEATURES

CLUSTERING FEATURES IN LIGHT NUCLEI

STRANGENESS AND STRANGELETS

STRANGENESS AND ANTIMATTER

Preface

S. Costa, A.Insolia, C.Tuvé

CRIS 2002 Local Organizers

CRIS 2002 is the fourth Conference in the CRIS series. Every other year, this series of topical Conferences allows scientists to debate some hot subject in Nuclear Physics and related fields.

The first topical CRIS Conference, CRIS '96, was devoted to discussing the still unsettled question of the liquid-gas phase transition in nuclei in connection with multifragmentation.

The second CRIS Conference, CRIS '98, was devoted to discussing intensity interferometry in different fields of physics, with the main emphasys on Nuclear Physics.

The third, CRIS 2000, was again devoted to discussing the liquid-gas phase transition question and the quark-gluon plasma phase transition, expected at higher energies.

CRIS 2002 had an even more ambitious goal: to draw a link from "standard" nuclear structure questions (proton and alpha emission, exotic cluster decay, clusters in light nuclei) and cluster production from the dynamical point of view to more "exotic" quark clusters, like the *uuddss* H0 particle, strangelets and strange neutron stars. It was quite a challenge putting together a program ranging from one field to the other.

Unlike previous meetings, which were held in nice hotels of the Acicastello township, near Catania, CRIS 2002 took place in the Department of Physics and Astronomy of the Catania University, then located in the Catania city center.

We thank all the speakers for their excellent talks. And we thank the about 100 participants, whose interested attendance was high through all sessions.

We are grateful to the members of the International Advisory Committee, whose opinions and suggestions were invaluable to setting up the program.

Coming to the organizational matters, as anyone who has organized a Conference knows very well, this was the result of the cooperative work of many people. We received financial and organizational support from many different Institutions. We will mention foremost the National Institute of Nuclear Physics (INFN) and the University of Catania.

In addition we have had remarkable support from the Sicilian Regional Government, from the Provincial Tourism Agency of Catania and from the Banca di Roma.

The City of Catania and Acicastello organized some of the social activities for the Conference: the farewell party on Friday night and the concert on Tuesday night.

In addition several companies supported our effort under many respects: Banca 121, GEOnext, Alitalia, Fratelli Russo, Caffè Torrisi, Shogun Travel Agency and, finally, the Teatro Massimo Bellini of Catania which offered a nice present to all participants.

Special mention deserve those people who worked hard every day for this Conference: Angela Torrisi, Anna Linda Magrì, Laura Romano and Tony Venasco, members of our local INFN Administrative Staff, who helped the Organizers for several months during the preparation of the Conference and all participants during the Conference week.

It is our pleasure to thank Alessia Tricomi who designed the CRIS logo, the black 'Etna' volcano erupting nuclei and particles from a 'quark-gluon plasma' lava bed, seen from the surfs of the Ionian sea. This logo is the distinctive graphic emblem of the CRIS Conferences.

CRIS 2002

International Advisory Committee
Juha Aysto *(Univ. Jyväskylä, Finland)*
Gordon Baym *(Univ. Illinois, Urbana-Champaign)*
Roberto Bonetti *(Milano Univ.)*
Hank Crawford *(Space Sciences Laboratory, UC Berkeley)*
Lidia Ferreira *(Univ. Tecnica de Lisboa)*
Robert L. Jaffe *(MIT)*
Emilio Migneco *(LNS-INFN, Catania)*
Grazyna Odyniec *(LBNL)*
Hans Georg Ritter *(LBNL)*

Local Organizing Committee

Salvatore Costa
Antonio Insolia
Cristina Tuvè

CRIS 2002

Funding Agencies and Sponsors

Istituto Nazionale di Fisica Nucleare
Universitá di Catania
Governo Regionale della Regione Sicilia
Comune di Catania
Comune di Acicastello
Azienda Provinciale Turismo di Catania
Banca di Roma, Catania
Banca121, Catania
GEOnext, Novara
Alitalia
Shougun Travel Agency, San Gregorio di Catania
Teatro Massimo Bellini, Catania
Distilleria F.lli Russo s.n.c., Santa Venerina (CT)
Torrisi, Compagnia Meridionale Caffè, Catania

Participant's Group Photo

Prof. Antonio Insolia, Conference Chairman, opens the Conference. Behind the desk, Prof. Ferdinando Latteri (left), Rector of the Catania University, and Prof. Renato Pucci (center), Dean of the Science Faculty.

Prof. Giuseppe Faraci, Vice-Director of the
Physics and Astronomy Department

Prof. Renato Pucci, Dean of the
Science Faculty

Prof. Ferdinando Latteri, Rector of the
Catania University

Prof. Francesco Catara, Director of INFN, Catania Section

Prof. Enrico Maglione, Padua University

Prof. Philip Woods, Edinburgh University

Audience

Audience

Audience

Audience

Proton and Alpha Cluster Emission

Proton emission from drip-line nuclei

Enrico Maglione* and Lídia S. Ferreira[†]

* Dip. di Fisica "G. Galilei" and INFN, Via Marzolo 8, I-35131 Padova, Italy,
email: maglione@pd.infn.it
[†] Centro de Física das Interacções Fundamentais, and Departamento de Física,
Instituto Superior Técnico, Av. Rovisco Pais, P-1049-001 Lisboa, Portugal,
email: flidia@ist.utl.pt

Abstract. Proton radioactivity from deformed drip–line nuclei is discussed. It is shown how it is possible to describe all experimental data currently available for decay from ground and isomeric states, and fine structure within a consistent theoretical approach.

PROTON RADIOACTIVITY

Proton emission from exotic nuclei lying beyond the proton drip-line has been observed in recent experiments [1, 2] with radioactive beams. Proton radioactive nuclei are in the mass region of heavy nuclei, where the Coulomb barrier is very high. The emitted proton is not bound to the nucleus, but is trapped by the barrier in a resonance state which will decay emitting a proton whose energy and half–life are measured. The tunnelling through the barrier is quite long, therefore the decay widths of these resonances are quite narrow of the order of 10^{-16}–10^{-22} MeV. The escape energy of the emitted proton is also very small, around 1 MeV, therefore these resonances lie very low in the continuum, and correspond essentially to single particle excitations, in contrast with what happens in stable nuclei.

From these observations, it is now possible to define almost completely the proton drip-line in the region of nuclear charges within $50 < Z < 82$, with the exception of Pm and Pr isotopes which are still missing. Different shapes for proton radioactive nuclei were observed, ranging from spherical [3] emitters, to nuclei with quite large deformations [4]. Most part of these decays occur from the ground state, but decay from isomeric excited states of the parent nucleus have also been observed. For deformed emitters, the daughter nucleus can have a first excited 2^+ rotational state very low in energy, to which decay is possible, and is usually referred as fine structure [5].

Due to the single particle character of this decays, and the large potential energy barrier a simple WKB calculations of the decay widths for spherical nuclei can already give a good estimate of the experimental data. Decay from deformed nuclei can be studied as decay from a Nilsson resonance in a deformed system [6, 7], or within the coupled-channel Green's function model [4], treating the parent nucleus in the strong coupling limit of the particle-rotor model [8]. The effect of Coriolis mixing was considered within the non–adiabatic coupled channel approach [9], and the coupled-channel Green's function [10] model, including the Coriolis interaction in the deformed

CP644, *Exotic Clustering: 4th Catania Relativistic Ion Studies,* edited by S. Costa, A. Insolia, and C. Tuvè
© 2002 American Institute of Physics 0-7354-0099-7/02/$19.00

Hamiltonian. A further development of the theory [11] has shown that the inclusion of Coriolis mixing requires a proper treatment of the pairing residual interaction.

The basic ingredient for all approaches are the energies and single particle wave functions, of the decaying nucleus. These quantities are determined from a phenomelogical single particle potential, and there are few possible interactions that can be used, that fit single particle properties of nuclei. In this work we will briefly review our theory for odd–even and odd–odd deformed proton emitters, based on the exact calculation of Nilsson resonances and discuss the importance of the choice of the single particle potential in the evaluation of decay widths. The importance of the Coriolis coupling and the pairing residual interaction are also addressed. It is shown that all experimental data currently available for decay from ground and isomeric states, and fine structure, can be consistently described within the model.

SINGLE PARTICLE POTENTIALS AND SINGLE PARTICLE LEVELS

Single particle potentials that describe the deformed nuclear mean field have pure phenomenological shapes depending on strength, radius and diffuseness parameters, adjusted to reproduce single particle properties of deformed nuclei. There are different choices for these potentials according to the sets of parameters known in the literature, that equally give reasonable fits of the data as seen in Ref. [12]. Modern fits to ground state spin and parity, on spherical and deformed odd–mass nuclei, and ground state equilibrium deformations were simultaneously taken into account and optimized to reproduce yrast states and proton s. p. gap. They gave origin to the "universal parameters" [13], valid throughout the periodic table including extensions to exotic nuclei. The most recent potential, "Davids" [10] set, is a compromise between geometrical parameters used for scattering states and nuclear structure calculations of high spin deformed states in the neighbourhood of Gd. The Becchetti–Greenlees [14] set comes from a different source of data, since it was fitted to low energy proton and neutron scattering data on medium size nuclei.

The relevance of the different parameterizations of the various s. p. potentials can become more explicit by comparing the s. p. energies evaluated from these interactions for deformed proton emitters. The emitted proton moves in a single particle Nilsson level, which corresponds to a resonance of the unbound core–proton system. The radial Schrödinger equation was then solved [15] for the mean field with deformed Saxon–Woods potential, deformed spin–orbit, and Coulomb interaction, to determine the bound states, and imposing outgoing wave boundary conditions to find the resonance states as a function of deformation.

The Nilsson levels for ^{141}Ho, a well deformed proton emitter, with a predicted quadrupole deformation $\beta_2 = 0.286$ and a small component of hexadecapole deformation $\beta_4 = -0.063$ [16], are shown in Fig. 1. Different interactions seem to produce a different ordering of levels, however, the position of the Fermi level is the most relevant property to describe decay. For other emitters [19], a similar behaviour was found for the Nilsson states according to the potential used.

4

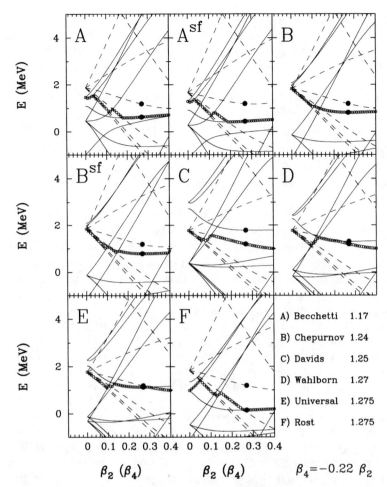

FIGURE 1. Proton Nilsson levels corresponding to ^{141}Ho for the different s. p. potentials listed from A to F. The label "sf" indicates that the spin orbit interaction is taken as spherical as in Ref. [9, 17], the open circles represent the Fermi surface, and the dashed lines the negative parity states coming from the $h11/2$ state. The full circle indicates the decaying state predicted in Ref. [16]. A hexadecapole deformation $\beta_4 = -0.22\beta_2$ was included. For easier comparison the radius of the various potentials are shown.

DECAY WIDTHS FOR DEFORMED PROTON EMITTERS

Odd-Z even-N emitters

Since nuclei on the drip–line have a Fermi level very close or even immersed in the continuum, decay of odd–Z even–N nuclei has been interpreted, as decay from a single particle Nilsson resonance of the unbound core–proton system. The states close to the Fermi surface are the most probable ones for decay to occur. The corresponding half–lives can be obtained from the imaginary part of the energy that solves the Schrödinger equation as discussed above, or simply by noticing that the partial decay width is the overlap between the initial and final states. Considering the wave function of the parent nucleus as the one of a particle plus rotor in the strong coupling limit [8], for decay to the ground state only the component of the s.p. wave function with the same angular momentum as the ground state contributes, and the width becomes,

$$\Gamma_{l_p j_p}(r) = \hbar/T = \frac{\hbar^2 k}{\mu(j_p + 1/2)} \frac{|u_{l_p j_p}(r)|^2}{|G_{l_p}(kr) + iF_{l_p}(kr)|^2} u_{K_i}^2, \tag{1}$$

where F and G are the regular and irregular Coulomb functions, respectively, and $u_{l_p j_p}$ the component of the wave function with momentum j_p, equal to the spin of the decaying nucleus. The quantity $u_{K_i}^2$ is the probability that the single particle level in the daughter nucleus is empty, evaluated in the BCS approach. Due to energy considerations, one expects proton decay to proceed mainly to the ground state of the daughter nucleus. In this way the proton has the largest possible energy and $J_d = 0$, so that j_p is equal to the total and initial momentum of the parent nucleus, $j_p = J_i = K_i$. This component of the wavefunction could be very small. In the case of decay to excited states, few combinations are permitted for $l_p j_p$ according to angular momentum coupling rules, and consequently different components of the parent wave function are then tested. It is possible that decays to different states are allowed, and a total width Γ_T should be defined as the sum of all partial widths, and branching ratios as the ratio between partial and total widths.

The decay width obtained from Eq. 1 depends on deformation, and is very sensitive to the wave function of the decaying state. Therefore, if it is able to reproduce the experimental value, will give clear information on the deformation and properties of the decaying state. Inserting in Eq. 1 the decaying functions corresponding to states lying close to the Fermi level, and obtained for the various parameterizations of the s. p. potentials as discussed before, the decay widths, and corresponding half–lives, can be determined and compared with the experimental data.

Let us consider again the example of [141]Ho. There is common agreement [4, 5, 6, 9, 10] that the [523]7/2⁻ level reproduces ground state decay and [411]1/2⁺ the decay of the isomeric state. As it can be seen from Fig. 1, these states are always close to the Fermi surface at deformations predicted by Möller and Nix [16], for all interactions. The branching ratio for decay from the ground and isomeric states of [141]Ho to the 2⁺ of [140]Dy are shown in Fig. 2. There is no experimental measurement for this fine structure, except an upper limit on the branching ratio of 1% [18]. For the purpose of the calculation, the energy of the 2⁺ was set equal to 160 keV [10] that should be reasonable, but of little influence on the total width due to the small branching ratio. In practice the half–life is the same as the one for decay to the ground state, known experimentally. The half–lives are perfectly described with a deformation close to the predictions of Ref. [16]. A small hexadecapole contribution $\beta_4 = -0.22\beta_2$ was included in the calculation.

TABLE 1. Total angular momentum and deformation that reproduce the experimental half–lives for the measured deformed odd–even proton emitters compared with the predictions of [16]. The theoretical results are from Refs. [6, 21, 22]. The label *m* refers to decays from isomeric states.

	Proton decay		Möller–Nix	
	J	β	J	β
^{109}I	1/2+	0.14	1/2+	0.16
^{113}Cs	3/2+	0.15 ÷ 0.20	3/2+	0.21
^{117}La	3/2+	0.20 ÷ 0.30	3/2+	0.29
^{117m}La	9/2+	0.25 ÷ 0.35		
^{131}Eu	3/2+	0.27 ÷ 0.34	3/2+	0.33
^{141}Ho	7/2−	0.30 ÷ 0.40	7/2−	0.29
^{141m}Ho	1/2+	0.30 ÷ 0.40		
^{151}Lu	5/2−	−0.18 ÷ −0.14	5/2−	−0.16
^{151m}Lu	3/2+	−0.18 ÷ −0.14		

It is reasonable to use any of the different parameterizations of the s. p. potentials to determine the decay rates, since they lead to quite similar results, inside the limits of experimental uncertainties, with the exception of the Becchetti–Greenlees potential. Calculations [19] made for other proton emitters have shown that the latter potential cannot describe the experimental data. It has a quite small radius parameter and interactions with larger radii were usually adopted uniformly for spherical and deformed systems in a consistent determination of the experimental spectroscopic factors [20]. The ordering of levels for this potential is very odd, therefore the Nilsson wave functions and decay widths become very unreasonable. It was fitted to scattering data around 15 and 40 MeV, mainly from medium weight nuclei, and it is difficult to expect that its extrapolation to low energies involved in proton radioactivity, should work well. Calculations with this interaction are then not very reliable.

We have applied [6, 21, 22] our model to all measured deformed proton emitters including isomeric decays. The results of the calculations are shown in Table 1, with the "universal" parameterization of the s. p. potential. The experimental half-lives are perfectly reproduced by a specific state, with defined quantum numbers and deformation, thus leading to unambiguous assignments of the angular momentum of the decaying states [6, 21]. Extra experimental information provided by isomeric decay observed in ^{117}La, ^{141}Ho and ^{151}Lu, and fine structure in ^{131}Eu can also be successfully accounted by the model. The experimental half–lives for decay from the excited states were reproduced in a consistent way with the same deformation that describes ground state emission.

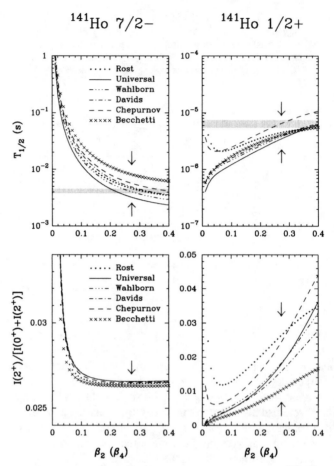

FIGURE 2. Total half–lives for decays from ground and isomeric states of ^{141}Ho, obtained from all single particle potential models discussed in this work (upper part of figure). The arrow indicates the deformation predicted by Ref. [16]. The experimental result [5, 4, 18] is within the shaded area. The branching ratios for the fine structure with the same potentials as in a) as shown in the lower part.

Decay from odd-Z odd-N nuclei

Emission from deformed systems with an odd number of protons and neutrons can be discussed in a similar fashion [7]. The decaying nucleus is described by a wave function of two particles–plus–rotor in the strong coupling limit. represented in terms of the single particle functions of the odd nucleons. However, in contrast with decay to ground state of odd-even nuclei where the proton is forced to escape with a specific

TABLE 2. As in Table 1 for the measured odd–odd proton emitters. Results from the calculation of Ref. [7]. The quantities K_p and K_n are the magnetic quantum numbers of the proton and neutron Nilsson wave functions, J the total angular momentum of the parent nucleus, and β and β_M the deformation coming from the proton decay calculation and the prediction of Ref. [16].

	K_p	K_n	J	β	β_M
^{112}Cs	3/2+	3/2+	0+,3+	0.12 ÷ 0.22	0.21
^{140}Ho	7/2−	9/2−,7/2+	8+,0−	0.26 ÷ 0.34	0.30
^{150}Lu	5/2−	1/2−	2+	−0.15 ÷−0.17	−0.16
^{150m}Lu	3/2+	1/2−	1−,2−	−0.22 ÷ 0.00	

angular momentum, many channels will be open due to the angular momentum coupling of the proton and daughter nucleus, $\vec{J}_d + \vec{j}_p$, giving the total width for decay as a sum of partial widths allowed by parity and momentum conservation,

$$\Gamma^{J_d} = \sum_{j_p=\max(|J_d-K_T|,K_p)}^{J_d+K_T} \Gamma^{J_d}_{l_p j_p} \tag{2}$$

where the width for decay in the channel $l_p j_p$ is given by,

$$\Gamma^{J_d}_{l_p j_p} = \frac{\hbar^2 k}{\mu} \frac{(2J_d+1)}{(2K_T+1)} <J_d,K_n,j_p,K_p|K_T,K_T>^2$$
$$\frac{|u_{l_p j_p}(r)|^2}{|G_{l_p}(kr)+iF_{l_p}(kr)|^2} u^2_{K_p}. \tag{3}$$

The factor $u^2_{K_p}$ is the probability that the proton single particle level is empty in the daughter nucleus, evaluated with the pairing interaction in the BCS approach. The quantity in brackets represents a Clebsch-Gordan resulting from the angular momentum coupling of the odd nucleons, and $K_T = |K_p \pm K_n|$ the spin of the bandhead state of the decaying nucleus. Since the neutron intrinsic state does not change during decay $K_d = K_n$. The total decay width depends on the quantum numbers of the unpaired neutron which cannot be considered only a spectator, but contributes significantly with its angular momentum to the decay.

As in the case of odd-even nuclei, there is a perfect description of experimental decay rates for deformations in agreement with predictions made by other models [16]. The same Nilsson state of the odd proton is used in the calculation of odd–odd and neighbour odd–even nuclei, as can be seen comparing Tables 1 and 2. Also similar deformations were found for the odd-odd and nearby odd-even nuclei. This represents a further consistency check of the model. The largest contribution of the residual interaction between the odd-neutron and odd-proton, i. e., the diagonal part, was taken into account exactly.

TABLE 3. Proton decay widths in units os 10^{-20} MeV of $7/2^-$ ground state at 1.190 MeV of ^{141}Ho, to the ground Γ_0 and $2^+(\Gamma_2)$ of ^{140}Dy, branching ratio and total half lives calculated in the adiabatic, with Coriolis mixing and with pairing residual interaction, compared with the calculations of Ref. [10].

Method	I^π	Γ_0	Γ_2	Γ_2/Γ_T	$t_{1/2}$
Adiab [10]	$7/2^-$	16.5	0.45	0.027	2.7 ms
Corl [10]	$7/2^-$	3.38	0.27	0.074	12.5 ms
Corl + pairing	$7/2^-$	17.7	0.21	0.012	2.6 ms
Exp. [18]	$7/2^-$	10.9 ± 1.0		< 0.01	4.2 ± 0.4 ms

EFFECT OF CORIOLIS MIXING AND PAIRING RESIDUAL INTERACTION

Calculations within the strong coupling limit were able to reproduce the experimental results. According to this model, the rotor had infinite moment of inertia, therefore the excitation spectrum was frozen and the nucleus had a degenerate ground state. No Coriolis mixing was considered. These calculations have been called adiabatic due to this collapse of the rotational spectrum into the ground state. The effect of a finite moment of inertia of the daughter nucleus on proton decay, was studied within the non-adiabatic coupled channel [9], and coupled-channel Green's function [10] methods but the excellent agreement with experiment found in the adiabatic context was lost. The results differ by factors of three or four from the experiment, and even the branching ratio for fine structure decay is not reproduced [9, 10, 23] as it can be seen from Table 3, for the decay of ^{141}Ho. Decay rates in deformed nuclei, are extremely sensitive to small components of the wave function. The Coriolis interaction mixes different Nilsson wave functions, and can be responsible for strong changes in the decay widths. We proved [11] that a correct treatment of the pairing residual interaction in the BCS approach, can modify this mixing of states which should be considered between quasi-particle states instead of particle ones as used in Refs. [9, 10, 23]. Such calculations bring back the perfect agreement with data of the strong coupling limit as seen from Table 3 and the good agreement with experiment, within experimental uncertainties. With the pairing residual interaction in the treatment of the Coriolis mixing, the decay width to the ground state increases by a factor of four, whereas the width for decay to the excited state is practically unchanged. The branching ratio is consequently reduced.

CONCLUSIONS

Proton radioactivity from deformed nuclei is by now well understood. A model was developed to describe decay from odd-even and extended to odd–odd nuclei. The resonance states and their corresponding half–lives for decay are evaluated exactly in a deformed nucleus described by realistic interactions. It was shown how it is possible to go beyond the description of the decaying nucleus as a particle–plus–rotor in the strong coupling limit and take into account the effect of Coriolis mixing. All available experimental data on even–odd and odd–odd deformed proton emitters from the ground and isomeric states, as well as the data on fine structure, were accurately and consistently reproduced, identifying the decay level and deformation of the decaying nucleus, and also supporting previous predictions made by other models on nuclear structure properties of the decaying nucleus.

For decay from odd–odd nuclei, the final decay width depends on the quantum numbers of the unpaired neutron state which cannot be considered only a spectator, but contributes significantly with its angular momentum to the decay.

ACKNOWLEDGMENTS

This work was supported by the Fundação de Ciência e Tecnologia (Portugal), Project: POCTI-36575/99.

REFERENCES

1.   P. J. Woods and C. N. Davids, Annu. Rev. Nucl. Part. Sci. **47** (1997) 541.
2.   A. A. Sonzogni, Nuclear Data Sheets **95** (2002) 1.
3.   S. Åberg, P. B. Semmes and W. Nazarewicz, Phys. Rev. **C56** (1997) 1762.
4.   C. N. Davids, *et al.,* Phys. Rev. Lett. **80** (1998) 1849.
5.   A. A. Sonzogni, *et al.,* Phys. Rev. Lett. **83** (1999) 1116.
6.   E. Maglione, L. S. Ferreira and R. J. Liotta, Phys. Rev. Lett. **81** (1998) 538; Phys. Rev. **C59** (1999) R589.
7.   L. S. Ferreira and E. Maglione, Phys. Rev. Lett. **86** (2001) 1721.
8.   V. P. Bugrov and S. G. Kadmenskii, Sov. J. Nucl. Phys. **49** (1989) 967 ; D. D. Bogdanov, V. P. Bugrov and S. G. Kadmenskii, Sov. J. Nucl. Phys. **52** (1990) 229.
9.   A. T. Kruppa, B. Barmore, W. Nazarewicz and T. Vertse, Phys. Rev. Lett. **84** (2000) 4549; B. Barmore, A. T. Kruppa, W. Nazarewicz and T. Vertse, Phys. Rev. **C62** (2000) 054315.
10.   H. Esbensen and C. N. Davids, Phys. Rev. **C63** 014315 (2001).
11.   G. Fiorin, E. Maglione and L. S. Ferreira, to be published.
12.   S. Cwiok, J. Dudek, W. Nazarewicz, J. Skalski and T. Werner, Comp. Phys. Comm. **46** (1987) 379.
13.   J. Dudek, Z. Szymanski, T. Werner, A. Faessler and C. Lima Phys. Rev. **C26** (1982) 1712.
14.   F. D. Becchetti and G. W. Greenlees, Phys. Rev. **182** (1969) 1190.
15.   L. S. Ferreira, E. Maglione and R. J. Liotta, Phys. Rev. Lett. **78** (1997) 1640.
16.   P. Möller, J. R. Nix, W. D. Myers and W.J. Swiatecki, At. Data Nucl. Data Tables **59** (1995) 185; P. Möller, R. J. Nix and K. L. Kratz, ibd. **66** (1997) 131.
17.   K. Rykaczewski *et al.*, Phys. Rev. **C60** (1999) 011301(R).
18.   D. Seweryniak *et al.*, Phys. Rev. Lett. **86** (2001) 1458.
19.   L. S. Ferreira, E. Maglione and D.E.P. Fernandes, Phys. Rev. **C65** (2002) 024323.
20.   E. Maglione and L. S. Ferreira, Eur. Journal of Phys. (2002) in press.
21.   L. S. Ferreira and E. Maglione, Phys. Rev. **C61** (2000) 021304(R); ibd. **C61** (2000) 47307.
22.   F. Soramel *et al.*, Phys. Rev. **C63** (2001) 031304(R).
23.   W. Królas *et al.*, Phys. Rev. **C65** (2002) 031303(R).

Anisotropic α - Decay in Deformed Nuclei

D. S. Delion*, A. Insolia† and R. J. Liotta **

*Institute of Physics and Nuclear Engineering, Bucharest, Romania
†Dept. of Physics and Astronomy, Univ. of Catania and INFN, Catania, Italy
**KTH - Institute of Physics, Stockholm, Sweden

Abstract. A miscroscopic description of the alpha decay of odd mass nuclei is given for axially deformed nuclei. Realistic mean field + pairing residual interaction in a very large single particle basis is used. Systematics for At and Rn isotopes, as well as for ^{221}Fr, ^{241}Am and Pa - isotopes. Within the model it suggested that a key role for the observed anisotropy is played by the penetration through the deofrmed barrier, as originally suggested by Hill and Wheeler. It is found that the approach gives predictions in good agreement with experimental data for well deformed nuclei. Theoretical predictions on α anisotropy in Am, Es and Fm isotopes are reported.

INTRODUCTION

It has been shown long time ago in nuclear orientation studies that in odd-mass actinides the alpha particles are emitted preferentially with respect to the direction of the total nuclear spin [1-5]. Recently, new experiments [6] have renewed the interest in this problem by reporting anisotropic emission in some At, near spherical isotopes, in connection with several theoretical descriptions of this effect.

Preferential emission of the alpha particles from deformed nuclei was first explained by Hill and Wheeler [7] and then by Bohr, Fröman and Mottelson [8] in terms of the penetration of the alpha cluster through the deformed Coulomb barrier. It was thus found that since for a prolate nucleus the barrier at the poles is thinner than at the equator, the probability to penetrate the barrier is larger along the nuclear symmetry axis. More recently, in order to explain observed anisotropies for almost spherical At isotopes, Berggren [9] proposed an alpha + core model. A quadrupole-quadrupole interaction acting between the already existing structureless alpha cluster and an odd-mass core was diagonalised in a weak couplig scheme. The strength of the interaction was adjusted to obtain the energy of the emitted alpha particle. Using this model several solutions with pronounced anisotropy were obtained [10]. No good comparison with the available data was obtained. Buck et al. [11] describe alpha decay from odd mass nuclei in a similar model, in which the depth of the alpha-core potential (taken as a square well), the alpha formation probability and the number of nodes in the radial wave function are fitted to the experimental data. Rowley et al. [12] followed the same philosophy, diagonalising the quadrupole-quadrupole interaction in an extreme cluster model basis. Stewart et al. [13], using either semiclassical or coupled - channels transmission matrices without any formation mechanism, calculated also anisotropic α emission.

In a series of papers [14,15] we followed the traditional Hill and Wheeler line, but using a realistic deformed mean field with a large configuration space + pairing residual

CP644, *Exotic Clustering: 4th Catania Relativistic Ion Studies,* edited by S. Costa, A. Insolia, and C. Tuvè
© 2002 American Institute of Physics 0-7354-0099-7/02/$19.00

interaction in computing the preformation amplitude of the alpha cluster inside the nucleus. We estimated the penetration through the deformed Coulomb barrier within the framework of the WKB approximation. The anisotropy was explained mainly by the effect of the deformed barrier (see ref. [16] for an overview on the microscopic approach to the alpha decay problem).

The aim of the present talk is to give a short account about our work on the anisotropic alpha particle emission from odd-mass nuclei at low temperature. We will discuss some predictions of ref.s [14,15] as well as some more recent calculations, yet unpublished, in connection with the experimental results obtained by Schuurmans et al. on anisotropy in At, Fr and Pa isotopes [17,18]. We will show that our predictions have been very well confirmed by the recent experimental findings in well deformed nuclei.

MICROSCOPIC DESCRIPTION OF THE ANISOTROPY

The mechanism describing the emission of the alpha particle used in refs. [14,15] consists in the classical two step process [19] : first the four nucleons cluster on a point at the nuclear surface with a given formation amplitude and afterwards this object penetrates the Coubomb barrier.

Let us consider the decay

$$B(I_i, K_i, M_i) \rightarrow A(I_f, K_f, M_f) + \alpha \tag{1}$$

where K_i, K_f and M_i, M_f are the projections of the initial and final total angular momenta in the intrinsic and laboratory frame, respectively [19].

We describe the mother and daughter nuclei within the BCS approximation. That is, the wave function of the nucleus X (A or B) is

$$|\phi^X\rangle = a^\dagger_{k\Omega} |(BCS)\rangle^X_\pi \otimes |(BCS)\rangle^X_\nu \tag{2}$$

where $\pi(\nu)$ label proton (neutron) degrees of freedom. The operator $a^+_{k\Omega}$ is the creation operator of the unpaired nucleon with projection Ω on the intrinsic symmetry axis and k denotes the other quantum numbers. In the case of favoured transitions of odd mass nucleus the quantum numbers $k\Omega$ of the odd nucleon are unchanged during the decay. The set of single particle levels was generated using a realistic axially deformed mean field of Woods-Saxon type [20]. The proton (neutron) pairing strength was used as parameter to reproduce the experimental proton (neutron) pairing gap [21] In this way the neutron-proton interaction, neglected within the model, is taken partially into account in some effective way. Details concerning the single particle potential are to be found in refs. [14,15,20].

The formation amplitude can be written as [14,15]

$$F(R, \vartheta, \varphi) = \sum_L F_L(R, \vartheta, \varphi)$$

$$= \sum_L \int d\xi_\alpha d\xi_A [\phi_\alpha(\xi_\alpha) \phi^A(\xi_A) Y_L(R, \vartheta, \varphi)]^*_{J_B M_B} \phi^B(\xi_B) \tag{3}$$

13

where R is the distance between cluster and daughter nucleus and ξ are the internal coordinates. The intrinsic wave function of the alpha particle has a standard Gaussian form. Additional details on the evaluation of the multidimensional integral in eq. (3) can be found in ref.s [14,15].

Experimentally, nuclei are tipically first produced, then separated, implanted on a foil of ferromagnet material (cooled down to few $10^{-2}K$) and, eventually, oriented by applying a strong magnetic field. Anisotropy is thus measured with respect to the direction of the applied magnetic field [17,18].

If full alignement is not achieved in the orientation process of the implanted isotopes, the conditions are such that one has to average on the initial distribution of the angular momentum projections M_i. The total width is given by

$$\Gamma(\vartheta,\varphi) = \frac{\hbar v}{4\pi}(\frac{R}{G_0(E,R)})^2 \sum_l F_l^2 W(\vartheta) \qquad (4)$$

where F_l is the partial formation amplitude of the emitted alpha particle, i. e.

$$F_l = exp\{-\frac{2l(l+1)}{\chi}\sqrt{(\frac{\chi}{kR}-1)}\} \times \sum_\Omega (-1)^\Omega < I_iK_il - \Omega|I_fK_f > \sum_{l'} K_{ll'}^\Omega a_{l'\Omega}(R) \qquad (5)$$

The matrix element $K_{ll'}$, as well as the quantities χ and $G_0(E,R)$ are defined as in ref. [14] (see also ref. [8] for additional detail). The microscopic formation amplitude enters into the calculation through the amplitude $a_{l'\Omega}(R)$ in eq.(4).

The function W in eq. (4) determines the angular distribution of the emitted particle. After recoupling l and l' to the angular momentum L of the emitted alpha particle and assuming an axially symmetric nucleus, one gets

$$W(\vartheta) = \sum_L A_L P_L(cos\vartheta) \qquad (6)$$

where the amplitudes A_L are given in terms of the F_l amplitudes of eq. (5) [15].

NUMERICAL CALCULATIONS AND DISCUSSION

We will apply the previous formalism to a few selected cases of anisotropic α decay from odd - even nuclei. In particular, we will discuss the ^{241}Am, the At - isotopes, the ^{221}Fr, and the ^{227,229}Pt - isotopes.

Application to the alpha-decay of ^{241}Am

We present in this section an application of the formalism developed above to the case of the favoured transition [15]

$$^{241}Am \rightarrow ^{237}Np + \alpha \qquad (7)$$

TABLE 1. Function $W(\vartheta)$ at $\vartheta = 0$ and $\vartheta = \pi/2$ compared with the measured anisotropy for the favored transition $^{241}Am \rightarrow ^{237}Np + \alpha(\Omega^\pi = 5/2^-)$

ϑ	W_{th}	W_{exp}
0	1.500	1.610
$\pi/2$	0.736	0.714

for which $K_i = K_f = I_i$ and the Nilsson quantum numbers of both the mother and daughter nuclei are $\frac{5}{2}^-$ [523]. The diagonalization of the deformed Woods-Saxon potential is done using 18 major shells [20]. The deformation parameters were chosen as $\beta_2 = 0.22$, $\beta_3 = 0$ and $\beta_4 = 0.08$. The total width was computed according to presented formalism. We obtained $\Gamma_{th} = 2.09 \times 10^{-34}$ MeV which is quite close to the experimental value $\Gamma_{exp} = 3.34 \times 10^{-34}$ MeV. As for the case of even - even nuclei [15,16], the absolute alpha decay widths, for odd nuclei, are given within the right order of magnitude. For instance, in the case of $^{243}Am \rightarrow ^{239}Np + \alpha$, we obtained $\Gamma_{th} = 1.17 \times 10^{-33}$ MeV, to be compared with the experimental value $\Gamma_{exp} = 1.96 \times 10^{-33}$ MeV. However, the main goal of our analysis is to determine the influence of deformation on the angular distributions of the emitted alpha particles.

The model predicts a large enhancement of the anisotropy versus quadrupole deformation. The role of deformations with multipolarities higher than the quadrupole one is much less important [15]. The influence of the intrinsic structure of the mother and daughter wave functions can be estimated by studying the dependence of the angular distribution as a function of the angular momentum transfer L. We found that including only the $L = 0$ component, the total width Γ versus θ shows a variation which is 10% smaller with respect to the case in which all L components are included. This result puts in evidence the important role of the barrier deformation. As a matter of fact, the $L = 0$ part of the formation amplitude is isotropic and in such a case the calculated anisotropy has to be ascribed entirely to the barrier. A similar feature was found in even - even axially deformed nuclei [14]. Although higher L contributions seem to give rise to a small effect, one should not forget that without the inclusion of deformations and without a large basis included in the calculation of the single particle states, one would fail to reproduce the absolute value of the width by many orders of magnitude [16].

The function $W(\vartheta)$ is the relevant quantity regarding the anisotropy in alpha decay processes. Once the microscopic formation amplitude is calculated one can easily expand the function W in terms of even order Legendre polynomials, as shown in eq. (6)

We found that in the case in which all nuclei are assumed to be aligned with the maximum projection of the total angular momentum in the laboratory frame, M_i, the coefficients A_L have all the same phase [15].

Applying the reduction procedure of ref. [3], the calculated W values can be well compared with the experimental data [3] at $1/T = 90.5^0K^{-1}$. The results are presented in Tab. I, where the function $W(\vartheta)$ at $\vartheta = 0$ and $\vartheta = \pi/2$ is compared with the measured anisotropy for the favored transition $^{241}Am \rightarrow ^{237}Np + \alpha(\Omega^\pi = 5/2^-)$.

The agreement can be considered remarkable, specially if one considers that the absolute normalization is given by the formation amplitude entering in the evaluation of the A_L coefficients.

At - isotopes

With the basis and the residual interaction previously discussed we proceed to evaluate the absolute decay widths and the W-coefficients as a function of the deformation parameters for At isotopes. An important motivation for this study is the comparison with recent experimental data taken at ISOLDE (CERN) [17].

For many At - isotopes (starting from odd-proton nucleus ^{207}At with $I_i = I_f = \frac{9}{2}^-$ we have calculated the dependence on the deformation parameter β_2 of the A_l amplitudes as well as of the anisotropy of the decay. We have found within our model [15] that the amplitudes A_l (as well as the $W(\vartheta)$ coefficients) are almost similar for the different isotopes. The only relevant parameter is the nuclear deformation.

We have already reported [15] that both the A_l amplitudes and the anisotropy are strongly dependent on the deformation. The results seem to suggest that the anisotropy increases as the prolate deformation increases.

This is a possible key to read the experimental data [17]. We agree that this interpretation brings into the problem the possibility that the deformation is increasing approaching ^{211}At. Some criticism has been raised against this interpretation [17]. Anyway the data of ref.[17] suggest (in term of the model) a sharp change of nuclear properties (deformation from prolate to oblate case, for instance) to justify the dramatic decrease of the measured anisotropic ratio [17] down to values smaller than 1.

It is worthwhile to mention that the model has no free parameter and that the anisotropic ratio is a strongly dependent on the deformation. See, for instance the results in ref.s [15], for the almost spherical Rn - isotopes, with appreciable differences between cases like ^{207}Rn and the nucleus ^{207}At, in which the mass number is the same and any difference should be attributed only to different properties of the odd neutron and proton orbital entering in the problem. As general comment, we can say that the coefficient A_2 has positive values (in phase with $A_0 = 1$) for the prolate deformations and negative values (opposite phase) for oblate ones. The other coefficients A_L with $L \neq 2$ are virtually negligible. The values of A_4 are one order of magnitude smaller than A_2. In spite of this, it is interesting to note that A_4 is positive and symmetric with respect to the deformation parameter β_2. A similar qualitative and even quantitative behaviour is found for the other At isotopes. Actually even for the odd-neutron case of ^{207}Rn ($I_i = I_f = \frac{5}{2}$) and the other Rn isotopes all the features discussed above are essentially the same.

The case of ^{221}Fr and Pa - isotopes

Finally, it is of great interest to refer the most recent results by Schuurmans et al. [17] for well deformed nuclei. For the ^{221}Fr, a ground state $K = 1/2$ was assumed with a prolate deformation [17], while a previous calculation [14,15] used an oblate ground state with $K = I = 5/2$ (referred in the first line of Tab. II as a theoretical predictions for

TABLE 2. Deformations, experimental and computed coefficients A_2, for the $^{221}_{87}Fr$ and the indicated Pa isotopes (adapted from ref. [18]).

	β_2	β_3	A_2^{th}	A_2^{exp}
$^{221}_{87}Fr$	-0.069	0.0	-0.215	-0.375
	0.120	0.15	-0.373	
$^{227}_{91}Pa$	0.168	0.0	0.649	0.696
	0.168	0.1	0.748	
$^{229}_{91}Pa$	0.185	0.0	0.733	1.13
	0.185	0.08	0.808	

the A_2^{th}).

For the Pa isotopes a prolate ground state was taken for the favoured $5/2^- \rightarrow 5/2^-$ ($^{227}_{91}Pa \rightarrow \alpha + ^{223}_{89}Ac$) and $5/2^+ \rightarrow 5/2^+$ ($^{229}_{91}Pa \rightarrow \alpha + ^{225}_{89}Ac$) transitions.

The results are reported in the Tab. II, where deformations, experimental and computed coefficients A_2, for the $^{221}_{87}Fr$ and the indicated Pa isotopes (adapted from ref. [18]) are reported. The agreement between the microscopic model and the experimental data is excellent. The calculation for the total widths gives $\Gamma = 0.64 \times 10^{-25}$ MeV and $\Gamma = 0.29 \times 10^{-28}$ MeV the $^{227}_{91}Pa$ and $^{229}_{91}Pa$, respectively. We found that the $L = 0$ part in the formation amplitude is the largest contribution. The neglect of the higher multipoles in the formation amplitude produces anisotropy only slightly smaller ($10 - 20\%$ in comparison with the case in which all multipolarities are included. This shows that the main role is played by the penetration through the deformed barrier. The effect is therefore expected to be larger for very well deformed nuclei.

New experimental data on α decay anisotropy should be soon available in the framework of a recently planned systematic investigation of the decay and fission in deformed nuclei [22]. In particular, new measurements have been planned for $^{241-243}$Am, $^{253-255}$Es and $^{255-257}$Fm isotopes [22].

We have just completed some preliminary calculations for the α anisotropy for those mentioned cases [23]. The results are reported in the next Table 3. Anyway, one should keep in mind that, to compare with the data available in the next future, one has to make proper corrections on the theoretical predictions. Without the mentioned filtering the comparison with experimental data would be meaningless. Thus, for instance, the reported anisotropy for ^{241}Am in Tab. 3 looks like much larger than that one, properly filtered for the comparison with the experimental data, reported in Tab. 1 for the same isotope.

In the calculations of Tab. 3, the new single particle basis of ref. [16] was used. The deformation parameters and K^π values were taken from ref. [22]. For all cases we deal with very well deformed nuclei. One should expect that the present model works well.

17

TABLE 3. Theoretical predictions for the α anisotropy for the indicated isotopes.

^AX	$W(0^o)$	$W(90^o)$
^{241}Am	2.183	0.464
^{243}Am	2.194	0.460
^{253}Es	2.700	0.318
^{255}Es	2.664	0.329
^{255}Fm	2.646	0.335
^{257}Fm	2.985	0.270

CONCLUDING REMARKS

In conclusion we have presented a realistic microscopic approach for the calculation of the formation amplitude for the alpha decay problem in axially symmetric deformed nuclei, within the well known approach by Mang and Rasmussen[17]. The main ingredients the use of a realistic deformed mean field, the large shell model space and the exact diagonalization of the deformed mean field. The predicted absolute values of the total alpha widths are reproduced within $10 - 30\%$. [16].

In the model we performed a systematic microscopic calculation of quantities related to alpha particle emission from oriented odd-mass nuclei and, in particular, for $_{85}At$, $_{86}Rn$, $_{87}Fr$ and $_{91}Pa$ isotopes. The probability of emitting an alpha particle in the polar direction (or in the equatorial direction) is strongly dependent on the emission angle. For prolate (oblate) deformations polar (equatorial) emission is preferred. We also found that deformations higher than quadrupole can play an important role in some cases. Even in the region of near spherical nuclei the anisotropy was found to be measurable. In addition, we have found that the main role in the observed anisotropy is due to the deformed barrier penetration. This has been recently confirmed experimentally in the case of well deformed nuclei [18].

We emphasised in this study the importance of anisotropies in alpha decay processes as a tool to extract intrinsic deformation parameters in nuclei.

New experimental data should be soon available on the anisotropic decay in the framework of a recently planned sistematic investigation of the decay and fission in deformed nuclei [22]. In particular, new measurements have been planned for $^{241-243}$Am, $^{253-255}$Es and $^{255-257}$Fm isotopes [22]. Theoretical unfiltered predictions have been reported for those cases. [23].

REFERENCES

1.   S.H.Hanauer, J.W.T.Dabbs, L.D.Roberts and G. W. Parker, Phys. Rev. **124** (1961)1512
     Q.O.Navarro, J. O Rasmussen ans D.A.Shirley, Phys. Lett. **2** (1962)353
2.   A.J.Soinski, R.B.Frankel, Q.O.Navarro and D.D.Shirley, Phys. Rev. **C2** (1970) 2379
3.   A.J.Soinski and D.D.Shirley, Phys. Rev. **C10** (1974) 1488
4.   D. Vandeplassche, E. van Walle, C. Nuytten and L.Vanneste, Phys. Rev. Lett. **49** (1982) 1390
5.   F. A. Dilmanian et al., Phys. Rev. Lett. **49** (1982) 1909

6. J. Wouters et al., Phys. Rev. Lett. **56** (1986) 1901
    J. Wouters et al., Nucl. Instr. and Meth. **B26** (1987) 463;
    N. G. Nicolis et al., Phys. Rev. **C41** (1990) 2118
7. D.L. Hill and J. D. Wheeler, Phys. Rev. **89** (1953) 1102
8. P. O. Fröman, Mat. Fys. Skr. Dan. Vid. Selsk. **1** (1957) no. 3;
    A. Bohr, P. O. Fröman abd B. Mottelson, Dan. Mat. Fys. Medd. **29** (1955) no. 10
9. T. Berggren, Phys. Lett. **197B** (1987) 1;
    T. Berggren, Hyperfine Interactions **43** (1988) 407
10. T. Berggren, Phys. Rev. **C50** (1994) 2494
11. B. Buck, A.C. Merchant and S.M. Perez, J. Phys. **G18** (1992) 143
12. N. Rowley, G.D. Jones and M.W. Kermode, J. Phys. **G18** (1992) 165
13. T.L. Stewart et al., Phys. Rev. Lett. **77** (1996) 36; T.L. Stewart et al., Nucl. Phys. **A611** (1996) 332
14. A. Insolia, P. Curutchet, R. J. Liotta and D. S. Delion, Phys. Rev. **C44** (1991) 545
    D.S.Delion, A.Insolia and R.J.Liotta, Phys. Rev **C46** (1992) 1346
15. D.S.Delion, A.Insolia and R.J.Liotta, Phys. Rev **C46** (1992) 884;
    D.S.Delion, A.Insolia and R.J.Liotta, Phys. Rev **C49** (1994) 3024
16. R. G. Lovas, R. J. Liotta, A. Insolia, K. Varga and D. S. Delion, Phys. Rep. **294** (1998) 265-362;
    D.S.Delion, A.Insolia and R.J.Liotta, Phys. Rev. **C 54** (1996) 292
17. P. Schuurmans et. al., (Nicole and Isolde Collaboration), Phys. Rev. Lett. **77** (1996) 4720
18. P. Schuurmans et. al., (Nicole and Isolde Collaboration), Phys. Rev. Lett. **82** (1999) 4787
19. R.G. Thomas, Progr.Theor.Phys. **12** (1954) 253
    H. J. Mang, Ann. Rev. Nucl. Sci. **14** (1964) 1
20. R. Bengtsson et al., Phys. Scrip. **39** (1989) 196;
    S. Cwiok et al., Comp. Phys. Comm. **46** (1987) 379.
21. A. Bohr and B. Mottelson, Nuclear Structure, (Benjamin, New York, 1975), vol. 2
22. *A study of spin depending phenomena in alpha decay and spontaneous fission of heavy actinide isotopes*; INTAS project "OPEN CALL 2000-0195"; and N. Severijns, private communication.
23. D.S.Delion, A.Insolia and R.J.Liotta, work in progress

Systematics of the Widths of
Alpha Decaying States of ^{12}C

D.V.Fedorov [1], A.S.Jensen [1], and H.O.U.Fynbo [2]

[1] *Department of Physics and Astronomy, University of Aarhus, DK-8000 Aarhus C, Denmark*
[2] *ISOLDE, EP division , CERN , CH-1213 Geneva , Switzerland*

Abstract. We attempt to describe the widths of the alpha decaying states of ^{12}C (0^+, 1^+, 1^-, 2^+, 2^-, 3^-, and 4^+) consistently within the three-alpha cluster model. We solve the Faddeev equations where we include short-range and Coulomb potentials between the alpha-particles and also a parametrized three-body potential. We estimate the widths of these states and compare with experimental data.

INTRODUCTION

Several low lying states of ^{12}C nucleus have only one non-electro-magnetic decay channel -- into three alpha particles. Experimentally the widths of these states differ irregularly by five orders of magnitude [1] and the energy distributions of the decay products exhibit very peculiar behavior.

The widths of these decaying states are largely determined by the properties of the wave-function at relatively large distances where the three alpha-clusters already are well formed. The three-cluster model must therefore be appropriate for a description of the large distance properties of these states, notably the widths [2,3].

In this contribution we attempt to consistently describe the alpha decaying states of ^{12}C (0^+, 1^+, 1^-, 2^+, 2^-, 3^-, and 4^+) within the three-alpha cluster model. We solve the Faddeev equations where we include short-range and Coulomb potentials between the alpha-particles and also a parametrized three-body potential. We use the hyper-spheric approach where the three-alpha decay is depicted as an effective motion of the three alpha-particles along the hyper-spheric coordinate through the effective hyper-spheric barrier. The transmission probability through this barrier then largely determines the width of the state. We calculate the widths of these states using the WKB approximation and compare with experimental data.

HYPER-SPHERIC APPROACH

We use the hyper-spheric coordinates where the hyper-radius ρ is the only coordinate with the dimension of length. It is a mass weighted average of the squares or the particle coordinates

CP644, *Exotic Clustering: 4th Catania Relativistic Ion Studies,* edited by S. Costa, A. Insolia, and C. Tuvè
© 2002 American Institute of Physics 0-7354-0099-7/02/$19.00

$$\rho^2 = \sum_{i=1}^{3} A_i r_i^2 \ , \tag{1}$$

where r_i and A_i are the coordinate of the i-th alpha particle and its mass in units of the nucleon mass m. The other five coordinates, denoted Ω, are dimensionless angles.

The Hamiltonian of the three-alpha system is written in these coordinates as

$$H = \frac{\hbar^2}{2m} \left(-\rho^{-5/2} \frac{d^2}{d\rho^2} \rho^{5/2} + \frac{15/4}{\rho^2} + \frac{1}{\rho^2} \Lambda^2 \right) + V(\rho, \Omega) \ , \tag{2}$$

where $V(\rho, \Omega)$ is the sum of the potential energy operators for the alpha-particles and Λ^2 is a differential operator which acts only upon the five angles Ω [2].

In the hyper-spheric adiabatic approach [2] the variables are divided into a slow hyper-radius ρ, and fast hyper-angles Ω. The eigenvalue problem for the fast variables,

$$\left[\frac{\hbar^2}{2m} \frac{1}{\rho^2} \Lambda^2 + V(\rho, \Omega) \right] \Phi_n(\rho, \Omega) = \epsilon_n(\rho) \Phi_n(\rho, \Omega) \ , \tag{3}$$

provides the eigenvalues, ϵ_n, and also the eigenfunctions, Φ_n, which are used as a basis for the expansion of the total wave-function Ψ

$$\Psi(\rho, \Omega) = \rho^{-5/2} \sum_n f_n(\rho) \Phi_n(\rho, \Omega) \ . \tag{4}$$

The eigenvalues $\epsilon_n(\rho)$ of the fast subsystem are the effective potentials for the slow subsystem in the hyper-radial equations

$$\left[\frac{\hbar^2}{2m} \left(-\frac{d^2}{d\rho^2} + \frac{15/4}{\rho^2} \right) + \epsilon_n(\rho) - E \right] f_n(\rho) = \sum_{n'} Q_{nn'} f_{n'}(\rho) \ , \tag{5}$$

where E is the total energy and Q is a non-adiabatic operator [2].

Already the decoupled lowest hyper-radial equation, often referred to as the hyper-spheric adiabatic approximation, provides an accurate quantitative description of the system. This equation reads

$$\left[-\frac{\hbar^2}{2m} \frac{d^2}{d\rho^2} + U_{eff}(\rho) - E \right] f(\rho) = 0 \ , \tag{6}$$

where the effective hyper-radial potential U_{eff} is defined as

$$U_{eff} = \frac{15/4}{\rho^2} + \epsilon(\rho) + W(\rho) \ , \tag{7}$$

where $\epsilon(\rho)$ is the lowest eigenvalue and $W(\rho)$ is a phenomenological three-body potential that is added to the hyper-radial equation in order to account for the compound state configurations which are beyond the scope of the three-alpha model. Here we for simplicity use a square-well potential with the range $\rho_0 = 4$ fm. The depth of the potential is then adjusted to reproduce the experimental energy of a given state.

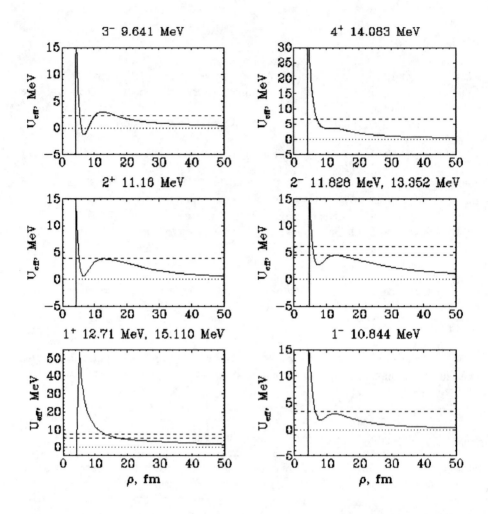

FIGURE 1. The effective hyper-radial potentials U_{eff} as functions of the hyper-radius ρ for the alpha-decaying states in ^{12}C with different spin-parities. The horizontal dashed lines indicate the energies of the states.

EFFECTIVE POTENTIALS

On figure 1 we show the effective potentials U_{eff} for the different alpha-decaying states of ^{12}C. The 0^+ case is not shown as it is has already been considered in [2] where also the width of this state was calculated.

The effective potentials exhibit attractive pockets at intermediate distances for all channels except for 1^+. The barrier for the 1^+ state is therefore very large and the widths of the 1^+ states should be correspondingly very small. The widths for the other states should be similar except for an anomaly for the 3^- channel where the energy of the state is lower than the second barrier (similar to the 0^+ case [2]). This second barrier decreases the width of the 3^- state compared to the 1^-, 2^+, 2^-, and 4^+ states.

These qualitative considerations correctly account for the gross features of the experimental data (see Table 1) where the widths of the 1^+ states are indeed extremely small, of the order of a dozen eV, the width of the 1^-, 2^+, 2^-, and 4^+ states are of the order of a hundred keV, while the 3^- state is somewhat narrower, of the order of a dozen keV.

WIDTHS IN THE WKB APPROXIMATION

We use the WKB approximation [4] to estimate the widths of the states. In this approximation the width Γ is given by the product of the transmission coefficient T and the "knocking rate" R,

$$\Gamma = T\,R, \quad R = \sqrt{\frac{2E}{m}}\,\frac{1}{2\hbar\rho_0}, \quad T = \frac{1}{1+\exp\left(2\int d\rho\sqrt{\frac{2m}{\hbar^2}\left(U_{eff}(\rho)-E\right)}\right)}, \quad (8)$$

where the integration in the transmission coefficient goes over the region where the system is under the barrier, $E < U_{eff}(\rho)$. The calculated widths are given in Table 1.

TABLE 1. Widths of the α-decaying states in ^{12}C

Excitation energy E*, MeV	J^π	Γ, keV	
		Experiment	Theory ($\rho_0 \approx 4.0$ fm)
7.6542	0^+	$8.5\pm1 \times 10^{-3}$	20×10^{-3} [2]
9.641	3^-	34 ± 5	50
10.844	1^-	315 ± 25	190
11.16	2^+	430 ± 80	390
11.828	2^-	260 ± 25	200
12.710	1^+	$18.1\pm2.8 \times 10^{-3}$	0.1×10^{-3}
13.352	2^-	375 ± 40	330
14.083	4^+	258 ± 15	50
15.110	1^+	$43.6\pm1.3 \times 10^{-3}$	3.3×10^{-3}

In conclusion the three-alpha model provides a good qualitative and in most cases also quantitative description of the width of the alpha-decaying states in ^{12}C. The only exception is the narrow 1^+ states where the three-alpha model gives too narrow widths.

REFERENCES

1. Aizenberg-Selov, Nucl.Phys. A506 (1990) 1
2. D.V.Fedorov and A.S.Jensen, Phys.Lett. B389 (1996) 631
3. R. Pichler, H. Oberhummer , Attila Csoto , S.A. Moszkowski, Nucl.Phys. A618 (1997) 55
4. P.J.Siemens and A.S.Jensen, *Elements of Nuclei*, Redwood City: Addison-Wesley, 1987, p. 259.

Alpha Emission and Clustering Features

Vortex Excitations in Light α Cluster Condensed Nuclei

Aiichi Iwazaki

Department of Physics, Nishogakusha University, Shonan Ohi Chiba 277-8585, Japan

Abstract. Vortex excitations arise in systems with infinite number of bosons condensed. Even in systems with finite number ($\sim 10^5$) of atomic bosons condensed the excitations have been observed. We point out that in system with smaller dynamical degrees of freedoms like finite nuclei there exist such vortex-like excitations. The nuclei are α cluster condensed nuclei. We show that the vortex-like excitations possess quantum numbers $J^P = n^{(-1)^n}$ and energies $\sim$ several MeV, measured from the energies of the $n\alpha$ condensed states.

Vortex excitations [1, 2] appear in many branches of physics.. Strictly speaking, they can arise in systems with infinite number of dynamical degrees of freedoms, namely, systems with infinite number of bosons condensed. Actually, the excitations are observed in superfluids such as superconductor. It involves almost infinite number of condensed Cooper pairs. Recent experimental techniques have made finite number ($\sim 10^5$) of bosonic atoms to be condensed[3] and have realized vortex excitations in the bose gas. Now , a natural question arises; do the topological excitations exist in a system with smaller number of dynamical degrees of freedom such as $10^3, 10^2$ or 10^1 ?. We expect that the vortex excitations or vortex-like excitations are present even in such a small system although the topological properties (topological stability, vorticity, etc.) of the excitations become obscure gradually as the number of degrees of freedom in the system decreases. In this report we show that vortex-like excitations may arise in nuclei with α cluster condensation.

Many nuclear states involve cluster structures[4] such as α, C, O etc. Among them, the states with the α cluster condensation have recently been argued[5] to exist around threshold energies at which the nuclei are dissociated into each cluster. For example, some excited states of ^{12}C or ^{16}O have been shown[6] to involve possibly α clusters interacting weakly with each other. These states have been shown to have small nuclear densities so that the clusters are almost free and condensed in an identical state. Furthermore, it has been shown theoretically that $n\alpha$ clusters condensed nuclear states with larger mass numbers $4n$ ($n \geq 3$) are possibly present.

Based on the possible existence of the α cluster condensed states in nuclei, we wish to find the vortex-like excitations in the nuclei. Our concern is whether or not their energies are accessible for observation in such small system. We also wish to clarify what quantum state does each cluster occupy in such a vortex-like nuclear state. Although the excitations lose their topological nature in such small systems, we expect that some nuclear states corresponding to vortices in infinite systems are still present.

First, we explain vortex excitations in a system with infinite number of dynamical

CP644, *Exotic Clustering: 4th Catania Relativistic Ion Studies*, edited by S. Costa, A. Insolia, and C. Tuvè

bosons. We suppose that the system is a nuclear matter composed of α clusters and that these alpha particles condense. We use a model of a complex boson field ϕ in order to describe the condensed states. The boson field represents quanta of α clusters which interacts with each other through a contact interaction in the model. The Hamiltonian is supposed to be given by[7]

$$H = \int d^3x \left(\frac{|\vec{\partial}\phi|^2}{2M} + \frac{\lambda(|\phi|^2 - \rho)^2}{2} \right) + V_c \tag{1}$$

where M (ρ) denotes mass of α particle (average number density of α clusters in the nuclear matter). The term $|\phi|^4$ represents the contact interaction between α clusters, i.e. $V(x-y) = \lambda \delta^3(x-y)$. The strength λ can be extracted from Gaussian potentials[8] and is given by $1.7 \times 10^{-4}/(\text{MeV})^2$. The term V_c represents Coulomb interaction: $V_c = e^2 \int (2|\phi(x)|^2 - 2\rho) \frac{1}{2|x-y|} (2|\phi(y)|^2 - 2\rho) d^3x d^3y$ where we have assumed neutrality of the matter.

Obviously, the ground state is given by $<\phi> \simeq \phi_0 = e^{i\theta_0}\sqrt{\rho}$ with a constant θ_0. Since the vortex is axial symmetric, the vortex solution does not depend on z axis. In order to find such vortex excitations, we take $\phi = \sqrt{\rho}f(r)e^{i\theta}$ and solve an equation derived from the Hamiltonian,

$$-\frac{1}{2M}\vec{\partial}^2\phi + \lambda(|\phi|^2 - \rho)\phi = \rho e^{i\theta} \left\{ -\frac{1}{2M}\left(f(r)'' + \frac{f(r)'}{r} - \frac{1}{r^2} \right) + \lambda\rho(f(r)^2 - 1)f(r) \right\} = 0, \tag{2}$$

where r (θ) denotes radial coordinate (azimuthal angle) in x-y plane and the prime denotes a derivative in r. We have assumed axial symmetric solutions which has no dependence on the coordinate z. The Coulomb interaction is taken into account perturbatively; after finding the solutions we calculate their Coulomb energies.

In order to avoid a singularity at $r = 0$, $f(r)$ has to satisfy the boundary condition, $f(r = 0) = 0$. $f(r)$ should also satisfy the boundary condition, $f(r = \infty) = 1$. The equation can be solved numerically to find the vortex soliton.

Here we should comment that in order to find the vortex solitons with finite energies, we need to include a gauge field (photon) coupled with the charge of the alpha particles. But, the effect is only important at the distance ($\sim 10^2$ fm) much larger than coherent length (~ 1 fm) of the field ϕ. Since our concern is only small nuclear matter with its radius much less than this distance, the gauge field can be neglected.

Obviously, the vortex represents an infinitely long object in the nuclear matter. The solution represents the fact that the α clusters rotate around the vortex center at $r = 0$. This is a vortex soliton in the infinite nuclear matter.

In order to find the bose condensed states in finite nuclei, we simply cut the finite matter out of this infinite matter. Thus, the mass number A of the nuclei is given by $A = 4n = \int d^3x|\phi_0|^2 = 4\pi R_0^3\rho/3$ where R_0 is the radius of the $n\alpha$ condensed nucleus with no vortex excitations; the radius R_0 is determined as a function as ρ and A. Here, the form of the nucleus is assumed to be spherical. Similarly, we cut the finite matter involving the vortex soliton from the infinite matter with the soliton. Namely, we assume that vortex state of nuclei is a spherical finite nuclear matter with axial symmetry, which is extracted from the infinite matter. Since the vortex excitation is an excited state of the nuclei with α condensation, the mass number of the state is the same as that of the alpha condensed state without vortex. Thus, its radius R_v is determined such that the mass number $\int_{\sqrt{r^2+z^2}<R_v} d^3x |\phi_v|^2$ is equal to $4n$.

This state of the finite nuclear matter is not stable vortex soliton, since the vorticity around the center ($r = 0$) of the matter is not conserved. The singular line at $r = 0$ in the finite matter can move and disappear, getting out of the surface of the matter. Namely, the factor $e^{i\theta}$ in the solution ϕ_v can be made to vanish without passing infinite potential barrier in its continuous deformation of the configuration $\phi_v(x)$. Although out treatment of the vortex-like excitations is very crude, this simple treatment can reveals significant properties of the states with vortices, as we will show below.

One of the properties in our concern is the energy of the vortex-like nuclear state. Although precise values of the energies can not be obtained in the crude approximation, we can find whether or not the energies are experimentally accessible or not. The energies of the vortex solutions in eq(2) are obtained by inserting the numerical solutions into the Hamiltonian,

$$ E_v = \frac{2\pi\rho}{M\sqrt{M\lambda\rho}}K, \quad K = \int_0^{S_v} dss\sqrt{S_v^2 - s^2}\left((f')^2 + f^2/s^2 + (f^2 - 1)^2\right) \quad (3) $$

$$ \simeq 6.2\,\text{MeV}\sqrt{\frac{n}{3}}\left(\frac{4\,\text{fm}}{R_0}\right)^{3/2}(S_v/7.1)(\log(S_v/7.1) + 1) \quad (4) $$

with $S_v = \sqrt{\lambda M\rho}\,R_v = 3.8\sqrt{n}(R_v/R_0)\sqrt{4\,\text{fm}/R_0}$, where we have not included their Coulomb energies since it is negligibly small. The radius R_v is given such that $n = \int |\phi|^2 d^3x \simeq \frac{0.05}{\sqrt{n}}(R_0/4\,\text{fm})^{3/2}\int_0^{S_v} sds\sqrt{S_v^2 - s^2}f[s]^2$. E_v represents the energy of the state with the vortex-like excitation measured from the energy of the α condensed state $<\phi> = \phi_0$ without the vortex. It is important to note that the energy is small enough to be experimentally accessible. This small energy results from the use of the small nuclear density, which has been discussed to be realized in the alpha condensed nuclei[6]. Namely, the alpha cluster condensed nuclei have small nuclear density (about a fifth of standard nuclear density) so that the vortex states have accessibly small energies.

The other properties are the angular momenta of the vortex-like state. The values of these quantities can be obtained without the use of the detail form of the solution. Thus, the present crude approximation is enough to find the angular momenta. Among others, the z component of the angular momentum can be obtained with only use of the fact, $\int d^3x |\phi_v|^2 = n$,

$$ J_z = \int d^3x(xP_y - yP_x) = \int d^3x |\phi_v|^2 = n \quad (5) $$

where the momentum P_i of the field is given by $P_i = \frac{1}{2i}(\phi_v^\dagger \partial_i \phi_v - \partial_i \phi_v^\dagger \phi_v) = |\phi_v|^2 \partial_i \theta$. We also find $J_x = J_y = 0$, using the property of the axial symmetry. The result indicates that all clusters in the finite nucleus take an identical state with angular momentum $l = 1$ and rotate around z axis, i.e. $l = l_z = +1$. Therefore, we expect that the states with the vortex-like excitations are such states with $J^P = 3^-$ for ^{12}C, $J^P = 4^+$ for ^{16}O, etc. Generally,

$J^P = n^{(-1)^n}$ for the nucleus $A = 4n$. All of the clusters occupy a p state with $l_z = 1$ so that their wave functions vanish at the center axis ($r = 0$) of the nucleus. This lack of clusters at the center axis of the nucleus is consistent with the form of the vortex solution, i.e. $f(r = 0) = 0$.

We should mention that although the vortex-like nuclei whose numbers of dynamical degrees of freedom are a few or around 10, are never coherent vortex solitons, they still keep some properties of the vortex solitons; wave functions vanish at $r = 0$ and z component of angular momenta are equal to the number of the alpha clusters involved in the states. Therefore, we may expect that these states are states in a limit of $n \to 2$ or 3 of the vortices, which have been observed in the bosonic atoms condensed states with $n \sim 10^5$.

Finally, we point out some candidates for the vortex-like nuclear states. First, we take ^{8}Be for which 2 alpha clusters play. The state $2^+(3\,\mathrm{MeV})$ is a candidate for the nuclear vortex-like state. The state is well known as a molecular state with large nuclear radius and decays into two alphas. This state would be the lightest vortex-like state, although its character as a vortex is rather obscure. Second, the state $3^-(9.6\,\mathrm{MeV})$ is a candiadate in ^{12}C. The state is also well known theoretically as a molecular state of equilateral triangle. Each cluster occupies a p state with $l_z = 1$. This molecular state is not so rigid that each cluster can move easily from its original position at the corner of the equilateral triangle. The molecular state is soft. This fact indicates the state being a vortex-like state rather than the molecular state. Finally, a candidate in ^{16}O is the state $4^+(16\,\mathrm{MeV})$, which is not yet fully understood. The state will be understood as a soft molecular state of equilateral square in the analysis by using standard cluster models. But, it is better to regard the state as a vortex-like state than the molecular state. Although topological properties in these small nuclei are obscure, the properties are made clearer as nuclei with bose condensation are heavier.

ACKNOWLEDGMENTS

We thank Y. Akaishi, O. Morimatsu and Y. Kanada-En'yo in KEK for fruitful discussions and H. Nemura for useful information of α particles. We also thank N. Itagaki for useful imformations of nuclear molecular states.

REFERENCES

1. R. Rajaraman. Soliton and Instantons (North Holland. Amsterdam 1982).
2. S.M. Girvin. The Quantum Hall Effect: Novel Excitations and Broken Symmeties in Les Houches Summer School 1998 (Springer Verleg 1999).
    F.Z. Ezawa M. Hotta and A. Iwazaki, Phys. Rev. B46, 7765.
3. M.R. Matthews, etal, Phys. Rev. Lett. 83, 2498 (1999).
4. Y. Fujiwara, H. Horiuchi, K. Ikeda, M. Kamimura, K. Kato, Y. Suzuki and E. Uegaki, Prog. Theor. Phys. Supplement No.68, 29 (1980).
5. G. Ropke, A. Schnell, P. Schuck and P. Nozieres, Phys. Rev. Lett. 80, 3177 (1998); M. Beyer, S.A. Sofianos, C. Kuhrts, G. Ropke and P. Schuck, Phys. Lett. B488, 247 (2000).
6. A. Tohsaki, H. Horiuchi, P. Schuck and G. Ropke, Phys. Rev. Lett. 87, 192501 (2001).
7. H.B. Nielsen and P. Olesen, Nucl. Phys. B61, 45 (1973).
8. S. Ali and A.R. Bodmer, Nucl. Phys. 80, 99 (1966).

Microscopic description of deformed α-particle condensation

A. Tohsaki[*], H. Horiuchi[†], Y. Funaki[†], P. Schuck[**] and G. Röpke[‡]

[*]Department of Fine Materials Engineering, Shinshu University, Ueda 386-8567, Japan
[†]Department of Physics, Kyoto University, Kyoto 606-8502, Japan
[**]Institut de Physique Nucléaire, 91406 Orsay Cedex, France
[‡]FB Physik, Universität Rostock, D-18051 Rostock, Germany

Abstract. A spatial deformation is introduced into the model wave function of the α-cluster condensate in order to study non-zero spin excitations of the α-particle condensate type. Application to ^{8}Be shows that the binding energies of the 0^+ and 2^+ states are nicely reproduced. Our 0^+ wave function is found to be exactly equal to the 0^+ wave function obtained by the full microscopic calculation of two α particles.

We have presented a conjecture that there exist excited states of rarefied density near the $n\alpha$ threshold in self-conjugate $4n$ nuclei, which can be considered as an $n\alpha$ condensate [1]. This conjecture was examined in ^{12}C and ^{16}O by using a new α-cluster wave function. Unfortunately this wave function can describe only the states of zero angular momentum. On the other hand, in ^{12}C there exist a few excited states with non-zero angular momenta above the 3α threshold. Actually various microscopic calculations gave the results that not only the second 0^+ state but also some excited states with non-zero spin have 3α structure of dilute density [3]. The purpose of this talk is to mention the possibility of spatial deformation of the α-cluster condensate. This is carried out introducing a wave function which is obtained by a natural extension of our prvious wave function of the spherical α-cluster condensate.

As a first step, we confirm the case of two α nuclei. For the purpose of our study we write down a new type of α-cluster wave function with deformation describing an α-particle Bose condensed state:

$$|\Phi_{n\alpha}\rangle = (C_\alpha^\dagger)^n |\text{vac}\rangle \tag{1}$$

where the α-particle creation operator is given by

$$
\begin{aligned}
C_\alpha^\dagger = {} & \int d^3R \, \exp(-R_x^2/R_{0x}^2 - R_y^2/R_{0y}^2 - R_z^2/R_{0z}^2) \int d^3r_1 \cdots d^3r_4 \\
& \times \varphi_{0s}(\mathbf{r}_1 - \mathbf{R}) a_{\sigma_1 \tau_1}^\dagger(\mathbf{r}_1) \cdots \varphi_{0s}(\mathbf{r}_4 - \mathbf{R}) a_{\sigma_4 \tau_4}^\dagger(\mathbf{r}_4)
\end{aligned} \tag{2}
$$

with $\varphi_{0s}(\mathbf{r}) = (1/(\pi b^2))^{3/4} e^{-\mathbf{r}^2/(2b^2)}$ and $a_{\sigma\tau}^\dagger(\mathbf{r})$ being the creation operator of a nucleon with spin-isospin $\sigma\tau$ at the spatial point $\mathbf{r}$. The total $n\alpha$ wave function therefore can be written as

$$\langle \mathbf{r}_1 \sigma_1 \tau_1, \cdots \mathbf{r}_{4n} \sigma_{4n} \tau_{4n} | \Phi_{n\alpha} \rangle$$

CP644, Exotic Clustering: 4th Catania Relativistic Ion Studies, edited by S. Costa, A. Insolia, and C. Tuvè

$$\propto \mathscr{A} \left\{ \exp\left(-\frac{2X_{1x}^2}{B_x^2} - \frac{2X_{1y}^2}{B_y^2} - \frac{2X_{1z}^2}{B_z^2} - \cdots - \frac{2X_{nx}^2}{B_x^2} - \frac{2X_{ny}^2}{B_y^2} - \frac{2X_{nz}^2}{B_z^2} \right) \right.$$
$$\left. \times \phi(\alpha_1) \cdots \phi(\alpha_n) \right\}, \tag{3}$$

where $B_k = (b^2 + 2R_{0k}^2)^{1/2}$ and $X_i = (1/4)\sum_n \mathbf{r}_{in}$ is the center-of-mass coordinate of the i-th α-cluster α_i. The internal wave function of the α-cluster α_i is $\phi(\alpha_i) \propto \exp[-(1/8b^2)\sum_{m>n}^4 (\mathbf{r}_{im} - \mathbf{r}_{in})^2]$. The wave function of Eq.(3) is totally antisymmetrized by the operator $\mathscr{A}$. It is to be noted that the wave function of Eqs.(1,3) expresses the state where $n\alpha$-clusters occupy the same deformed $0s$ harmonic oscillator orbit with B_k an indepedent variational width parameter. For example if B_k is of the size of the whole nucleus whereas b remains more or less at the free α-particle value, then the wave function (3) describes an $n\alpha$ cluster condensed state in the macroscopic limit $n \to \infty$. For finite systems we know from the pairing case that such a wave function still can more or less reflect Bose condensation properties. Of course the total center-of-mass motion can and must be separated out of the wave function of Eq.(1) for finite systems. In the limiting case of $B_k = b$ (i.e. $R_{0k} = 0$), Eq.(3) describes a Slater determinant of deformed harmonic oscillator wave functions. We also would like to point out that for $B_k \neq 0$ the wave function (1,3) is different from Brink's α-cluster state [2].

We calculate 2α state employing Volkov No.1 force with $M = 0.56$ [4]. The binding energy of α particle is 27.08 MeV for the size parameter $b = 1.37$ fm. The lowest binding energy in ^{8}Be takes the value 54.33 MeV for spherical parameters of R_{0k} which is only about 0.17 MeV lower than the theoretical two-α threshold energy, in good agreement with the experimet. It is known that this parameter set gives a good reproduction of the α-α scattering phase shift. The energy minimum with 0^+ is located at $R_{0k} = 3$ fm. However, the deformed configuration is not yet stable in a spherical condensate, and namely is around $R_{0x} = R_{0y} = 1.8$ fm and $R_{0z} = 7.8$ fm in the prolate region. The energy gain is only about 0.12 MeV from the spherical case. Reflecting this small energy gain, there is a valley with an almost flat bottom running from this energy minimum to the second energy minimum in the oblate region. We can also point out that the states with respect to two minima are almost equivalent from the overlap between wave functions. As for 2^+ state we do not see energy minimum but plateau region whose height is approximately 3 MeV measured from the energy of the 0^+ state. Altough, without imposing the correct boundary condition of the resonance, we cannot make a definite statement, we safely conjecture that we will have a 2^+ excitation energy around 2.9 MeV which is in good agreement with experiment.

From the results for 0^+ and 2^+ states, in using the deformed condensate wave function, we have two conclusions: (1) For the 0^+ state, the introduction of the deformation plays no essential role and the description with spherical condensate is justified. (2) The deformed 2α-state yields a nice description of the 2^+ state, which gives a new understanding about the character of the 2^+ state. Namely, the 2^+ state can be considered to be a gas-like state of two α-particles in a relative d-wave. The merit of the introduction of deformation is just this point. We have succeeded in obtaining the picture that not only the ground state but also the excited state can be interpleted in terms of a gas-like structure. This conclusion encourages us to introduce the deformed wave function of the

α-condensate in the other self-conjugate $4N$ nuclei with non-zero spin excitations for a future study.

Our wave functions are approximations to the RGM wave functions. Since the numerical calculation of the 2α RGM is not difficult with present-day computors, one may ask why we do not discuss the comparison of our wave functions and the RGM ones. The answer is apparent: Our study in this talk is to extract and to elucidate th α-condensate character of the ^{8}Be of which the RGM study until now has not shown even though it has given numerical values. For 0^+ state, we solved the RGM equation in order to compare with our results. We obtained the binding energy of 54.446 MeV which absolutely coincides with our result. The squared overlap between the RGM wave function and ours is 0.9980. We can easily say that we have obtained almost converged value for the lowest eigenvalue of the RGM equation. It is surprising but it clearly shows that our model wave function is very much suited to the ground state of ^{8}Be.

We conclude for ^{8}Be that (1) the spherical α condensate reproduces the character of ground state in relative s-wave, (2) the deformed wave function gives us only a slight energy gain of 0.12 MeV for the ground 0^+ state, (3) the wave function is almost equivalent to the RGM one, and (4) the deformed wave function nicely reproduces 2^+ state around 2.9 MeV. It is worth mentioning that our wave function has so far reproduced quite accurately 0^+ threshold states in ^{12}C and ^{16}O and the rotational state in ^{8}Be. The fact that these results are obtained without any adjustable parameter is surprisingly remarkable and gives strong credit to the correctness of the physics contained in our wave function.

REFERENCES

1. A. Tohsaki, H. Horiuchi, P. Schuck, and G. Röpke, Phys. Rev. Letters **87**, 192501(2001).
2. D. M. Brink, Proc. Int. School Phys. Enrico Fermi **36** (Academic Press, New York, 1966);
3. Y. Fujiwara, H. Horiuchi, K. Ikeda, M. Kamimura, K. Kato, Y.Suzuki, and E. Uegaki, Prog. Theor. Phys. Supplement No.68, 29 (1980).
4. A. B. Volkov, Nucl. Phys. **74**, 33 (1994).

Alpha and light nucleus emission within a generalized liquid drop model

G. Royer, C. Bonilla and R. Moustabchir

Laboratoire Subatech, 4 rue A. Kastler, 44 - Nantes, France

Abstract. The potential energy governing the spontaneous α, C, O, F, Ne, Mg and Si emissions has been determined within a generalized liquid drop model including the proximity effects between the emitted light nucleus and the daughter one and taking into account empirically the experimental Q value. The decay path has been described by a quasi-molecular shape sequence leading rapidly to two spherical touching nuclei before crossing the barrier. The partial half-lives deduced from the WKB barrier penetration probability are in very good agreement with experimental data and accurate analytical expressions are proposed. The partial half-lives of the Be, Li, He and H sub-barrier emissions have been calculated by adding an excitation energy to the Q value and new formulae are given.

INTRODUCTION

The spontaneous α, C, O, F, Ne, Mg and Si decays as well as the emission of Be, Li, He and H isotopes from slightly excited nuclei have been studied in the fusion-like deformation path within a generalized liquid drop model previously developed to study the fusion and fission processes [1, 2, 3]. In these exit channels the separation in two fragments occur rapidly and the one-body shapes play a minor role. The proximity forces in the neck between the fragments smooth strongly the unrealistic pure Coulomb peak and the potential barrier top corresponds to two well separated spheres maintained in unstable equilibrium by the balance between the repulsive Coulomb forces and the attractive nuclear proximity forces. Only excitation energies E^* lower than the potential barrier height are considered.

PARTIAL HALF-LIVES

The decay is viewed as a tunneling process and within a very asymmetric fission mode for which the preformation probability at the surface is one. Consequently the decay constant of the light nucleus emission is simply given by $\lambda = v_0 P$. The assault frequency v_0 has been taken as $v_0 = 10^{20} \ s^{-1}$.

The barrier penetrability P is calculated within the action integral

$$P = exp[-\frac{2}{\hbar} \int_{r_{in}}^{r_{out}} \sqrt{2B(r)(E(r) - E^*)}dr], \tag{1}$$

CP644, *Exotic Clustering: 4th Catania Relativistic Ion Studies*, edited by S. Costa, A. Insolia, and C. Tuvè
© 2002 American Institute of Physics 0-7354-0099-7/02/$19.00

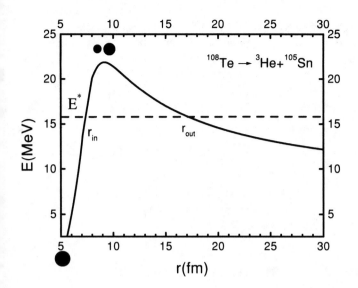

FIGURE 1. Potential barrier governing the ^{3}He emission from ^{108}Te. E^* is the excitation energy and r_{in} and r_{out} the turning points. r is the centre-of-mass distance.

where r_{in} and r_{out} correspond to the points where $E(r) = E^*$ (see Fig. 1).
The inertia B(r) is related to the reduced mass by [4]

$$B(r) = \mu(1 + 1.3f(r)) \tag{2}$$

with

$$f(r) = \begin{cases} \sqrt{\frac{R_{cont}-r}{R_{cont}-R_{in}}}, r \leq R_{cont} \\ 0, r \geq R_{cont} \end{cases} \tag{3}$$

R_{cont} is the distance between the mass centres at the contact point.
The partial half-life is finally obtained by $T_{1/2} = ln2/\lambda$.

SPONTANEOUS C, O, F, NE, MG AND SI EMISSIONS

The theoretical partial half-lives for the ^{14}C, ^{20}O, ^{23}F, $^{24-26}$Ne, $^{28-30}$Mg and ^{32}Si emissions are in good agreement with the experimental data. Predictions for other possible decays meeting two criteria : partial half-life $\leq 10^{30}s$ and branching ratio relative to α emission $\geq 10^{-24}$ have been given [4]. An analytic formula based on a fitting procedure applied on a set of 144 cluster decay half-lives predicted by the GLDM is proposed for the partial half-life of the spontaneous light nucleus decay.

$$Log_{10}T_{1/2}(s) = -11.035 + R_{cont}\sqrt{A_{12}Q_{exp}}[-4.311 - \frac{1.17277}{\sqrt{Q_{exp}}}$$

$$-0.054618\sqrt{\frac{Q_{exp}}{A_{12}R_{cont}}} + \frac{1.62744}{\sqrt{x}} + 4.4956\sqrt{x} - 1.7526x], \tag{4}$$

where $x = R_{cont}Q_{exp}/e^2 Z_1 Z_2$ and $A_{12} = A_1 A_2/(A_1 + A_2)$. The input data are the charge product $Z_1 Z_2$, the experimental Q value and the mass numbers A_1 and A_2 which allow to determine the dimensionless reduced mass A_{12} and R_{cont}.

ALPHA DECAY BELOW THE BARRIER

The predictions for the partial half-lives of the spontaneous α emission have been compared with a recent experimental data set on 373 α emitters. The rms deviation between the theoretical and experimental values of $log_{10}[T_{1/2}(s)]$ is only 0.63 and 0.35 for the subset of even-even nuclides (see Fig. 2). Predictions for the heaviest and superheavy elements have also been provided [5].

The following new formulae give $log_{10}[T_{1/2}(s)]$ as functions of the mass and charge numbers of the α emitter, the Q value and the excitation energy [6]. In addition to the experimental data at $E^* = 0$ a whole set of theoretical data given by the GLDM has been built to take into account the excitation energy, the experimental data being absent.

For the even Z- even N nuclei the following formula has been obtained with a rms deviation of 0.28 between the data set given by the GLDM and the values given by the formula

$$log_{10}\left[T_{1/2}(s)\right] = (-25.31 - 1.1629A^{\frac{1}{6}}\sqrt{Z} + \frac{1.5864Z}{\sqrt{Q_\alpha + E^*}})$$

$$\times (1 - 4.5182 \times 10^{-4}E^{*2}). \tag{5}$$

For the even-odd nuclei the formula is, with a rms deviation of 0.41

$$log_{10}\left[T_{1/2}(s)\right] = (-26.65 - 1.0859A^{\frac{1}{6}}\sqrt{Z} + \frac{1.5848Z}{\sqrt{Q_\alpha + E^*}})$$

$$\times (1 + 1.1170 \times 10^{-2}E^* - 1.4903 \times 10^{-3}E^{*2}). \tag{6}$$

For the odd-even nuclei the rms deviation is 0.27

$$log_{10}\left[T_{1/2}(s)\right] = (-25.68 - 1.1423A^{\frac{1}{6}}\sqrt{Z} + \frac{1.592Z}{\sqrt{Q_\alpha + E^*}})$$

$$\times (1 + 8.9617 \times 10^{-3}E^* - 1.3446 \times 10^{-3}E^{*2}). \tag{7}$$

36

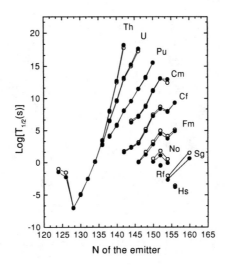

FIGURE 2. $log_{10}[T_{1/2}(s)]$ for the spontaneous α emission from Th, U, Pu, Cm, Cf, Fm, No, Rf, Sg and Hs isotopes. The open and full circles correspond to the experimental and theoretical data.

For the odd-odd nuclei the following formula leads to a rms deviation of 0.5

$$log_{10}\left[T_{1/2}(s)\right] = (-29.48 - 1.113A^{\frac{1}{6}}\sqrt{Z} + \frac{1.6971Z}{\sqrt{Q_\alpha + E^*}})$$
$$\times (1 - 8.8806 \times 10^{-3}E^*). \qquad (8)$$

For the available experimental data on spontaneous α decay the rms is respectively 0.285, 0.39, 0.36 and 0.35. The introduction of the excitation energy does not diminish the accuracy of the formulae.

BE, LI, HE, AND H SUB-BARRIER EMISSION FROM SLIGHTLY EXCITED NUCLEI

The transmutation of nuclear waste and the production of energy and radio-isotopes by accelerator-driven systems is under consideration. The knowledge of all the nuclear reactions which constitute a non negligible part of the cross section is needed and the energy range is from thermal up to a GeV. To contribute to this study, here the above-mentioned procedure has been applied to determine formulae giving the Be, Li, He and H decay half-lives for excitation energies below the potential barrier height [6].

For $^9_4$Be the formula is, with a rms deviation of 0.48,

$$log_{10}\left[T_{1/2}(s)\right] = (-31.69 - 2.238A^{\frac{1}{6}}\sqrt{Z} + \frac{4.47183Z}{\sqrt{Q+E^*}})$$
$$\times(1 + 2.384 \times 10^{-3}E^* - 9.556 \times 10^{-5}E^{*2}). \tag{9}$$

For $^7_3$Li, with $\sigma = 0.38$,

$$log_{10}\left[T_{1/2}(s)\right] = (-27.55 - 1.796A^{\frac{1}{6}}\sqrt{Z} + \frac{3.0016Z}{\sqrt{Q+E^*}})$$
$$\times(1 + 2.665 \times 10^{-3}E^* - 1.109 \times 10^{-4}E^{*2}). \tag{10}$$

For $^6_3$Li, with $\sigma = 0.36$,

$$log_{10}\left[T_{1/2}(s)\right] = (-27.45 - 1.647A^{\frac{1}{6}}\sqrt{Z} + \frac{2.8051Z}{\sqrt{Q+E^*}})$$
$$\times(1 + 3.107 \times 10^{-3}E^* - 1.213 \times 10^{-4}E^{*2}). \tag{11}$$

For $^6_2$He, with $\sigma = 0.35$,

$$log_{10}\left[T_{1/2}(s)\right] = (-24.85 - 1.424A^{\frac{1}{6}}\sqrt{Z} + \frac{1.8773Z}{\sqrt{Q+E^*}})$$
$$\times(1 + 2.873 \times 10^{-3}E^* - 1.288 \times 10^{-4}E^{*2}). \tag{12}$$

For $^3_2$He, with $\sigma = 0.24$,

$$log_{10}\left[T_{1/2}(s)\right] = (-23.60 - 1.003A^{\frac{1}{6}}\sqrt{Z} + \frac{1.3665Z}{\sqrt{Q+E^*}})$$
$$\times(1 + 3.896 \times 10^{-3}E^* - 1.662 \times 10^{-4}E^{*2}). \tag{13}$$

For $^3_1$H, with $\sigma = 0.14$,

$$log_{10}\left[T_{1/2}(s)\right] = (-22.65 - 0.7187A^{\frac{1}{6}}\sqrt{Z} + \frac{0.6775Z}{\sqrt{Q+E^*}})$$
$$\times(1 + 3.079 \times 10^{-3}E^* - 1.795 \times 10^{-4}E^{*2}). \tag{14}$$

For $^2_1$H, with $\sigma = 0.16$,

$$log_{10}\left[T_{1/2}(s)\right] = (-22.02 - 0.6039A^{\frac{1}{6}}\sqrt{Z} + \frac{0.5626Z}{\sqrt{Q+E^*}})$$
$$\times(1 + 3.06 \times 10^{-3}E^* - 1.871 \times 10^{-4}E^{*2}). \tag{15}$$

For $^1_1$H, with $\sigma = 0.08$,

$$log_{10}\left[T_{1/2}(s)\right] = (-21.32 - 0.4214A^{\frac{1}{6}}\sqrt{Z} + \frac{0.3961Z}{\sqrt{Q+E^*}})$$
$$\times(1 + 1.208 \times 10^{-3}E^* - 9.972 \times 10^{-5}E^{*2}). \tag{16}$$

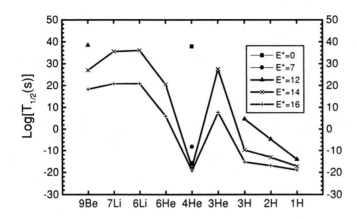

FIGURE 3. $log_{10}[T_{1/2}(s)]$ as functions of the excitation energy E^*(MeV) and the emitted light particle for an excited ^{209}Pb nucleus.

As an example, the dependence of the partial half-lives on the emitted light nucleus and the excitation energy is shown in Fig. 3 for a ^{209}Pb nucleus formed and excited by the absorption of one neutron. The Q value is positive only for the α emission.

REFERENCES

1. Royer, G., and Remaud, B., *Nucl. Phys. A*, **444**, 477-497 (1985).
2. Moustabchir, R., and Royer, G., *Nucl. Phys. A*, **683**, 266-278 (2001).
3. Royer, G., and Zbiri, K., *Nucl. Phys. A*, **697**, 630-638 (2002).
4. Royer, G., and Moustabchir, R., *Nucl. Phys. A*, **683**, 182-206 (2001).
5. Royer, G., *J. Phys. G*, **26**, 1149-1170 (2000).
6. Bonilla, C., and Royer, G., *Heavy-ion Physics*, submitted (2002).

Formation and Decay of Super Heavy Composite Systems

Toshiki Maruyama*, Aldo Bonasera†, Massimo Papa** and Satoshi Chiba*

*Japan Atomic Energy Research Institute, Tokai, Ibaraki 319-1195, Japan
†Istituto Nazionale di Fisica Nucleare Laboratorio Nazionale del Sud, Via S. Sofia 44, Catania
95123, Italy
**Istituto Nazionale di Fisica Nucleare - Sezione di Catania, Via S. Sofia 64, Catania 95123, Italy

Abstract. We investigate the formation and the decay of heavy systems which are above the fission barrier. By using a microscopic simulation of constraint molecular dynamics (CoMD) on Au+Au collision, we observe composite states stay for very long time before decaying by fission.

INTRODUCTION

Collisions between heavy nuclei at relatively low energy region have attracted strong interests of heavy-ion physicists for 3 different reasons, namely the creation of super heavy elements (SHEs), fission dynamics of very heavy systems, and creation of electron-positron pair due to the strong Coulomb field of the composite heavy nuclei as a verification of the QED process. In these processes, the lifetime of the composite system, created by the fusion of the projectile and the target, which decays eventually by fission is the key issue to understand the underlying reaction mechanisms and to estimate the probability of occurrence of these processes.

SHEs are produced in two ways: one is "cold fusion" which is complete fusion below the classical barrier, and the other is "hot fusion" which allows several neutrons to be emitted. Even though the name is "hot", such reactions are still at very low energy near the barrier and the total mass number is very close to the aimed one. As far as the formation of SHE is concerned, the "fusion" of very heavy nuclei where the fission barrier no more exists is found to be ineffective [1, 2].

In the study of fission dynamics of heavy systems including the spontaneous fission and the fusion-fission of heavy composite, the competition of neutron emission between the fission and the fission delay have been discussed intensively. However almost all the discussion are done for mass regions where the classical fission barrier exists.

Sometime ago the low energy collision of very heavy nuclei has been paid much attention regarding the spontaneous positron emission from strong electric fields [3]. If the binding energy of an electron can exceed the electron mass due to the strong electric field in a very heavy system, an electron-positron pair might be created by a static QED process. Although no clear evidence of static positron creation was observed below Coulomb energy region, they have pointed out [4] the importance of nuclear reaction which causes the time delay of separation of two nuclei. The reaction mechanism of very heavy nuclei, however, has not been discussed by fully dynamical models.

CP644, *Exotic Clustering: 4th Catania Relativistic Ion Studies,* edited by S. Costa, A. Insolia, and C. Tuvè
© 2002 American Institute of Physics 0-7354-0099-7/02/$19.00

In this paper we discuss the possibility of molecule-like states of heavy nuclei and the time scale of very heavy composite system formed by the fusion-fission or deep inelastic processes. To investigate these problems theoretically we use a recently developed constraint molecular dynamics (CoMD) model [5]. This model has been proposed to include the Fermionic nature of constituent nucleons by a constraint that the phase space distribution should always satisfy the condition $f \leq 1$. In this paper we apply CoMD to the ^{197}Au+^{197}Au collisions at low energies where fusion-fission or deep-inelastic process may occur. In the following we give a brief review of the model [5].

COMD SIMULATION OF AU+AU SYSTEM

The CoMD model mainly consists of two parts: classical equation of motion (EOS) of many-body system, and stochastic process which includes constraint of Pauli principle and the two-body collisions. The effective interaction used for the classical EOS consists volume term, density-dependent term, surface term and the Coulomb term. The parameters in the above interaction is explained in Ref. [6].

The Pauli principle is taken into account in two ways: One is the Pauli blocking of the final state of two-body collision and the other is the constraint which brings into the system the Fermi motion in a stochastic way. The starting point of the constraint is the requirement that the phase-space occupation for each particle should be less than 1. At each time step and for each particle i the phase space occupation $\overline{f}_i$ is checked. If $\overline{f}_i$ has a value greater than 1 an ensemble K_i of nearest particles (including the particle i) is determined within the neighborhood in the phase space. Then we change randomly the momenta of the particles belonging to the ensemble K_i in such a way that for the newly generated sample the total momentum and the total kinetic energy is conserved ("many-body elastic scattering"). The new sample is accepted only if it reduces the phase space occupation $\overline{f}_i$ [5].

To simulate the collision of two ^{197}Au nuclei, we prepare the ground state by applying the frictional cooling method together with the constraint of CoMD. The ground states we obtain have binding energy of 8.4 MeV/nucleon and the root mean square radius of 5.34 fm. They are rather stable for 1000 fm/c. For instance our ^{197}Au ground states evaporates 3.1 nucleons during 1000 fm/c. The collision events are performed for impact parameter b of 0 and 6 fm for incident energy in laboratory system of $E_{lab} = 5 \sim 35$ MeV/nucleon.

Figure 1 shows a typical event of CoMD calculation with incident energy $E_{lab} = 10$ MeV/nucleon with impact parameter $b = 6$ fm. The two nuclei form a quite deformed compound system, they keep such a deformation almost 2500 fm/c and finally fission takes place. The reaction mechanism we are observing here may be in-between the deep inelastic and molecular resonance.

Figure 2 shows the time dependences of the largest cluster mass for the impact parameters $b = 0$ and 6 fm calculated by CoMD and QMD. In CoMD calculations we see at the beginning the largest cluster mass $A_{max} = 197$ which corresponds to projectile and target mass number. Within about 50 fm/c, A_{max} becomes 394 except for the incident energy $E_{lab} = 5$ MeV which is below the barrier where two nuclei never contact. At

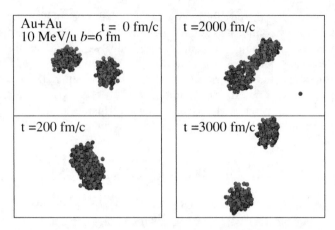

FIGURE 1. Snapshot of the ^{197}Au+^{197}Au reaction at $E_{\text{lab}} = 10$ MeV/nucleon and $b = 6$ fm. The time indicated in each panel is not from the contact of two nuclei but indicates only that of the simulation.

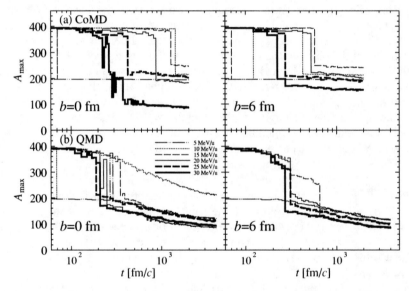

FIGURE 2. The time-dependence of the largest fragment mass A_{max}. From the top (a) CoMD and (b) QMD. The left panels show cases of head-on collision and the right $b = 6$ fm.

incident energies above the barrier, the formed large system will decay into smaller fragments by different modes according to the energy and angular momentum. At higher incident energies ($E_{\text{lab}} \geq 30$ MeV/nucleon) the largest cluster mass changes suddenly at the early stage and continuously decreases in time. This indicates multifragmentation for head-on collisions and deep inelastic reaction for peripheral collisions followed by the emission of nucleons and small fragments. At lower incident energies ($E_{\text{lab}} \leq 20$

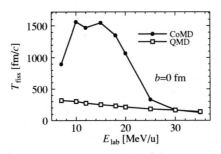

FIGURE 3. The "fission" life time of composite system for CoMD and QMD calculation.

MeV/nucleon) there is a sudden change of the largest cluster mass at very late time, which indicates a fission of the system. One should note that in our calculation of Au+Au system there is almost no event where the system decays only by emitting particles or light fragments, i.e., pure incomplete fusion. The instability due to the Coulomb repulsion plays the major role in the decay process.

Here we should note that the plotted largest fragment mass are obtained by only one event for each incident energy and impact parameter. Therefore the fission time includes large amount of statistical error. In Fig. 2 (b) the same quantity as Fig. 2 (a) is displayed for QMD calculation (using CoMD code switching off the constraint). The difference between the CoMD and the QMD is clear and dramatic. At low energy collisions there are no fission process and the system decays only by emitting nucleons and light fragments. At higher energies there is some sudden change of the largest fragment mass even in QMD calculation, which is the passing-through process or deep inelastic process.

LIFE TIME OF PRODUCED COMPOSITE SYSTEMS

Assuming a very simple form of the time-dependent fission width $\Gamma(t) = \Gamma_f \, \theta(t - T_d)$, the averaged fission time T_{fiss} can be obtained by the survival probability of the compound system against two-body process P_{surv} as

$$P_{surv} = \exp\left[-(t - T_d)\Gamma_f/\hbar\right], \qquad (1)$$
$$T_{fiss} \equiv T_d + \hbar/\Gamma_f \qquad (2)$$

where T_d is the delay time and Γ_f is the "fission width" after the delay time. The probability $P_{surv}(t)$ is obtained directly by the simulation. For all the calculations the fitting works well, particularly the effect of delay time. The assumption of constant fission width after the delay time, on the other hand, is not completely supported because of poor statistics and still existing dynamical effects. One should note that the fitting by (1) and (2) is just to extract the "fission" time of the super heavy composite. Especially the time scale of QMD results is obviously different from that of fission process.

The extracted fission time T_{fiss} are plotted in Fig. 3. With effective interaction used in Reference [5] the calculated fission time is about an order longer than this calculation. However in both cases of CoMD calculations the longest life time of heavy composite is found at $E_{lab} = 10$ MeV/nucleon. Just above the Coulomb barrier, the system might not form a fully thermalized single composite but might be quasi separated in the phase space, which makes the system split easily. For higher energies, the fully thermalized system needs some fluctuations to reseparate even though there is no classical barrier for fission. Therefore the fission time gets shorter with increase of the incident energy.

In QMD calculation what is in marked contrast to the CoMD calculation is that the "fission time" has no maximum energy and shows monotonic decrease. This is due to the lack of the Pauli principle which suppresses two nuclei from overlapping at very low energies (above the Coulomb barrier).

For peripheral collision ($b = 6$ fm), the life time of very heavy composite is shorter than the head-on collisions. But the incident-energy dependence is very similar to the $b = 0$ fm cases. Though the mechanism is much more dynamical, (1) and (2) fit well again. *Nevertheless, the super heavy composite system formed by the head-on collision of Au+Au may survive rather long time of $10^3 \sim 10^4$ fm/c.*

An interesting aspect of the long-lived very heavy system is, as mentioned before, the spontaneous positron-electron production from the strong electric field as a static QED process. The total charge of Au+Au system may be still smaller than the necessary charge ($Z \sim 170$) for this process. However, the nuclear reaction of, e.g. U+U system, should be qualitatively the same as what we observe in Au+Au system. One can get longest life time of strong electric field (stronger than the case of Rutherford or molecular trajectory) around $E_{lab} = 10$ MeV/nucleon at some impact parameter and the production of positrons from the static QED process should be largest around that energy.

ASYMMETRIC FISSION

As mentioned above, production of SHE is one of the most important subject in the heavy-ion collision problem. Besides cold- and hot-fusion, mass transfer in collision of very heavy nuclei was tried before. One could produce, e.g. up to Fm ($Z = 100$) in U+U system, or Md ($Z = 101$) in U+Cm system, by such a mechanism [1, 2]. The incident energy, however, was very close to the Coulomb barrier and the reaction was rather gentle with the transfer of ~ 20 nucleons. In our CoMD calculation for $E_{lab} \geq 7$ MeV/nucleon, the reaction mechanism is more violent and there happens the transfer of much more nucleons though the mass loss from the system is also large. In Fig. 4 plotted is the mass-asymmetry $(A_1 - A_2)/(A_1 + A_2)$ of the fission process in CoMD calculation and the fission fragment mass A_1 and $A_2 (< A_1)$. The mass-asymmetry increases with the incident energy. For $b = 0$ the asymmetry amounts to about 0.1 at $E_{lab} = 7$ MeV/nucleon and almost 0.4 at 30 MeV/nucleon. As a result, the largest fission fragment can be much larger than the initial projectile and target nuclei as shown at the lower panel of the figure. Of course we should consider the thermal mass loss and subsequent fission due to the excitation of fragments. However, such kind of fusion-fission mechanism at around 10 MeV/nucleon should be taken into account for the SHE production. The new 4π detectors can accumulate lots of statistics plus they can make coincidence studies to see if the fragments come from fission.

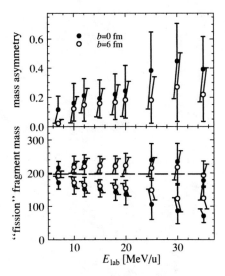

FIGURE 4. The mass asymmetry of the fission fragments and the fragment mass. The dashed line indicates the initial mass of projectile and target nuclei. Error bars indicate statistical standard deviation.

SUMMARY

In summary, we have discussed the formation and decay of super heavy composite in the Au+Au collisions. The CoMD calculation which takes into account the Fermionic nature of the nucleon many-body system can describe well the low-energy dynamics including fusion, fission, deep inelastic, emission of nucleons and small fragments, and multifragmentation. Although there are still some ambiguities on the effective interaction, the life time of super heavy composite is found to be rather long up to $10^3 \sim 10^4$ fm/c. Some experimental explorations such as detection of e^+e^- formation at around 10 MeV/nucleon and measurement of the energy averaged angular distribution and/or excitation function for binary processes are encouraged.

REFERENCES

1. H. Gäggeler et al., Proc. 4th Int. Conf. on Nuclei Far From Stability, L.O. Skolen, Helsinger 1981, CERN 81-09, p 763.
2. G. Herrmann, Proc. 4th Int. Conf. on Nuclei Far From Stability, L.O. Skolen, Helsinger 1981, CERN 81-09, p 772.
3. "Quantum Electrodynamics of Strong Fields", ed. W. Greiner, Plenum Press New York and London, 1983; R. Bär et al, Nucl. Phys. A583, 237 (1995); S.M Ahmad et al, Nucl. Phys. A583, 247 (1995); A. Lépine-Szily et al, Nucl. Phys. A583, 263 (1995).
4. J. Reinhardt, in "Quantum Electrodynamics of Strong Fields", ed. W. Greiner, Plenum Press New York and London, 1983.
5. M. Papa, T. Maruyama and A. Bonasera, Phys. Rev. C 64, 024612 (2001).
6. T. Maruyama and A. Bonasera, M. Papa and S. Chiba, Eur. Phys. J. A 14, 191 (2002).

Clustering Features in Light Nuclei

Theoretical prediction of three-body cluster resonances in ^{11}Li and ^{10}He

S. Aoyama*, T. Myo†, K. Katō† and K. Ikeda**

* *Information Processing Center, Kitami Institute of Technology, Kitami, 090-8507 Japan*
† *Division of Physics, Graduate School of Science, Hokkaido University, Sapporo 060-0810, Japan*
** *The Institute of Physical and Chemical Research (RIKEN), Wako 351-0198, Japan*

Abstract. With the core+n+n three-body cluster model, we investigated ^{11}Li and ^{10}He. By using a ^{9}Li-n interaction which reproduce the degeneracy of single particle $s_{1/2}$- and $p_{1/2}$-orbital states above the ^{9}Li+n threshold, a low-lying excited resonance is predicted in ^{11}Li together with the ground state, and they are found to be partners in mixed configurations of $(0p_{1/2})^2$ and $(1s_{1/2})^2$. The observed three-body resonance in ^{10}He is considered to be the two valence neutrons mainly in the p-orbit, $(0p_{1/2})^2$. In addition to a solution of the two valence neutrons in the p-orbit, we obtained a solution of the two valence neutrons mainly in the s-orbit, $(1s_{1/2})^2$, as the ground state. Experimentally, such a state has not yet been observed. This newly predicted state of ^{10}He with the main component of $(1s_{1/2})^2$ would corresponds to the ground state of ^{11}Li with a halo structure.

INTRODUCTION

The mechanism of the large root-mean-squared radii for the drip-line nuclei is explained as a result of the halo structure. As a typical halo nucleus, ^{11}Li has been studied by using the three-body cluster model of ^{9}Li+n+n. By using a ^{9}Li-n potential, which makes ^{10}Li not only having the p-state but also the s-state near to the ^{9}Li+n threshold (virtual state), it is shown that the ground state property of ^{11}Li is explained well [1]. Though the physical origin of the virtual state is not clear, several mechanisms have been proposed. Recently, by using a coupled channel model taking into account the ^{9}Li-core excitation, Katō *et al.* makes a new ^{9}Li-n potential [2] based on the shell model picture of the so called pairing blocking effect proposed by Sagawa *et al.*[3] By using this ^{9}Li-n potential, we get a conclusion that the ground state has the large s-wave component of $(1s_{1/2})^2$ in addition to $(0p_{1/2})^2$.

Considering the case of the large s-wave component, we have newly two questions at least: i)where is the pairing excited resonance in ^{11}Li? ii)where is the ground state in ^{10}He? The reason of the former is that the ground state of ^{11}Li has a similar weight of of $(1s_{1/2})^2$ and $(0p_{1/2})^2$. When the main component of the ground state is $(1s_{1/2})^2$, we can expect a paring excited state having the $(0p_{1/2})^2$ main component. Experimentally, an excited state of ^{11}Li, which is suggested as the pairing excited state, is observed at E_x=1.02 MeV[4]. The reason of the latter is that the observed ground state of ^{10}He is considered to have the configuration of the $(0p_{1/2})^2$ because of its decay width.

CP644, *Exotic Clustering: 4th Catania Relativistic Ion Studies*, edited by S. Costa, A. Insolia, and C. Tuvè

Since neutron number is same both for ^{11}Li and for ^{10}He, we expect them to a similar configuration of valence two neutrons. However, the observed resonance in ^{10}He exhibit a large component of $(0p_{1/2})^2$, though the observed ground state of ^{11}Li exhibit the large component of $(1s_{1/2})^2$. Then, it is natural to think that the observed resonance in ^{10}He corresponds to the excited resonance in ^{11}Li rather than the ^{11}Li ground state. And the missing ground state corresponds to the ^{11}Li ground state.

As far as methods of solving three-body resonance are concerned, we use the complex scaling method (CSM) [5] and the method of analytical continuation in the coupling constant (ACCC)[6][7]. The former has been widely used in the analyses of three-body cluster resonance. The later has recently been apply to the practical calculation of the three-body cluster resonances [8], and show the usefulness in the study of three-body S-wave resonance because the CSM can not treat the three-body S-wave resonance in many cases.

MODEL AND METHOD

Hybrid-TV model

We used the variation method with the so-called Hybrid-TV model [9]. Using this model, we could accurately treat both the core-n and n-n correlations within a small base number, as discussed in Refs.[9, 10]. The variation function for the Hybrid-TV model is expressed by the superposition of the V-type function and the T-type function as

$$\Psi_{JM} = \Phi_{JM}(V) + \Phi_{JM}(T). \tag{1}$$

The details are given in Ref.[10].

We employ the core+n+n Hamiltonian, which is given in Refs.[2] and [11]:

$$H = \sum_{i=1}^{2} [\frac{1}{2\mu} p_i^2 + V_{core-n}^F(\eta_i) + V^{Pseud.}(\eta)]$$

$$+ V_{nn}(|\eta_1 - \eta_2|) + \frac{1}{\mu_{core}} p_1 \cdot p_2. \tag{2}$$

The details of each term are given in Refs.[2] and [11], though the Hamiltonian is extended from a simple core+n+n model for ^{11}Li in order to take into account of the core excitation in Ref.[2]. As a neutron-neutron interaction, V_{nn}, we use the Minnesota potential[12]. The core-n interaction, V_{core-n}^F, is constructed microscopically by using the effective nucleon-nucleon interaction of the modified Hasegawa Nagata potential (MHN)[13]. In the present calculation, we introduced the parameter $\lambda = 1 + \delta$ to the mid-range attractive part of the core-n potential as $V_{core-n}^F(\eta) = V^{(1)}(\eta) + \lambda V^{(2)}(\eta) + V^{(3)}(\eta)$.

Complex Scaling Method

The complex scaling is defined by the following transformation:

$$U(\theta)f(x) = \exp(i\frac{3}{2}\theta)f(\exp(i\theta)x). \tag{3}$$

Under this transformation, the Schrödinger equation is given as

$$H(\theta)|\Psi_\theta> = E(\theta)|\Psi_\theta>, \tag{4}$$

where
$$H(\theta) = U(\theta)HU(\theta)^{-1}, \qquad |\Psi_\theta> = U(\theta)|\Psi>. \tag{5}$$

The ABC-theorem [5] indicates that the solutions of the transformed Schrödinger equation have the following properties: 1) The resonant solutions are also described by square-integrable functions in addition to normalizable bound states. 2) The energies of bound states are not changed by scaling. 3) When we choose the scaling parameter θ larger than the angle $tan^{-1}(\Gamma/2E_r)$ corresponding to the resonant position (resonance energy E_r and width Γ), E_r and $\Gamma/2$ are obtained as real and imaginary parts of the complex eigenvalue $E(\theta)$, e.g., $E(\theta) = E_r - i\Gamma/2$. 4) The continuum spectra are obtained along lines on the complex energy plane, which start at the threshold energies of decays of the system into sub-systems and have an angle -2θ from the positive real axis. The details of applying the CSM to the core+n+n model are given in Ref.[14].

Complex Scaling Method

We consider an unbound state with Hamiltonian H. In the ACCC, we introduce a parameter λ (a coupling constant) in the Hamiltonian as $H(\lambda) = H_0 + \lambda V$. We employ the attractive part of the potential as V. Then, $H(\lambda = 1)$ is the original Hamiltonian. By increasing λ, we can obtain a bound-state solution, because V is the attractive part of the potential. For a two-body system, it is known that the square root of the energy behaves as $k_l(\lambda) \sim \sqrt{\lambda - \lambda_0}$ for $l \neq 0$ and $k_0(\lambda) \sim (\lambda - \lambda_0)$ for $l = 0$ around the branching coupling constant λ_0[7]. Here, l is the relative angular momentum. In the case of $l \neq 0$, λ_0 is easily obtained as a coupling constant which gives the threshold energy (E(λ_0)=0, $k_l(\lambda_0) = 0$). In the case of $l = 0$, it is known that $k_0(\lambda_0)$ is not zero ($k_0(\lambda_0) = i\chi_0$, $\chi_0 < 0$). For a three-body system, it is discussed by Tanaka $et\ al.$ that the branching energy of the S-wave for the Borromean system is nearly equal to zero due to the presence of an effective barrier of the three-body system [8]. The calculated result shows that the branching energy E(λ_0) is nearly equal to zero within the numerical error. Using the obtained momentum of k as a function of λ in the bound-state region, we carried out analytical continuation to the unbound region with the Padé approximation. The Padé approximation is given as

$$k_l^{MN}(x) = i\frac{c_0 + c_1 x + c_2 x^2 + \cdots + c_M x^M}{d_0 + d_1 x + d_2 x^2 + \cdots + d_N x^N}, \tag{6}$$

where $x = \sqrt{\lambda - \lambda_0}$. For the bound-state region, since $x = \sqrt{\lambda - \lambda_0} > 0$, $k_l^{MN}(x)$ is purely imaginary on the positive axis. For the unbound region, since $x = \sqrt{\lambda - \lambda_0}$ is imaginary, $k_l^{MN}(x)$ is complex. We use δ as a substitute for λ. The details are given in Ref.[7].

RESULTS AND DISCUSSIONS

Firstly, we investigate the excited resonance of ^{11}Li by using the CSM. In Fig. 1, we display typical calculated $3/2^-$ energy levels and the experimental low-lying states in the ^{9}Li+n+n threshold energy region. The potential parameters, δ, are determined that the ground state energy of ^{11}Li is reproduced. In the present calculation, an excited $3/2^-$ resonance is obtained. The ground state is described as the mixed configurations of $(1s_{1/2})^2$ and $(0p_{1/2})^2$. And its main component is $(1s_{1/2})^2$. Then, the excited resonance is considered to be a pairing excitation from the ground state which has the main component of $(0p_{1/2})^2$. Such a state is also obtained in the shell model calculation.

For example, by using the WBP interaction [15], a low-lying excited $3/2^-$ state in ^{11}Li is predicted at $E_x = 1.49 \pm 1$MeV MeV [16]. The present energy gap depends on the employed nucleon-nucleon interaction in the ^{9}Li-n potential, being 0.56 MeV for HN No. 1 and 0.91 MeV for MHN. The difference essentially originates from the odd-state component of the microscopic nucleon-nucleon interaction [2]. For the nucleon-nucleon interaction between the valence neutrons, we used the Minnesota potential [12], which is one of the most commonly employed effective nucleon-nucleon interactions used in microscopic calculations of neutron-rich nuclei.

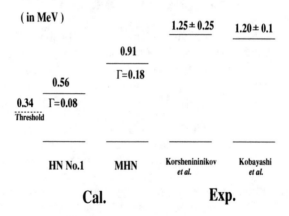

FIGURE 1. The calculated $3/2^-$ energy levels and the experimental low-lying levels.[17][18]

Next, we investigate the excited resonance of ^{10}He by using the ACCC. In Fig. 2, we give the calculated complex energy as a function of the potential strength parameter, δ, within a single channel. Here, we show a plot of $E^{55}(\delta)$ ($=E_r - i\Gamma/2$) with the 55 Padé approximation (M=5, N=5), where the energy was calculated with the relation of $E = \frac{(\hbar k)^2}{2M}$. The trajectory of the solid curve with circles is that of two neutrons in the

$p_{1/2}$-orbit, $(0p_{1/2})^2$. We also plot the calculated complex energy of the CSM (crosses: δ=0, 0.02,$\cdots$, 0.18). The extrapolated energy of the ACCC (the solid curve with circles) exhibits a good correspondence with a more accurate value, which was calculated with the CSM. For example, in the case of $\delta = 0.102$, E^{55}=2.05−i0.97 MeV and E ^{CSM} =2.067−i0.925 MeV [11].

The trajectory of the solid curve with squares is that of two neutrons in the $s_{1/2}$-orbit, $(1s_{1/2})^2$. The solution is not given in the case of the CSM, because we cannot solve it due to the divergent property of the complex scaled Gaussian potential, as mentioned above. This is the main reason why we used the ACCC in the present analyses. Considering the ^{8}He+n subsystem, we expect to be $\delta \sim 0.25$, because the s-state of the ^{8}He+n system is observed as a virtual state [19].

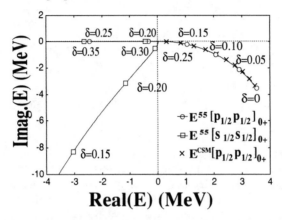

FIGURE 2. Trajectory of the complex energy by changing the potential strength parameter (δ). The trajectory of the solid curve with circles is that of two neutrons in the $p_{1/2}$-orbit. The trajectory of the solid curve with squares is that of two neutrons in the $s_{1/2}$-orbit. The crosses indicate the calculated complex energy for two neutrons in the $p_{1/2}$-orbit with the CSM (δ=0, 0.02,$\cdots$, 0.18).

Finally, in Fig. 3, the observed resonance in ^{10}He [20] (first column) and in ^{11}Li (second column) [17] are given. Also, the calculated 0^+ sates in ^{10}He [21] (third column) and the calculated $3/2^-$ states in ^{11}Li [22] (fourth column) are also given. For ^{10}He, the potential strength, δ=0.25, is employed. The ground state energy is E_r=0.05 MeV and the decay width is Γ=0.21 MeV. This configuration of the ground state of ^{10}He would correspond to the ground state of ^{11}Li. For the excited 0^+ state, we use the CSM because of the accuracy of the solution and the computational time. The calculated resonance energy is E_r=1.68 MeV and the decay width is Γ=1.12 MeV. This excited state of ^{10}He would correspond to our predicted state [22] of ^{11}Li. For ^{10}He, the observed resonance has been analysed to have mainly the component of $(0p_{1/2})^2$. Also, for ^{11}Li, the observed ground state has been analysed to have the main component of $(1s_{1/2})^2$. Thus, an observation of the corresponding state in each nucleus is important in order to consistently understand the binding mechanism of the three-body S-wave.

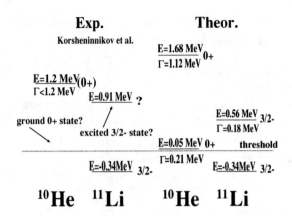

FIGURE 3. Observed resonance in ^{10}He [20] (first column) and the observed states in ^{11}Li [17] (second column). The calculated 0^+ sates in ^{10}He (third column) [21] and the calculated $3/2^-$ states in ^{11}Li [22] (fourth column) are shown.

SUMMARY AND CONCLUSION

In this work, with the core+n+n three-body cluster model, we investigated ^{11}Li and ^{10}He.

For ^{11}Li, we studied the excited $3/2^-$ resonance with the CSM. This state is understood as a pairing excitation of the valence neutrons from the ground state, because the ground state has the large mixed configuration of s- and p-waves. The excited $3/2^-$ state is obtained in the very low energy region $E_x = 0.5$-1 MeV. Since the low-lying state in ^{11}Li observed by Gornov et al., which is suggested as the second $3/2^-$ state, has $E_x = 1.02$ MeV,[4] the newly predicted state may have already been observed. However, the calculated energy has a small ambiguity, because the present model has parameters that should be determined from the experimental data of the subsystems (^{9}Li+n, n+n). Thus, the experimental determination of the p-wave resonant state and the s-wave virtual state in ^{10}Li should make the binding mechanism of ^{11}Li clearer. Furthermore, it is expected that a more realistic pairing interaction increases the energy gap between the $3/2_1^-$ and $3/2_2^-$ states.

For ^{10}He, we studied the ground 0^+ state with the ACCC. The calculated energy of unbound states is very near to the ^{8}He+n+n threshold energy within a reasonable ^{8}He-n potential ($\delta = 0.24 \sim 0.27$). Furthermore, by using any parameter of the potential strength, we could not obtain a ground-state solution which corresponds to the observed resonance energy around $E_r=1.2$ MeV (the calculated maximum energy of the ground state is 0.05 MeV). On the contrary, the solution of the first excited state was obtained at $E_r =1.68$ MeV, where the ^{8}He+n subsystem was also reproduced at the same time. Then, the observed resonance in ^{10}He is considered to have the main component of $(0p_{1/2})^2$, which is the same conclusion as in a previous paper [11]. Therefore, we conclude that the ground state of ^{10}He has not yet been observed. Also, it should exist in the energy region of the ^{8}He+n+n threshold. This conclusion is also supported by an analogy from the neighboring nucleus, ^{11}Li.

ACKNOWLEDGMENTS

This research was supported by the Ministry of Education, Culture, Sports, Science and Technology (Japan), Grant-in-Aid, No.14740142, 2002.

REFERENCES

1. I. J. Thompson and M. V. Zhukov, Phys. Rev. C **49**, 1904 (1994).
2. K. Katō, T. Yamada and K. Ikeda, Prog. Theor. Phys. **101**, 119 (1999).
3. H. Sagawa, B. A. Brown and H. Esbensen, Phys. Lett. B **309**, 1 (1993).
4. M. G. Gornov, Yu. Gurov, S. Lapushkin, P. Morokhov, V. Pechkurov, T. K. Pedlar, K. Seth, J. Wise and D. Zhao, Phys. Rev. Lett. **81** 4325 (1998).
5. J. Aguilar and J. M. Combes, Commun. Math. Phys. **22**, 269 (1971). E. Balslev and J. M. Combes, Commun. Math. Phys. **22**, 280 (1971).
6. V.I. Kukulin and V.M. Krasnopol'sky, J. Phys. **A 10**, 33 (1977), V.I. Kukulin, V.M. Krasnopol'sky and M. Miselkhi, Sov. J. Nucl. Phys. **29**, 421 (1979).
7. V.I. Kukulin, V.M. Krasnopol'sky and Horacek, Theory of Resonances: Principles and Applications (Kluwer Academic Publisher, Dordrecht, Netherlands, 1989), 219.
8. N. Tanaka, Y.Suzuki, K.Varga and R.G. Lavas, Phys. Rev.**C59**, 1391 (1999).
9. K. Ikeda, Nucl. Phys. **A538**, 355c (1992).
10. S. Aoyama, S. Mukai, K. Katō, K.Ikeda, Prog. Theor. Phys. **93**, 99 (1995).
11. S. Aoyama, K. Katō and K. Ikeda, Phys. Rev.**C55**, 2379 (1997).
12. D. R. Thompson, M. LeMere and Y. C. Tang, Nucl. Phys. **A286**, 53 (1977); Y. C. Tang, M. LeMere and D. R. Thompson, Phys. Rep. **47**, 167 (1978).
13. F. Tanabe, A. Tohsaki and R. Tamagaki, Prog. Theor. Phys,**53**, 677 (1975).
14. S. Aoyama, K. Katō and K. Ikeda, Prog. Theor. Phys. Suppl **142**, 35 (2001).
15. E. K. Warburton and B. A. Brown, Phys. Rev. C **46** 923, (1992).
16. S. Karataglidis, P. G. Hansen, B. A. Brown, K. Amos and P. J. Dortmans, Phys. Rev. Lett. **79** 1447, (1997).
17. A.A. Korsheninnikov, E.Yu. Nikolskii, T. Kobayashi, A. Ozawa, S. Fukuda, E.A. Kuzmin, S. Momota, B.G. Novatskii, A.A. Ogloblin, V. Pribora, I.Tanihata, K. Yoshida, Phys. Rev. **C53**, R537 (1996).
18. T. Kobayashi, Nucl.Phys. A **538** 343c, (1992).
19. L. Chen, B. Blank, B.A. Brown, M. Chartier, A. Galonsky, P.G. Hansen, M. Thoennessen, Phys. Lett. **B505**, 21 (2001).
20. A.A. Korsheninnikov, K. Yoshida, D.V. Aleksandrov, N. Aoi, Y. Doki, N. Inabe, M. Fujimaki, T. Kobayashi, H. Kumagai, C.-B. Moon, E.Yu. Nikolskii, M.M. Obuti, A.A. Ogloblin, A. Ozawa, S. Shimoura, T. Suzuki, I. Tanihata, Y. Watanabe and M. Yanokura, Phys. Lett. **B326**, 31 (1994).
21. S. Aoyama, Phys. Rev. Lett. **89**, 052501 (2002).
22. S. Aoyama, K. Katō and K. Ikeda, Prog. Theor. Phys. **107**, 543 (2002).

Polarization of ^{4}He in $_\Lambda^5$He

H. Nemura*, Y. Akaishi* and Y. Suzuki†

*Institute of Particle and Nuclear Studies, KEK, Tsukuba 305-0801, Japan
†Department of Physics, Niigata University, Niigata 950-2181, Japan

Abstract. Variational calculations for s-shell hypernuclei are performed by explicitly including Σ degrees of freedom. Two sets of YN interactions (D2 and SC97e(S)) are used. The bound-state solution of $_\Lambda^5$He is obtained by using each of YN potentials, and a large energy expectation value of the tensor $\Lambda N - \Sigma N$ transition part is found by using the SC97e(S). The internal energy of the ^{4}He subsystem changes a lot by the presence of a Λ particle with the strong tensor $\Lambda N - \Sigma N$ transition potential.

INTRODUCTION

Few-body calculations for s-shell Λ hypernuclei with mass number $A = 3 - 5$ are important not only to explore exotic nuclear structure, including the strangeness degrees of freedom, but also to clarify the characteristic features of the hyperon-nucleon (YN) interaction. Although several interaction models have been proposed[1, 2, 3, 4], the detailed properties (e.g., $^1S_0$ or $^3S_1 - ^3D_1$ phase shift, strength of $\Lambda N - \Sigma N$ coupling term) of the YN interaction are different among the models. The observed separation energies (B_Λ's) of light Λ hypernuclei are expected to provide important information on the YN interaction, because the relative strength of the spin-dependent term or of the $\Lambda N - \Sigma N$ coupling term is affected from system to system.

^{4}He is one of the most tightly bound nuclei in light mass region. It is well-known that the tensor correlation plays an important role for the binding of the ^{4}He. More than a third, or about one half, of the interaction energy comes from the tensor force for the ^{4}He[5, 6, 7]. Considering the fact that the experimental Λ separation energy (B_Λ) of $_\Lambda^5$He is only about 3 MeV and that both of ^{4}He and Λ have isospin $I = 0$, one may think that $_\Lambda^5$He is a weakly coupled system of Λ and α particle, and effects of the tensor ΛN interaction or of the $\Lambda N - \Sigma N$ coupling would be suppressed.

On the other hand, considering the fact that the one-pion-exchange mechanism, which is the origin of the tensor NN interaction in part, induces the $\Lambda N - \Sigma N$ transition for the YN sector, the tensor $\Lambda N - \Sigma N$ transition potential as well as the tensor NN potential could play an important role for the α particle with a Λ particle attached, i.e., $_\Lambda^5$He. If this is the case, it means that the structure of the core nucleus, ^{4}He, would be strongly influenced by the presence of a Λ particle. Therefore, a genuine five-body calculation is crucial for the study of the structural aspects of $_\Lambda^5$He.

If one constructs a phenomenological central ΛN potential, which is consistent with the experimental $B_\Lambda(_\Lambda^3\text{H})$, $B_\Lambda(_\Lambda^4\text{H})$, $B_\Lambda(_\Lambda^4\text{He})$, $B_\Lambda(_\Lambda^4\text{H}^*)$ and $B_\Lambda(_\Lambda^4\text{He}^*)$ values as well as the Λp total cross section, that kind of potential would overestimate the $B_\Lambda(_\Lambda^5\text{He})$

CP644, *Exotic Clustering: 4th Catania Relativistic Ion Studies*, edited by S. Costa, A. Insolia, and C. Tuvè
© 2002 American Institute of Physics 0-7354-0099-7/02/$19.00

value[8, 9]. This is known as an anomalously small binding of $^5_\Lambda$He[10]. In view of the aim to pin down a reliable YN interaction, a systematic study for the complete set of the s-shell Λ hypernuclei is desirable.

The purpose of this study is twofold: First is to describe an *ab initio* calculation of $^5_\Lambda$He as well as $A = 3, 4$ hypernuclei explicitly including Σ degrees of freedom. Second is to conduct a new view of the $^5_\Lambda$He, due to taking account of explicit Σ admixture, beyond $\alpha + \Lambda$ model.

FORMALISM

In the present study, we use the G3RS potential[11] for the NN interaction, and two kinds of YN interactions, D2 and SC97e(S), are used. Both YN interactions include the $\Lambda N - \Sigma N$ coupling. The D2 potential consists of only the central component, while the SC97e(S) has tensor and spin-orbit components in addition to the central one. These YN interactions have Gaussian form factors where parameters are set to reproduce the low energy S matrix of corresponding original Nijmegen YN interactions[12] (model D[1] and soft-core model SC97e[3]).

The trial function is given by a combination of basis functions:

$$\Psi_{JMTM_T} = \sum_{k=1}^{N} c_k \varphi_k, \quad \text{with} \tag{1}$$

$$\varphi_k = A \left[\exp\left\{ -\frac{1}{2} \sum_{i,j=1}^{A-1} (A_k)_{ij} \mathbf{x}_i \cdot \mathbf{x}_j \right\} v_k^{2K_k + L_k} \left[Y_{L_k}(\hat{\mathbf{v}}_k) \times \chi_{S_k} \right]_{JM} \eta_{kTM_T} \right]. \tag{2}$$

Here, A is an antisymmetrizer acting on nucleons, and $\chi_{S_k} \left(\eta_{kTM_T} \right)$ is the spin (isospin) function. The abbreviation $\mathbf{x} = (\mathbf{x}_1, \cdots, \mathbf{x}_{A-1})$ is a set of relative coordinates. A set of linear variational parameters $(c_1, \cdots, c_N)$ is determined by the Ritz variational principle. The angular momentum part is expressed by the global vector representation, and the global vector, $\mathbf{v}_k$, is given by a linear combination of the relative coordinates:

$$\mathbf{v}_k = \sum_{i=1}^{A-1} (u_k)_i \mathbf{x}_i. \tag{3}$$

The A_k and u_k are sets of nonlinear parameters which characterize the spatial part of the basis function. Allowing the factor $v_k^{2K_k}$ ($K_k \neq 0$) is useful to improve the short-range behavior of the trial function. The value of K_k is assumed to take 0 or 1. The variational parameters are optimized by a stochastic procedure. Recently, we performed a five-body calculation of $^5_\Lambda$He[13] by using the stochastic variational method[6]. The reader is referred to Refs.[6, 13] for details of the calculation.

RESULTS

Table 1 lists the results of the Λ-separation energies. Both D2 and SC97e(S) YN interactions reasonably well reproduce the experimental B_Λ values for the $A = 3, 4$ hypernuclei. Particularly, the order of the spin doublet structure of the $A = 4$ system is correctly reproduced; the ground (excited) state has spin-parity, $J^\pi = 0^+(1^+)$ for both isodoublet hypernuclei $^4_\Lambda$H and $^4_\Lambda$He. Although the energy difference of the spin doublet structure is well described by each of these YN potentials, the property of the spin-dependent part for the YN potentials is different from each other. For example, the scattering lengths of the D2 potential in spin singlet (triplet) is $a_s = -1.99$ fm ($a_t = -2.09$ fm). On the other hand, the scattering lengths of the SC97e(S) is $a_s = -2.37$ fm ($a_t = -1.83$ fm).

Though the calculated $B_\Lambda(^5_\Lambda$He) is slightly larger or smaller than the experimental value, the discrepancies are rather small in view of the anomalous binding problem [10]. This is a *first ab initio* calculation to produce the bound state of $^5_\Lambda$He with explicit Σ degrees of freedom. The present result implies that the key to resolve the anomalous binding problem is to include explicitly Σ degrees of freedom.

Table 2 lists the probability, P_Σ (in percentage), of finding a Σ particle in the system. Relatively larger P_Σ values are found for the $^4_\Lambda$H and the $^4_\Lambda$He. These results are consistent with the concept of the coherent $\Lambda - \Sigma$ coupling[12]. For the excited state of $A = 4$ system or the ground state of $A = 5$ system, the P_Σ value obtained by the SC97e(S) is larger than that obtained by the D2. Particularly, the sizable amount of $P_\Sigma(^5_\Lambda$He) is obtained with the SC97e(S). This implies that the $\Lambda - \Sigma$ coupling plays an important role, even for the $^5_\Lambda$He, despite a large excitation energy of the core nucleus, ^{4}He (with the isospin 1), in the Σ-component.

Figure 1 displays the energy expectation values of the kinetic and potential energy terms for $^5_\Lambda$He. The contributions from the spin-orbit and the Coulomb potentials are not included in the figure, though the calculations include them. The most left-side step (T_c) is the energy expectation value of the kinetic energy of the core nucleus (c) subtracted by the center-of-mass (CM) energy of c. The T_c is defined by

$$T_c = \sum_{i=1}^{A-1} \frac{\mathbf{p}_i^2}{2m_N} - \frac{\left(\sum_{i=1}^{A-1} \mathbf{p}_i\right)^2}{2(A-1)m_N}. \tag{4}$$

The next three steps take the spin-singlet central, the spin-triplet central, and the tensor parts of the NN interaction (V_{NN}) into account successively. The V_{NN} takes account of a summation over appropriate particle pair. The internal energy of the core nucleus is given by $T_c + V_{NN}$. The solid line is the case of no presence of Λ particle, and the energy level at $E \approx -28$ MeV gives the energy of ^{4}He $+ \Lambda$ threshold. The energy expectation

TABLE 1. Λ separation energies, given in units of MeV, of $A = 3 - 5$ Λ-hypernuclei.

	$B_\Lambda(^3_\Lambda$H)	$B_\Lambda(^4_\Lambda$H)	$B_\Lambda(^4_\Lambda$He)	$B_\Lambda(^4_\Lambda$H*)	$B_\Lambda(^4_\Lambda$He*)	$B_\Lambda(^5_\Lambda$He)
D2	0.05	2.16	2.11	1.17	1.15	3.39
SC97e(S)	0.10	2.06	2.02	0.92	0.90	2.75
Experiment	0.13(5)	2.04(4)	2.39(3)	1.00(4)	1.24(4)	3.12(2)

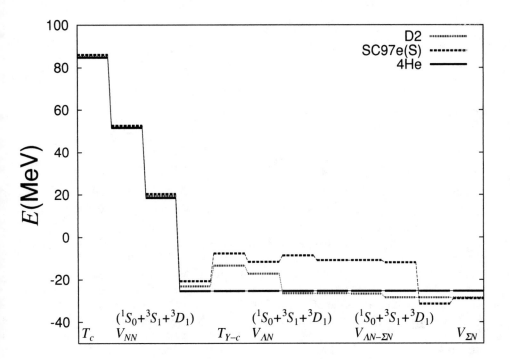

FIGURE 1. Energies of $^5_\Lambda$He.

value obtained with the SC97e(S) (D2) is shown by broken (dotted) line. The presence of a Λ particle reduces the internal energy of the ^{4}He subsystem:

$$\Delta E_c = \left(\langle T_c \rangle + \langle V_{NN} \rangle\right)_{^5_\Lambda \text{He}} - \left(\langle T_c \rangle + \langle V_{NN} \rangle\right)_{^4\text{He}}$$

$$\approx \begin{cases} 2.3\text{MeV} & \text{(D2)} \\ 4.7\text{MeV} & \text{(SC97e(S))}. \end{cases} \tag{5}$$

The change of the energy, ΔE_c, for the SC97e(S) is considerably large (it is about two times bigger than that for the D2), despite the fact that the rms radius of N from the CM of ^{4}He for the $^5_\Lambda$He hardly changes from that for ^{4}He. (Both radii are 1.5 fm.) The difference of ΔE_c between SC97e(S) and D2 is mainly due to the difference of

TABLE 2. Probabilities of finding a Σ particle, given in percentage, of $A = 3 - 5$ Λ-hypernuclei.

	$P_\Sigma(^3_\Lambda\text{H})$	$P_\Sigma(^4_\Lambda\text{H})$	$P_\Sigma(^4_\Lambda\text{He})$	$P_\Sigma(^4_\Lambda\text{H}^*)$	$P_\Sigma(^4_\Lambda\text{He}^*)$	$P_\Sigma(^5_\Lambda\text{He})$
D2	0.14	1.93	1.85	0.42	0.42	0.61
SC97e(S)	0.15	1.49	1.45	0.98	0.96	1.55

the reduction of the tensor NN interaction,

$$\left(\langle V_{NN}(\text{tensor})\rangle\right)_{^5_\Lambda\text{He}} - \left(\langle V_{NN}(\text{tensor})\rangle\right)_{^4\text{He}} \approx \begin{cases} 1.5\text{MeV} & \text{(D2)} \\ 2.9\text{MeV} & \text{(SC97e(S))}. \end{cases}$$

The next step of the energy levels, T_{Y-c}, includes the kinetic energy of the relative motion between the Y and the CM of c, which is given by

$$T_{Y-c} = \frac{\pi^2_{Y-c}}{2\mu_Y} + (m_Y - m_\Lambda)c^2, \tag{6}$$

where $\mu_Y = \frac{(A-1)m_N m_Y}{(A-1)m_N + m_Y}$ is the reduced mass for the $Y + c$ system and π_{Y-c} is the canonical momentum of the relative coordinate between Y and c $(Y = \Lambda, \Sigma)$. T_{Y-c} also counts the difference in the rest-mass energy between Λ and Σ.

To include the spin-triplet (central) part of the D2 $\Lambda N - \Lambda N$ diagonal potential, the energy level reaches across the $^4\text{He} + \Lambda$ threshold. Therefore, for the $^5_\Lambda\text{He}$, the main part of the D2 YN potential is the (spin triplet) $\Lambda N - \Lambda N$ diagonal component, and the $\Lambda N - \Sigma N$ transition part gives small attractive energy. On the other hand, the $\Lambda N - \Lambda N$ diagonal part of the SC97e(S) is rather repulsive. The energy levels including $\Lambda N - \Lambda N$ diagonal part of the SC97e(S) are extremely (more than 10 MeV) higher than the $^4\text{He} + \Lambda$ threshold. The tensor $\Lambda N - \Sigma N$ transition part brings a surprisingly large attractive energy (about -20 MeV). This large coupling energy makes $^5_\Lambda\text{He}$ bound for the SC97e(S).

The calculated wave function is divided into orthogonal components according to the total orbital angular momentum (L), the total spin (S), the core nucleus spin (S_c) and the core nucleus isospin (I_c). Table 3 displays the probability of each component for $^5_\Lambda\text{He}$. The table also lists the probability of S-state or of D-state for ^4He. The sizable amount of probability of the Σ-component is found in the D-state by using the SC97e(S), while the Σ probability with the D2 potential appears in the S-state and is rather small. Moreover, the sum of D-state probabilities in the Λ-component with the SC97e(S) is slightly

TABLE 3. Probability, given in percentage, of each component with the total orbital angular momentum (L), total spin (S), core nucleus spin (S_c) and core nucleus isospin (I_c) in Λ- or in Σ-component for $^5_\Lambda\text{He}$. The probability in S- or in D-state for ^4He is also listed.

| | $L=0$ | | $L=2$ | | |
| | $S=\frac{1}{2}$ | | $S=\frac{3}{2}$ | | $S=\frac{5}{2}$ |
	$S_c=0$	$S_c=1$	$S_c=1$	$S_c=2$	$S_c=2$
$^5_\Lambda\text{He}$ $(YN:\text{D2})$					
$(I_c=0)\otimes\Lambda$	89.54	0.03	~ 0	3.90	5.91
$(I_c=1)\otimes\Sigma$	0.27	0.31	0.02	~ 0	0.02
$^5_\Lambda\text{He}$ $(YN:\text{SC97e(S)})$					
$(I_c=0)\otimes\Lambda$	89.14	0.03	0.19	3.74	5.36
$(I_c=1)\otimes\Sigma$	0.10	0.09	1.34	~ 0	0.01
^4He	89.56			10.44	

smaller than that with the D2 or the D-state probability for ^{4}He. Therefore, though the presence of a Λ in ^{4}He with the strong tensor $\Lambda N - \Sigma N$ transition potential (e.g., SC97e(S)) influences the structure of the D-state component and reduces the energy expectation value of the tensor NN interaction, the large coupling energy $\langle V_{\Lambda N - \Sigma N} \rangle$ of the tensor part bears the bound state of $^5_\Lambda$He instead.

SUMMARY

We have made a systematic study of all s-shell hypernuclei based on ab initio calculations using YN interactions with an explicit Σ admixture. Both YN interactions, D2 and SC97e(S), reasonably well reproduce the experimental separation energies for $A = 3, 4$ systems. Regarding the anomalous binding problem of $^5_\Lambda$He, the present calculation convinces us that one of the keys to resolve the problem is to include the explicit Σ degrees of freedom. The SC97e(S) seems to be more favorable YN interaction than the D2, since it is natural to include the tensor component in the $\Lambda N - \Sigma N$ transition. A sizable amount of P_Σ was obtained for $^5_\Lambda$He by using the SC97e(S). The contribution of the energy from the tensor $\Lambda N - \Sigma N$ coupling is quite large, and this coupling is considerably important to make $^5_\Lambda$He bound. The present study for $^5_\Lambda$He is a first step toward a detailed description of light strange nuclear systems. The core nucleus, ^{4}He, is no longer rigid in interacting with a Λ particle. A similar situation can occur for strangeness $S = -2$ systems. Investigations into the strength of the $\Lambda\Lambda$ interaction based on the experimental data of the binding energy for double Λ hypernuclei (e.g., $^{\;\;6}_{\Lambda\Lambda}$He [14]) should take account of the energy reduction of the core nucleus (ΔE_c for $^{\;\;6}_{\Lambda\Lambda}$He is expected to be larger than the present ΔE_c for $^5_\Lambda$He).

REFERENCES

1. Nagels, M. M., Rijken, T. A., and de Swart, J. J., *Phys. Rev.* D15, 2547-2564 (1977); *ibid.* D20, 1633-1645 (1979).
2. Maessen, P. M. M., Rijken, Th. A., and de Swart, J. J., *Phys. Rev.* C 40, 2226-2245 (1989).
3. Rijken, Th. A., Stokes, V. G. J., and Yamamoto, Y., *Phys. Rev.* C 59, 21-40 (1999); Stokes, V. G. J., and Rijken, Th. A., *ibid.* C 59, 3009-3020 (1999).
4. Fujiwara, Y., Nakamoto, C., and Suzuki, Y., *Phys. Rev. Lett.* 76, 2242-2245 (1996); *Phys. Rev.* C 54, 2180-2200 (1996).
5. Akaishi, Y., "Few-Body System in Realistic Interaction," in *Cluster Models and Other Topics*, edited by T. T. S. Kuo and E. Osnes World Scientific, Singapore, 1986, pp.259-393.
6. Suzuki, Y., and Varga, K., *Stochastic Variational Approach to Quantum-Mechanical Few-Body Problems*, Lecture Notes in Physics, Vol. m54, Springer-Verlag, Berlin Heidelberg, 1998.
7. Pieper, S. C., *et al.*, *Phys. Rev.* C 64, 014001 (2001).
8. Dalitz, R. H., Herndon, R. C., and Tang, Y. C., *Nucl. Phys.* B47, 109-137 (1972).
9. Nemura, H., Suzuki, Y., Fujiwara, Y., and Nakamoto, C., *Prog. Theor. Phys.* 103, 929-958 (2000).
10. Gibson, B. F., and Hungerford III, E. V., *Phys. Rep.* 257, 349-388 (1995).
11. Tamagaki, R., *Prog. Theor. Phys.* 39, 91-107 (1968).
12. Akaishi, Y., Harada, T., Shinmura, S., and Khin Swe Myint, *Phys. Rev. Lett.* 84, 3539-3541 (2000).
13. Nemura, H., Akaishi, Y., and Suzuki, Y., nucl-th/0203013, *Phys. Rev. Lett.* (to be published).
14. Takahashi, H., *et al.*, *Phys. Rev. Lett.* 87, 212502 (2001).

Exotic Cluster Structure of Light Neutron-rich Nuclei

N. Itagaki*, T. Otsuka*, S. Okabe† and K. Ikeda**

*Department of physics, University of Tokyo, Hongo, Tokyo 113-0033, Japan
†Center for Information and Multimedia Studies, Hokkaido University, Sapporo 060-0810, Japan
**The Institute of Physical and Chemical Research (RIKEN), Wako, Saitama 351-0198, Japan

Abstract. The molecule-like structure of the C isotopes is investigated using a microscopic $\alpha+\alpha+\alpha+n+n+\cdots$ model. The valence neutrons are classified based on the molecular-orbit (MO) model, and both π-orbit and σ-orbit are introduced around three α-clusters. The excited states of ^{16}C is one of the most promising candidates for the linear-chain structure. The equilateral-triangular shape of 3α surrounded by valence neutrons is also suggested for ^{14}C. The cluster around the ^{10}Be+α threshold energy corresponds to the experimentally observed band built on top of the second 3^- state.

INTRODUCTION

It has been a long-standing dream of the nuclear structure physics to find an linear-chain configuration and equilateral-triangular shape. Recently, various cluster structures have been studied extensively, and especially, α-cluster structure is promising for such an exotic shape. For example, fragments of He isotopes (^{4}He, ^{6}He, ^{8}He) have been observed from the excited states of ^{10}Be[1] and ^{12}Be[2, 3], and the presence of molecular structure of $\alpha+\alpha$ surrounded by valence neutrons is suggested. On the theoretical side, these states are studied by molecular-orbital (MO) model and the nature of weakly bound neutrons around the two α-clusters has been manifested [4, 5]. Also, this α-α clustering is related to the lowering of a $1/2^+$ neutron orbit in energy, and is an important mechanism to enhance the disappearance of the $N = 8$ magic number in the Be isotopes[6].

Recently, the discussions of the well-developed cluster structure are extended to the neutron-rich nuclei, and the role of valence neutrons which stabilize the linear-chain structure has been pointed out. For example, von Oertzen has extended his analyses for the molecular structure in Be isotopes [7] to C isotopes, and the linear-chain state consisting of 3α and valence neurons around it has been speculated. Even if the 3α-system without valence neutrons (^{12}C) does not have a linear-chain structure, the valence neutrons around it are expected to increase the binding energy and stabilize the linear-chain state[8].

CP644, *Exotic Clustering: 4th Catania Relativistic Ion Studies*, edited by S. Costa, A. Insolia, and C. Tuvè
© 2002 American Institute of Physics 0-7354-0099-7/02/$19.00

FRAMEWORK

A microscopic $\alpha+\alpha+\alpha+n+n \cdots$ model is introduced: the total wave function is fully antisymmetrized and is given by a superposition of basis states (with coefficients $\{c_i\}$) with different relative distances among the α-clusters and various configurations of the valence neutrons around the α-clusters:

$$\Phi(^{14}\text{C}) = \sum_i c_i P^J_{MK} \mathscr{A} [\phi_1^{(\alpha)} \phi_2^{(\alpha)} \phi_3^{(\alpha)} (\psi_1^{(n)}\chi_1)(\psi_2^{(n)}\chi_2)\cdots]. \tag{1}$$

The projection onto a good angular momentum is performed by P^J_{MK}, and the coefficients $\{c_i\}$ are obtained after this projection by diagonalizing Hamiltonian matrix. Each of the three α-clusters ($\phi_j^{(\alpha)}$, $j = 1,2,3$) consists of two protons and two neutrons:

$$\phi_j^{(\alpha)} = G^{p\uparrow}_{\vec{R}_{\alpha j}} G^{p\downarrow}_{\vec{R}_{\alpha j}} G^{n\uparrow}_{\vec{R}_{\alpha j}} G^{n\downarrow}_{\vec{R}_{\alpha j}} \chi_{p\uparrow}\chi_{p\downarrow}\chi_{n\uparrow}\chi_{n\downarrow}, \tag{2}$$

$$G_{\vec{R}_{\alpha j}} = \left(\frac{2v}{\pi}\right)^{\frac{3}{4}} \exp[-v(\vec{r} - \vec{R}_{\alpha j})^2]. \tag{3}$$

Here, χ represents the spin-isospin eigen function, and the oscillator parameter ($\beta = \frac{1}{\sqrt{2v}}$) is set equal to 1.46 fm. The wave function of each valence neutron ($\psi_k^{(n)}\chi_k$ $k = 1,2$) is expressed by a linear combination of local Gaussians $\{G_{\vec{R}_l}\}$ with coefficients $\{d_l\}$,

$$\psi_k^{(n)}\chi_k = \sum_l d_l^k G_{\vec{R}_l} \chi_k, \tag{4}$$

here, $G_{\vec{R}_l}$ represents a Gaussian wave function centered at $\vec{R}_l$. We introduce π-orbit and σ-orbit for the valence neutrons shown in Refs.[5, 6].

The adopted effective nucleon-nucleon interaction is the Volkov No.2 [9] with the exchange parameters $M = 0.6$ ($W = 0.4$), $B = H = 0.125$ for the central part, and the G3RS spin-orbit term [10] for the spin-orbit part as in Refs. [5, 6]. All the parameters of this interaction were determined from the $\alpha + n$ and $\alpha + \alpha$ scattering phase shifts and the binding energy of the deuteron[11].

RESULT

We show the calculated results for the stability of the linear-chain state of α-clusters for various neutron configurations. The isotopes and configurations which we take into account are ^{12}C, ^{14}C$(3/2_\pi^-)^2$ (two n's in the π-orbits), ^{14}C$(1/2_\sigma^-)^2$ (two n's in the σ-orbits), ^{16}C$((3/2_\pi^-)^2(1/2_\pi^-)^2)$ (four n's in the π-orbits) and ^{16}C$((3/2_\pi^-)^2(1/2_\sigma^-)^2)$ (two n's in the π-orbits and two n's in the σ-orbits). Two variational paths are introduced corresponding to the breathing-like and the bending-like degrees of freedom. The parameters d and θ stand for the α-α distance and the bending angle of the 3α-core, respectively.

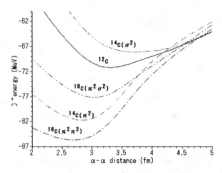

FIGURE 1. The 0^+ energy curves with respect to the α-α distance (d) for ^{12}C (solid curve), ^{14}C$((3/2_\pi^-)^2)$ (dashed curve), ^{14}C$((1/2_\sigma^-)^2)$ (dotted curve), ^{16}C$((3/2_\pi^-)^2(1/2_\pi^-)^2)$ (dash dotted curve), and ^{16}C$((3/2_\pi^-)^2(1/2_\sigma^-)^2)$ (dash two-dotted curve). The coefficients of local Gaussians $\{g_j\}$ describing the valence neutrons are treated as variational parameters to take into account deviations of the original MO.

Firstly, we show the 0^+ energy curves for the linear-chain structure with respect to the breathing-path in Fig. 1. It is found that the energy pocket around $d = 3$ fm becomes deeper as the increase of number of valence neutrons in the π-orbit (^{12}C $\rightarrow$ ^{14}C$(3/2_\pi^-)^2$ $\rightarrow$ ^{16}C$((3/2_\pi^-)^2(1/2_\pi^-)^2)$). The 3α-system (^{12}C) has minimal energy around $d = 3.5$ fm, however, this is too shallow to conclude the stability of the linear-chain state. On the contrary, in ^{14}C$(3/2_\pi^-)^2$, there appears evident minimal energy around $d = 3$ fm. The energy (~ -82 MeV) is lower than ^{12}C by 11 MeV and the energy pocket is much deeper. This energy corresponds to the excitation energy of 18 MeV from the ground state calculated with an equilateral-triangle configuration for the 3α-core. ^{16}C$((3/2_\pi^-)^2(1/2_\pi^-)^2)$ is most stable among these states and it has an energy pocket of ~ -86 MeV, and corresponding α-α distance is $d = 2.5$ fm, shorter than ^{12}C and ^{14}C$(3/2_\pi^-)^2$. Therefore, the π-orbit is found to stabilize the linear-chain structure as the increase of valence neutrons and to prevent a breathing-like break-up of the system.

The stability of these linear-chain states with respect to the bending-like path is examined. The 0^+ energy curves of ^{12}C, ^{14}C$((3/2_\pi^-)^2)$, ^{14}C$((1/2_\sigma^-)^2)$, ^{16}C$((3/2_\pi^-)^2(1/2_\pi^-)^2)$, and ^{16}C$((3/2_\pi^-)^2(1/2_\sigma^-)^2)$ with respect to the θ-value are shown in Fig. 2. Except for the case of ^{16}C$((3/2_\pi^-)^2(1/2_\sigma^-)^2)$, the curvature of these states is rather monotonic and the energy minimum does not clearly appear. This feature is much different from ^{12}C, ^{14}C$((3/2_\pi^-)^2)$, ^{14}C$((1/2_\sigma^-)^2)$, and ^{16}C$((3/2_\pi^-)^2(1/2_\pi^-)^2)$ cases.

Here, ^{16}C$((3/2_\pi^-)^2(1/2_\sigma^-)^2)$ is found to be only the case which is stable with respect to both the breathing and the bending-like path. The 0^+ energy increases by 15.7 MeV from $\theta = 0^o$ to $\theta = 30^o$, in which the kinetic energy part is 10.3 MeV. To understand the energy increase with the increase of bending angle θ of this case, we calculate and compare the overlap between the wave functions with $\theta = 0^o$ and $\theta = 30^o$ for various configurations. In ^{12}C, the wave functions with $\theta = 0^o$ and $\theta = 30^o$ have the squared overlap of 0.91, and ^{14}C$((3/2_\pi^-)^2)$ has almost the same value. ^{14}C$((3/2_\sigma^-)^2)$ has the value of 0.85, smaller than ^{14}C$((3/2_\pi^-)^2)$ by only 6 %, and ^{16}C$((3/2_\pi^-)^2(1/2_\pi^-)^2)$ has almost the same value as the ^{14}C$((1/2_\sigma^-)^2)$ case. This result shows that the overlaps

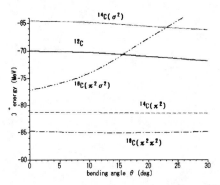

FIGURE 2. The 0^+ energy curves with respect to the bending angle (θ) for ^{12}C (solid curve), ^{14}C$((3/2_\pi^-)^2)$ (dashed curve), ^{14}C$((3/2_\sigma^-)^2)$ (dotted curve), ^{16}C$((3/2_\pi^-)^2(1/2_\pi^-)^2)$ (dash dotted curve), and ^{16}C$((3/2_\pi^-)^2(1/2_\sigma^-)^2)$ (dash two-dotted curve). The coefficients for the linear combination of Gaussians describing MO are optimized at $\theta = 0^o$ and fixed. The α-α distance (d) is fixed to 3 fm.

additionally decreases a little for the σ-orbital neutrons, and also for the π-orbital neutrons as the increase of the valence neutrons. In spite of these, the overlap between $\theta = 0^o$ and $\theta = 30^o$ for ^{16}C$((3/2_\pi^-)^2(1/2_\sigma^-)^2)$ case shows a significantly large decrease to 0.60. ^{16}C$((3/2_\pi^-)^2(1/2_\sigma^-)^2)$ is only the configuration which shows drastic decrease of the overlap between $\theta = 0^o$ and $\theta = 30^o$. It can be known that the drastic decrease as the increase of the bending angle is due to the increase of overlap between two neutrons in the π-orbit and two neutrons in the σ- orbit. When there arises the overlap between them, the overlap component in the total wave function is diminished due to Pauli exclusion principle, that is, so-called Pauli blocking.

Next, we discuss the equilateral-triangular configuration in ^{14}C using $\alpha+\alpha+\alpha+n+n$ model. For the two valence neutrons, we introduce two kinds of the basis states: shell-model-like and molecular-orbital basis states. In the shell-model-like basis state, the valence neutrons are described as shell-model orbits around the center of mass of the 3α. The 3α-cluster ($\{\phi_j^{(\alpha)}\}$ $j = 1,2,3$) is introduced on the xy-plane with the equilateral-triangular shape, and the valence neutrons are introduced as the p-orbits in the z-direction (ψ_z, $\psi_z \propto ze^{-vr^2}$), since these orbits are not Pauli-blocked by the 3α-core located on the xy-plane. For the molecular-orbital basis state, the model space is $\alpha+^{10}$Be$(\alpha+\alpha+2n)$, and the 3α-core is assumed to form an equilateral-triangular configuration. Because of the antisymmetrization and angular momentum projection, two valence neutrons between two α's are placed between other pairs of α's.

The energy levels of the negative parity states of ^{14}C are presented in Fig. 3. Here, we superpose both shell-model-like and molecular-orbital basis basis states with various distance between α-clusters. The energy is measured from the ^{10}Be$+\alpha$ threshold energy (calculated as -88.1 MeV), which corresponds to $E_x = 12.01$ MeV, experimentally. There appear two 3^- states below the threshold energy. The first one mainly has the $K = 3$ component which reflects the equilateral triangular symmetry, and the obtained spectra well fit into the $J(J+1)$ rule. The rotational band structure is shown in the 4-th column. The absolute energy of the two 3^- states are -94.8 MeV and -91.0 MeV. The

65

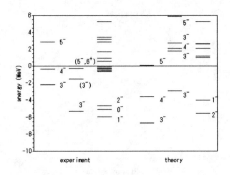

FIGURE 3. The energy levels of ^{14}C (negative parity states). The energy is measured from the ^{10}Be+α threshold energy.

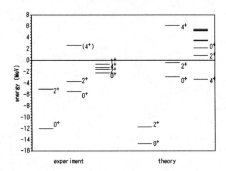

FIGURE 4. The energy levels of ^{14}C (positive parity states). The energy is measured from the ^{10}Be+α threshold energy.

levels classified in the 5-th column are other 3^-, 4^-, and 5^- states calculated.

Furthermore, it is significant that these states are experimentally observed around the ^{10}Be+α threshold energy as shown in the 1st column. The situation is much different from ^{12}C. In ^{12}C, 3^- is already above the ^{8}Be+α threshold by 2 MeV and 4^- and 5^- have not been observed. On the contrary, in ^{14}C, these states are calculated below the threshold, and candidates have been observed. Experimentally, the candidate for the band head of this cluster rotational band structure is the second 3^- state. Unfortunately, in our calculation, this band head corresponds to the first 3^- state. To improve this discrepancy, it would be necessary to incorporate the α-breaking effect to get the correct energy for this more shell-model-like 3^- state and make the state lower than the 3^- state of the cluster-band.

The energy levels of the positive parity states with respect to the ^{10}Be+α threshold energy are presented in Fig. 4. The calculated ground and second 0^+ energies are -102.8 MeV and -91.0 MeV, respectively. It is shown that three states (0_2^+, 2_2^+, 4^+ state at 6.12 MeV) around the ^{10}Be+α threshold energy fit into the $J(J+1)$ rule, which are plotted in the 5-th column. In this energy region, corresponding states are also experimentally observed, as shown in the 2nd column. The calculated electro-magnetic transition probabilities prove that these states are members of a rotational band. The B(E2: $2_2^+ \rightarrow 4^+$ (6.12 MeV)) and the B(E2: $0_2^+ \rightarrow 2_2^+$) values are calculated as 16.4 e^2fm^4 and 39.1 e^2fm^4, respectively, and this ratio of 0.42 almost agrees with the well known value of 0.51 for a rigid-rotor. This result suggests that the positive-parity rotational band with the cluster structure also appears around the ^{10}Be+α threshold energy.

In Fig. 4, the calculated ground 0^+ level and the first 2^+ level are presented in the 4-th column. The energy spacing between them is calculated to be too small (3.0 MeV), in comparison with the observed value of 7.01 MeV (1st column). To improve the level spacing, it would be necessary to take into account other effects, such as α-breaking effect. For example, it has been shown in Refs. [12, 13] that in ^{12}C, the coupling effect between the 3α state and the α-breaking state increases the level spacing between 0^+ and 2^+. This is because, the spin-orbit interaction strongly acts for the ground 0^+ state by taking into account the component of the closed $p_{3/2}$ sub-shell for the protons and the neutrons. Similar mechanism is expected to work also in ^{14}C.

We have shown that both of the positive-parity (band head is 0^+) and the negative-parity rotational band (band head is 3^-) appear around the ^{10}Be+α threshold energy. Although the neutron configurations are slightly different, these two rotational bands are considered to be the inversion doublet structure.

The inversion doublet structure, which is one of the most strong evidence for the presence of the cluster structure, has been first proposed by Horiuchi and Ikeda for ^{16}O (^{12}C+α) and ^{20}Ne (^{16}O+α)[14]. If there is cluster structure, the normal parity and opposite parity (higher nodal) rotational bands appear close in energy. In the present case, the positive-parity and the negative-parity rotational bands appear around the ^{10}Be+α threshold energy, and corresponding states are experimentally observed. These results mean that the picture of inversion doublet structure also can be seen in systems with 3 clusters and valence neutrons. The band head energy is below the α-threshold contrary to the ^{12}C case, and this may suggest a new mechanism for the appearance of cluster structure in neutron-rich nuclei in addition to the so-called threshold rule.

REFERENCES

1. N. Soić *et al*, Eurohys. Lett, 34(1), 7 (1996).
2. A. A. Korscheninnikov *et al*. Phys. Lett. B **343**, 53 (1995).
3. M. Freer *et al*., Phys. Rev. Lett. **82**, 1383 (1999)
4. Y. Kanada-En'yo, H. Horiuchi and A.Doté, Phys. Rev. C**60**, 064304 (1999).
5. N. Itagaki and S. Okabe, Phys. Rev. C **61**, 044306 (2000).
6. N. Itagaki, S. Okabe, and K. Ikeda, Phys. Rev. C **62**, 034301 (2000).
7. W. von Oertzen, Z. Phys. A**354**, 37 (1996); A357, **355** (1997).
8. N. Itagaki, S. Okabe, I. Ikeda, and I. Tanihata, Phys. Rev. C **64**, 014301 (2001).
9. A.B. Volkov Nucl. Phys. 74, 33 (1965).
10. N. Yamaguchi, T. Kasahara, S. Nagata and Y. Akaishi, Prog. Theor. Phys. 62, 1018 (1979).
11. S. Okabe and Y. Abe, Prog. Theor. Phys. 61, 1049 (1979).
12. Y. Kanada-En'yo, Phys. Rev. Lett. **81**, 5291 (1998).
13. S. Okabe and N. Itagaki, *Proc. of the XVII RCNP Int. Symp.on Innovative Computational Methods in Nuclear Many-body Problems, Osaka*, 120 (1998).
14. H. Horiuchi and K. Ikeda, Prog. Theor. Phys. **40**, 277 (1968).

Study of the neutron-rich molecular resonances

Makoto Ito[*†], Kiyoshi Kato[*] and Kiyomi Ikeda[†]

*Divison of Physics, Grad. School of Sci., Hokkaido Univ., 060-0810 Sapporo, Japan
†RI Beam Science Laboratory, RIKEN(The institute of Physical and chemical Research), Wako,
Saitama 351-0198, Japan

Abstract. New cluster model based on the Generator Coordinate Method (GCM) is proposed for the studies on the molecular resonances recently observed in the neutron-rich Be isotope. The new model is applied to the ^{10}Be=^{4}He+^{6}He system and its adiabatic energy surface is calculated. It is found that both the weak-coupling states and the strong-coupling ones are quite naturally generated depending on the ^{4}He–^{6}He relative distance. The stability of both states are also discussed.

INTRODUCTION

Recently, the pronounced resonant structures have been observed through the breakup reactions of the neutron-rich ^{12}Be nucleus into the ^{6}He+^{6}He and ^{4}He+^{8}He decay channels [1]. The observed resonance states are distributed in the 10 to 25 MeV excitation energy interval, with spins in the range of $4\hbar$ to $8\hbar$ [1]. According to the energy-spin characteristics of the observed ^{6}He+^{6}He breakup states, the moment of inertia, $\hbar^2/2\mathscr{I}$, of the observed resonances is approximately 0.15 ($\pm$0.04) MeV, which is much smaller than that of spherical ^{12}Be (0.36 MeV) and almost the same as that of two touching ^{6}He nuclei (0.165 MeV). Therefore, this result strongly indicates that the breakup into ^{6}He+^{6}He occurs primarily through a deformed "^{6}He+^{6}He molecule-like states", in which two ^{6}He nuclei keep their identities and contact each other.

In not only the ^{12}Be nucleus but also other Be isotope such as ^{10}Be and ^{14}Be, similar molecule-like states with the ^{4}He+^{6}He and ^{6}He+^{8}He configurations have also been observed in recent experiments [1, 2]. The ^{6}He and ^{8}He nuclei themselves are known as soft nuclei with a neutron halo (or skin) around the α core due to their small neutrons separation energy. Therefore, it is very interesting that the observed resonance states can be interpreted in terms of the ^{6}He and ^{8}He cluster states or the so-called "weak-coupling states" of these clusters, in which individual clusters keep their identities and they are weakly coupled to each other.

On the other hand, the recent studies suggest that the low-lying states of the Be isotope have much complicated structure compared with those of the normal $N \sim Z$ nuclei. In such a low-lying states, the α–α clustering is well developed, which leads to the molecular orbital formation of the valence neutrons [3]. In contrast to the weak-coupling states, the molecular-orbital states correspond to the so-called "strong-coupling states" having a certain kind of the intrinsic configuration.

The orbitals of the valence neutrons are drastically changed from the ground state to the excited one if the ^{6}He or ^{8}He cluster correlation will strongly occur in the

CP644, Exotic Clustering: 4th Catania Relativistic Ion Studies, edited by S. Costa, A. Insolia, and C. Tuvè
© 2002 American Institute of Physics 0-7354-0099-7/02/$19.00

observed resonance states. The weak-coupling correlation will be generated at a large $\alpha-\alpha$ distance compared with that in the low-lying states. Therefore, we need to consider the new cluster model which can cover the weak-coupling states at a large $\alpha-\alpha$ distance and the strong-coupling states at a small $\alpha-\alpha$ distance. In the present study, therefore, we propose new cluster model based on the Generator Coordinate Method (GCM) [4] and discuss its application to the simple neutron-rich system, $^{10}Be=^4He+^6He$.

FRAMEWORK

Coupled-Channel Generator Coordinate Method. In this section, we briefly describe the framework of new cluster model based on the coupled-channel Generator Coordinate Method (CC-GCM). Here, we consider the $\alpha+^6He$ system as a simple example. In this model, the GCM wave function of the total system $\Psi_R^{JK\pi}(^{10}Be)$, which localized at a certain distance $\mathbf{R}$, is expressed in terms of the linear combination of various channel wave functions,

$$\Psi_R^{JK\pi}(^{10}Be) = \sum_\beta C_\beta \psi_{\beta R}^{JK\pi}(\alpha+^6He) \ , \tag{1}$$

where $\psi_{\beta R}^{JK\pi}(\alpha+^6He)$ denotes the channel wave function in which the ^6He nucleus is in the internal state "β". The channel wave function is an eigenstate of the total angular momentum (J) and parity (π), which can be generated from the intrinsic wave function based on the standard projection technique,

$$\psi_\beta^{JK\pi}(\alpha+^6He) = \mathscr{P} \int d\Omega \mathscr{D}_{MK}^{J*}(\Omega)R(\Omega)\mathscr{A}\{\varphi(\alpha)\phi_\beta(^6He)\} \ . \tag{2}$$

Here, $\mathscr{P}, R(\Omega)$, and $\mathscr{D}_{MK}^{J*}(\Omega)$ mean the parity-projection operator, the rotation operator, and Wigner's D–function, respectively. $\varphi(\alpha)$ and $\phi_\beta(^6He)$ show the internal wave function of the α particle and that of the ^6He nucleus, respectively, which are described by the lowest shell model configuration in the harmonic-oscillator (HO) potential. We take b=1.44 fm for the width parameter of HO. The internal wave functions are fully anti-symmetrized and its explicit form is given by

$$\mathscr{A}\{\varphi(\alpha)\cdot\phi_\beta(^6He)\} = \mathscr{A}\{\varphi(\alpha)\cdot\varphi(\alpha)\varphi_n(p_1S_1)\varphi_n(p_2S_2)\} \ . \tag{3}$$

In this equation, $\varphi_n(pS)$ shows the single-particle wave function of the valence neutron in the ^6He nucleus, which is specified by the direction of the p–orbitals (p) and the spins of the neutrons (S) and hence, $\beta=(p_1S_1p_2S_2)$.

The energy eigenvalues which depend on the $\alpha-^6He$ relative distance can be calculated by solving the following eigenvalue equation,

$$< \psi_{\gamma R}^{JK\pi}(\alpha+^6He) \,|\, H-E \,|\, \Psi_R^{JK\pi}(^{10}Be) >= 0 \ . \tag{4}$$

Therefore, the mixing of the different channels are optimized depending on the α–^{6}He relative distance, which leads to the dynamical change from the strong-coupling states to the weak-coupling ones depending on the distance. As for the nucleon–nucleon interaction, we adopted Volkov No.2 with the Wigner and Majorana parts.

Internal states of ^{6}He.   We need to consider only the four kinds of the internal states in the case that we treat the compound $\alpha+^6$He system with $J^\pi=0^+$. When we take the z–axis to be parallel to the α–α relative coordinate, the states of valence two neutrons in ^{6}He can be expressed as

(a) (Symmetric) $\otimes$ (S=0) ; $(xx + yy)(\uparrow\downarrow - \downarrow\uparrow)$

(b) (Anti-symmetric) $\otimes$ (S=1,S_z=0) ; $(xy - yx)(\uparrow\downarrow + \downarrow\uparrow)$

(c) (Symmetric) $\otimes$ (S=0) ; $zz(\uparrow\downarrow - \downarrow\uparrow)$

(d) (Anti-symmetric) $\otimes$ (S=1,S_z=±1) ; $\mathscr{A}\{(yz - izx)\uparrow\uparrow - (yz + izx)\downarrow\downarrow)\}$,

where the x, y, and z show the direction of the p–orbitals, while the $\uparrow$ and $\downarrow$ do that of the intrinsic spins. In this notation, for example, xy means that the first neutron occupies the p–orbital of the x-direction and the second one does that of the y-direction and so on.

The above states have the connection with the intrinsic deformation of the ^{6}He nucleus. In the $\alpha+^6$He system, both (a) and (b) correspond to the α–pole configuration with the oblately deformed ^{6}He nucleus, while (c) and (d) do to the α–pole one with the prolate ^{6}He and the α–equator one with the oblate ^{6}He, respectively. Namely, these basis states correspond to the strong-coupling states having a definite geometrical configuration.

RESULTS

In Fig. 1, we show the energy surface of individual basis states shown in the previous section. That is, they correspond to the energy surfaces of the α–^{6}He system with strong-coupling configuration. At a small distance, the energy of (a) (the lower solid curve) is the lowest of all the states, because, in the overlapping limit, this state goes to the lowest shell model configuration $(0s)^4(0p)^6$ of the compound system ^{10}Be. On the other hand, the energy of (c) (the lower dashed curve) becomes quite high at a small distance due to the anti-symmetrization effect, while it becomes the lowest at a distance between 4 fm to 7 fm, which leads to the energy inversion between (a) and (c). This is due to the difference of the intrinsic shape between them. Namely, the latter state, having the α–pole configuration with prolate ^{6}He, can contact to each other at a large distance compared with the former state, which leads to the gain of the interaction energy between α and ^{6}He.

We also show the energy surface obtained by the coupled-channel calculation in the same figure. In comparison with the solid and dashed curves, one can clearly confirm that the results of the coupled-channel calculation are completely same as those with-

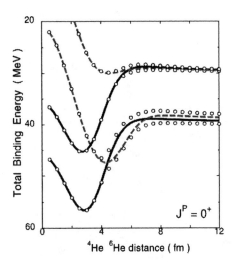

FIGURE 1. α–^{6}He energy surfaces of individual basis states ($J^\pi=0^+$). The lower and upper solid curves show the surfaces of the state (a) and (b), respectively, while the lower and upper dashed ones do those of the state (c) and (d), respectively. The white dots show the results of the coupled-channel calculation which includes all the states.

out channel coupling at a small distance $R < 4$ fm. This means that the strong-coupling states becomes appropriate bases for describing the total system at such a small relative distance. In the asymptotic distance, however, the results of the coupled-channel calculation deviate from those without the channel coupling. Therefore, another kind of states is generated at an asymptotic region by the channel-coupling effect.

In order to confirm the asymptotic states generated by the channel-coupling effects, we artificially constructed the weak-coupling states in which the ^{6}He nucleus had a definite intrinsic spins and couples to the α particle with a relative orbital momentum L. The weak-coupling state can be expressed by the linear combination of the strong-coupling bases shown in the previous section.

In Fig. 2, we show the comparison between the energy surfaces of the coupled-channel calculation and those of the weak-coupling states. Both two solid curves show the surfaces of the weak-coupling states in which ^{6}He has the 0^+ intrinsic spins. In the lower solid curve, the p waves of the valence two neutrons in ^{6}He couple to zero orbital-angular-momentum ($L=0$) and the intrinsic spins of the neutrons also couple to the singlet states ($S=0$). Namely, in the lower surfaces, ^{6}He has the $L=S=0$ configuration in the L–S coupling scheme. On the other hand, in the upper solid curve, both the spatial part L and the intrinsic spin S form the triplet configuration, $L=S=1$ and they totally couple to the 0^+ states. The former one can be constructed by the linear combination of the basis (a) and (c), while the latter one can be constructed from the basis (b) and (d). In the similar way, the two dashed curves correspond to the weak-coupling states in which the ^{6}He nucleus forms the 2^+ states. In the lower dashed surface, the ^{6}He nucleus has the $L=2$, $S=0$ configuration, while it takes the $L=S=1$ configuration, which totally couples to 2^+ states, in the upper surface. The former and latter 2^+ states can also be constructed

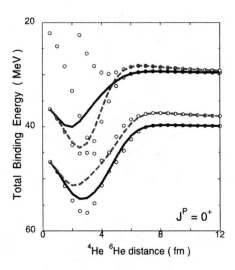

FIGURE 2. The comparison between the energy surfaces of the coupled-channel calculation and those of the weak-coupling states which were artificially constructed. The solid and dashed curves show the energy surfaces of the weak-coupling states, while the white dots do those of the coupled-channel calculation. The lower and upper solid curves show the states in which ^{6}He forms the 0^+_1 and 0^+_2 states, respectively, while the lower and upper dashed ones do the states in which the ^{6}He nucleus are excited to the 2^+_1 and 2^+_2 states, respectively. See text for details.

from the linear combination of (a) and (c) and from that of (b) and (d), respectively.

It is important to notice that the energy surfaces of the coupled-channel calculation are completely same as those of the weak-coupling states. This means that the channel coupling effect exactly generates the weak-coupling states in an asymptotic region. In the small distance, however, the energies of the weak-coupling states are different from the coupled-channel solution. Therefore, the ^{6}He nucleus hardly keeps their asymptotic states with the definite spins in the case that the ^{6}He nucleus and the α particle becomes much closer. When we switched on the channel-coupling among the constructed weak-coupling states, the energy surfaces are changed to the white dots shown in Fig. 2, which are obtained by the coupled-channels with the strong-coupling basis.

According to the results shown in Figs. 1 and 2, we can interpret the energy surfaces obtained by the coupled-channel calculation. The result is shown in Fig. 3. In each curve, the internal states of ^{6}He are dynamically changed depending on the α–^{6}He distance. In the lowest surface, for example, the ^{6}He nucleus has the 0^+ spin-parity states with the spherical shape at an asymptotic region ($R > 6$ fm). The ^{6}He nucleus begins to be deformed to a prolate shape in order to gain the interaction energy between two clusters when ^{6}He approaches the α particle (4 fm$< R <$7 fm). However, the energy eigenvalue becomes quite high due to the anti-symmetrization effect if two cluster becomes closer with the same configuration kept. To avoid the energy loss due to the anti-symmetrization effect, the ^{6}He nucleus begins to be deformed to the oblate shape ($R <4$ fm). Finally, α and ^{6}He completely overlap, which goes to the lowest shell model

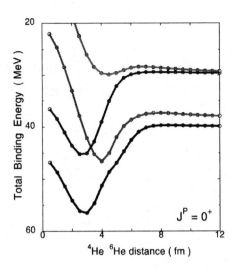

FIGURE 3. The interpretation of the energy surfaces obtained by the coupled-channel calculation. See text for details.

states of the compound system. The similar interpretation can be done for other energy surfaces. Therefore, the present coupled-channel calculation generates the dynamical change from the weak-coupling states to the strong-coupling ones depending on the α–^{6}He relative distance.

SUMMARY AND CONCLUSION

In the present study, we have proposed the new cluster model based on the coupled-channel Generator Coordinate Method and applied it to the ^{4}He+^{6}He system as a simple example. The individual channels are specified by the intrinsic configuration of the strong-coupling scheme and their dynamical coupling are solved at each ^{4}He–^{6}He relative distance. We have found that the strong-coupling states are favored at a small distance, while the weak-coupling states are done at a large distance. Therefore, the present model will be powerful tool for studies on the molecular resonances in the neutron-rich Be isotope because the model quite naturally generates both the weak-coupling states at an asymptotic distance and the strong-coupling ones at a small distance.

REFERENCES

1. M. Freer *et al.*, Phys. Rev. Lett. **82**, 1383 (1999); Phys. Rev. C **63**, 034301 (2001).
2. A. Saito *et al.*, Prog. Theor. Phys., in press.
3. N. Itagaki *et al.*, Phys. Rev. C **61**, 044306 (2000); Phys. Rev. C **62**, 034301 (2000).
4. H. Horiuchi *et al.*, Suppl. Prog. Theor. Phy. 62, 1(1977) and references therein.

Coupled-channel study for O-isotopes with the core plus valence neutrons model

H. Masui*, T. Myo*, K. Katō* and K. Ikeda†

*Graduate School of Science, Hokkaido University, Sapporo 060-0810, Japan
†The Institute of Physical and Chemical Research(RIKEN), Wako 351-0198, Japan

Abstract. We study the ^{17}O by using the core+n model. As a new approach, we perform the coupled-channel and RGM calculation not only for bound states but also resonant states of ^{17}O. We obtain the state dependent core-n folding potential and reasonable SPE in the ^{16}O core.

INTRODUCTION

In recent years, much interest for the unstable nuclei has been concentrated on [1, 2, 3]. Neutron rich nuclei have a particular properties, A halo structure is the one of them. Among these nuclei, ^{23}O and ^{24}O have a possibility to be the halo nuclei, one- and two-neutron's halo, respectively. In neutron rich side of the O-isotopes, as the neutron number goes up, the orbitals for valence neutrons $1s_{1/2}$, $0d_{5/2}$ and $0d_{3/2}$ come closer. Hence the neutrons can easily make a pair and have a pairing-correlation in the core. Such orbital degeneracy possibly make a orbital inversion for last one neutron in ^{21}O and ^{23}O.

To investigate such effect and dynamics to the core-n interaction microscopically, we work on the "core plus valence neutrons" model. The procedure of the calculation has been developed and improved in lighter nuclei He, Li, and Be. Investigation for resonance states have been done by using the complex scaling method (CSM) [4]. Recently we successfully applied the other calculational method, the Jost function method (JFM) [5]. By using these two methods, we can study the properties of unstable nuclei including the unbound states precisely.

In this work, we investigate the O-isotopes in the core plus valence neutrons model. Here the dynamics of the core, which is the pairing-excitation of neutrons inside the core and other mechanisms are taken in this model. We examine the effect of the pairing-correlation in the core, a tensor-renormalization for the central part of the effective interaction and the NN-interaction dependence for the core-n folding potential. As a first example, we perform a coupled-channel calculation for $^{16}O+n$ system.

MODEL AND METHOD

As the neutron number goes higher, the isotope becomes more unstable. Therefore, we have to study properties of such nuclei not only for bound states, but also resonant states.

CP644, Exotic Clustering: 4th Catania Relativistic Ion Studies, edited by S. Costa, A. Insolia, and C. Tuvè
© 2002 American Institute of Physics 0-7354-0099-7/02/$19.00

To investigate the resonant states, we use the complex scaling method (CSM) [4] and the Jost function method (JFM) [5].

Core + n model

The core-n potential is made from the effective NN-interaction by folding way. The NN-interactions which we use here are, Hasegawa-Nagata(HN) [6] and Volkov [7] interactions. We only apply the central-part of these potentials and make folding potentials by using the harmonic oscillator wave function of the ^{16}O core. The LS-potential which is necessary to be introduce for making the spin-orbit splitting between $d_{5/2}$ and $d_{3/2}$ is taken as a density-derivative type. We choose the strength of LS-potential so that the splitting of $d_{5/2}$ and $d_{3/2}$ is produced. The other contributions, for example, the tensor force are neglected in this stage.

Complex scaling method and Jost function method

The complex scaling method.   In CSM [4], to solve the resonant states the following transformation is performed for a radial coordinate (momentum):

$$r \to re^{i\theta} \quad (p \to pe^{-i\theta}) . \tag{1}$$

Here θ is a rotational angle parameter in the complex coordinate (momentum) plane. We choose the angle θ so as to cover the resonant pole:

$$0 \le \tan^{-1}(\Gamma/2E_r) \le 2\theta . \tag{2}$$

With this condition, the rotated resonant wave function converges to zero in the asymptotic region, and the resonant poles can be found in the wedge region between the positive real axis and a 2θ line

The Jost function method.   The JFM is quite useful to obtain the position of poles both on the complex energy and on the momentum planes. Details of the procedure for an actual calculation are given in the original papers of Sofianos and Rakityansky [5] and also in our previous papers [8, 9]. The essential procedures of JFM are as follows: Introducing unknown functions $\mathscr{F}^{(\pm)}(k,r)$ as coefficients of the incoming $(-)$ and outgoing $(+)$ waves, we express the regular solutions of the Schrödinger equation as

$$\phi(r) \equiv \frac{1}{2} \left[H_l^{(+)}(kr)\mathscr{F}^{(+)}(k,r) + H_l^{(-)}(kr)\mathscr{F}^{(-)}(k,r) \right] . \tag{3}$$

To solve the equation for $\mathscr{F}^{(\pm)}(k,r)$, we introduce the additional constraint condition, which is usually chosen in the variable-constant method:

$$H_l^{(+)} \left[\partial_r \mathscr{F}^{(+)} \right] + H_l^{(-)} \left[\partial_r \mathscr{F}^{(-)} \right] = 0 . \tag{4}$$

Using this condition, the equation for $\mathscr{F}^{(\pm)}(k,r)$ becomes a first-order differential equation:

$$\frac{\partial \mathscr{F}^{(\pm)}(k,r)}{\partial r} = \pm \frac{\mu}{ik\hbar^2} H_l^{(\mp)}(kr)V(r)\left\{H_l^{(+)}(kr)\mathscr{F}^{(+)}(k,r) + H_l^{(-)}(kr)\mathscr{F}^{(-)}(k,r)\right\}. \tag{5}$$

In the asymptotic region, where the potential goes zero faster than r^{-1}, the $\mathscr{F}^{(\pm)}(k,r)$ converge into constant $\mathscr{F}^{(\pm)}(k)$, and are equivalent to the "Jost functions" of the original Schrödinger equation. By using the Jost functions $\mathscr{F}^{(\pm)}(k)$, the S-matrix is expressed as

$$S(k) = \mathscr{F}^{(+)}(k)/\mathscr{F}^{(-)}(k). \tag{6}$$

The numerical convergence of the Jost functions $\mathscr{F}^{(\pm)}(k)$ is quite well. Physical quantities concerned with the S-matrix can be obtained very easily and accurately. And we can obtain the s-wave poles of the S-matrix for virtual states in the complex-momentum plane by solving $\mathscr{F}^{(-)}(k) = 0$.

RESULTS

We show the calculated results for ^{17}O. First, we study the positive parity states, $5/2^+$, $1/2^+$ and $3/2^+$. The $3/2^+$-state is a resonant state, therefore, we calculate such the resonant state by using JFM. Next, we calculate the negative parity state, $1/2^-$.

Positive parity states

We investigate the positive parity states $5/2^+$, $1/2^+$ and $3/2^+$. These three states can be considered as the single-particle states of $0d_{5/2}$, $1s_{1/2}$ and $0d_{3/2}$, respectively. To reproduce the states, we change the potential strength in a state dependent way as $V \rightarrow (1 + \delta_c)V$ for HN forces and change w parameter for Volkov forces. Table 1 shows the calculated results for positive parity states. The result indicates that the state-dependence between s- and d-waves are strong.

TABLE 1. The strength parameters and V_{ls} of the ^{16}O-$n$folding potential for a state dependent approach.

NN-int.	s-wave	d-wave	V_{ls}
HN-1 : δ_c	0.015	0.122	84.0
HN-2	0.085	0.190	76.0
Volkov-1 : w	0.493	0.542	91.5
Volkov-2	0.480	0.519	98.5

Coupled-channel approach. To reduce such the strong state dependence, we perform the coupled-channel calculation by taking into account the core-polarization (pairing-correlation in the core). If the $1s_{1/2}$-orbit is occupied, the strength of the s-wave potential can be taken a larger value than that of the case for an absence of pairing-correlation. This pairing-correlation mechanism has an important effect for ^{9}Li-n interaction shown in Ref.[10].

We perform the coupled-channel calculation for the s-wave ($1/2^+$) state. However, the result show that the state-dependence still remains for s- and d-wave states. The reason is follows: The coupling strength of each configuration in the core Hamiltonian is relatively small to the single particle energy (SPE). Especially, $(0p_{1/2})^2$ to $(1s_{1/2})^2$ is much smaller than the SPE from $0p_{1/2}$ to $1s_{1/2}$, that is less than one order of magnitude.

Therefore, it is necessary to introduce other mechanisms for solving the system without the state dependence.

Inclusion of RGM-kernel. Next, we apply the exchange kernel of the RGM (the resonating group method) to this calculation. The exchange effect is stronger for d-wave than that for s-wave. We use the knock-on exchange kernel of RGM (model-K kernel) [11] for our calculation, since the model-K kernel has a simple form compared to the full RGM-kernel and also has enough information for the effect of exchange between the valence particle and the core.

TABLE 2. The strength parameters and V_{ls} of the ^{16}O-$n$folding potential for consistent parametrization.

NN-int.	δ/w	δ_{RGM}	V_{ls}
HN-1	-0.0829	0.729	102
HN-2	-0.0569	0.798	93.5
Volkov-1	0.427	0.104	96.9
Volkov-2	0.418	0.0169	99.7

The model-K kernel is a truncated RGM-kernel. Therefore, we multiply the factor $(1+\delta_{RGM})$ to the kernel as

$$K(r,r') \rightarrow (1+\delta_{RGM})K(r,r') . \tag{7}$$

The calculated results are listed in Table 2. We can obtain the three positive parity states, $5/2^+$, $1/2^+$ and $3/2^+$ without any state dependence.

Negative parity state

By using the model-K kernel and the parameter sets for the positive parity states, next, we investigate the negative parity state $1/2^-$, which can be considered the one neutron-hole state in the $0p_{1/2}$-orbit. For the study of this negative parity state, we perform a coupled-channel calculation of two different configuration in the core nucleus. The ground state configuration is $(0p_{1/2})^2_v$ and the first excited one is $(0d_{5/2})^2_v$ for the last two neutrons in ^{16}O. In this calculation, we fix the parameter for the potential but change the single-particle energy (SPE) from $0p_{1/2}$ to $0d_{5/2}$.

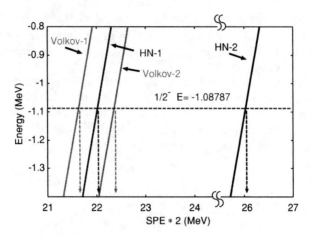

FIGURE 1. The SPE dependence for the coupled-channel calculation of the $1/2^-$ state.

The SPE and the ratio of the core-configurations are listed in Table 3. Here ψ_1 and ψ_2 are channel wave functions, which correspond to two configurations in the core as $(0p_{1/2})^2$ and $(0d_{5/2})^2$, respectively. The SPE dependence to the binding energy is shown in Fig. 1. The Volkov-1, 2 and HN-1 forces give comparable results to that of the SPE in Ref. [12], in which the single-particle energy is estimated as 11.63MeV.

TABLE 3. Deduced single-particle energy (SPE) from the coupled-channel calculation. The third column shows the level of $p_{1/2}$ in the core and the forth column shows the energy level of the second-channel.

NN-int.	SPE (MeV)	$(0p_{1/2})$	level of ψ_2	$|\psi_1|^2 : |\psi_2|^2$
HN-1	11.01	−22.66	−1.21	0.6 : 99.4
HN-2	13.04	−26.63	−1.18	0.5 : 99.5
Volkov-1	11.18	−22.87	−1.27	1.0 : 99.0
Volkov-2	10.81	−22.06	−1.29	1.1 : 98.9
WB*	11.63			

* taken from Ref.[12]

SUMMARY

We study the low-lying states of ^{17}O not only for the bound states but also the resonant states in the core+n model space. For the positive parity states, the folding potentials have strong state dependence. Even we take into account the pairing-correlation in the core, the state dependence still remains. On the other hand, the RGM-kernel, which we introduce here is model-K one [11], can give the state independent parameter sets for the low-lying positive parity states.

By using these state independent potentials and kernels, we calculate the lowest negative parity state $1/2^-$. The coupled-channel calculations give the comparable SPE to Ref. [12].

As the next step, we study the ^{18}O by using the core+n+n three-body model, and the knowledge from these calculation will be reflected in the construction of the ^{18}O-n potential. Moreover the information about the tendency to the higher neutron number nuclei will give us interesting results for the drip-line nuclei.

ACKNOWLEDGMENTS

The authors are grateful to the laboratory members of nuclear theory group at Hokkaido University for helpful discussions.

REFERENCES

1. I. Tanihata et al., Phys. Lett. **B160**, 380 (1985). I. Tanihata et al., Phys. Rev. Lett. **55**, 2676 (1985).
2. I. Tanihata, J. Phys., **G22**,157 (1996).
3. Ogawa et al., Nucl. Phys. **A691**, 599 (2001). Ogawa et al., Nucl. Phys. **A693**, 32 (2001).
4. J. Aguilar and J. M. Combes, Commun. Math. Phys. **22**, 269 (1971). E. Balslev and J. M. Combes, Commun. Math. Phys. **22**, 280 (1971).
5. S. A. Sofianos and S. A. Rakityansky, J. of Phys. **A30**, 3725 (1997),3725.; **A31**, 5149 (1998).
6. A. Hasegawa and S. Nagata, Prog. Theor. Phys. **45**, 1786 (1971).
7. A. B. Volkov, Nucl. Phys. **74**, 33 (1965).
8. H. Masui, S. Aoyama, T. Myo and K. Katō, Prog . Theor. Phys., **102**, 1119 (1999).
9. H. Masui, S. Aoyama, T. Myo, K. Katō and K. Ikeda, Nucl. Phys. **A673**, 207 (2000).
10. K. Katō, T. Yamada and K. Ikeda, Prog. Theor. Phys. **101**, 119 (1999).
11. T. Kaneko, M. LeMere and Y. C. Tang, Phys. Rev. **C44**, 1588 (1991).
12. E. K. Warburton and B. A. Brown, Phys. Rev. **C46**, 923 (1992).

Search for 4-n cluster in ^{8}He+^{4}He collision and 4-n system

R. Wolski[1,2], S .I. Sidorchuk[1], D. D. Bogdanov[1], A. S. Fomichev[1], A. M. Rodin[1], S. V. Stepantsov[1], G. M. Ter-Akopian[1], W. Mittig[3], P. Roussel-Chomaz[3], H. Savajols[3], N. Alamanos[4], V. Lapoux[4], M. Matos[5], R. Raabe[6]

[1]*Flerov Laboratory of Nuclear Reactions, JINR, Dubna, 141980 Russia,*
[2]*Institute of Nuclear Physics, Cracow, Poland,*
[3]*GANIL BP5027, 14021 Caen, France,*
[4] *DAPNIA/SPhN/Saclay, 91191 Gif-sur-Yvette Cedex, France,*
[5]*Comenius University, Mlynska Dolina, 84215 Bratislava, Slovakia,*
[6]*Instituut voor Kern- en Stralingsfysica, University of Leuven, B-3001 Leuven, Belgium*
e-mail: wolski@nrsun.jinr.ru,

Abstract. A search for the transfer of 4-neutron cluster in the elastic scattering of ^{8}He on a gaseous helium target at a beam energy of 26 A·MeV is reported. The upper limits of the cross section at the backward scattering angles were obtained. DWBA calculations show that the two-step 2n transfer will contribute more than the one-step 4n transfer. The d (^{8}He,4n) ^{6}Li reaction for investigation of the 4-neutron system is discussed.

INTRODUCTION

The 4-neutron cluster is definitely an exotic one. The existence of bound or quasi-bound tetra neutron system is a long standing problem for theoretical and experimental investigation, e.g. see compilation [1]. Numerous attempts to produce tetra-neutron have been undertaken. In spite of all efforts no convincing evidence of bound or resonant 4-n system has been found. Diverse theoretical estimations of 4n binding energy are rather in accord that 4n is unbound but are inconclusive concerning a presence of low lying resonance, [2] and references therein. The authors of Ref. [2] suggest a resonance in 4-n system at the excitation energy of 3.0MeV. They have calculated 4-body scattering phase shifts using either Volkov 1 or Tang nucleon-nucleon potential within hyperharmonic basis. The main conclusion is that the phase shifts became more energy dependent if more hyperharmonics are included.

The subject became recently even more exciting after authors of Ref. [3] in their study of break-up of ^{14}Be observed, in coincidence with ^{10}Be fragments, few events of neutrons interacting with protons in a liquid scintilator. For these events, the proton recoiling energies were significantly larger than neutron energies measured by the time of flight. Such events could be attributed to the bound 4-n cluster produced in the reaction.

If confirmed, the existence of a bound 4-n system is of great importance. Recently the author of Ref [4] extensively discusses various calculations and concludes that, to

CP644, Exotic Clustering: 4th Catania Relativistic Ion Studies, edited by S. Costa, A. Insolia, and C. Tuvè
© 2002 American Institute of Physics 0-7354-0099-7/02/$19.00

get tetra-neutron bound, one needs nucleon-nucleon potential parameters which lead to large overbinding of the stable light nuclei.

The availability of ^{8}He beam offers a new opportunity for the study the 4-neutrons final state interaction (FSI).

In the present paper we briefly report on our ^{8}He + ^{4}He elastic scattering experiment and discuss the use of ^{8}He beam for an investigation of 4-n system by a α-particle transfer reaction.

^{8}He+^{4}He ELASTIC SCATTERING

^{8}He nucleus, notable for its large neutron-to-proton ratio N/Z =3, could be treated as a candidate for one possessing a neutron-skin. In a simple model COSMA [5], convenient due to its transparency, the ^{8}He is considered of having an α-particle, as an almost inert core, together with a full $0p_{3/2}$ sub-shell of four neutrons a thick skin.

The elastic scattering is the simplest direct reaction which can be used for studying properties of nuclei involved. To describe the angular distribution and the total absorption one introduces the optical model (OM) potential. Sometime, at backward scattering angles, a pronounce enhancement over the OM prediction is observed. The effect is usually interpreted as the presence of exchange reactions leading to the exit channel identical to the entrance one. In the case of the ^{8}He + ^{4}He system, such process is the transfer the 4-neutron cluster. However, one should be aware of various combinations of sequential processes also.

Experimental Layout

The secondary ^{8}He beam was obtained by the fragmentation of ^{11}B primary beam, obtained from the U-400M cyclotron, on a beryllium production target (255 mg/cm^2). The magnetic separator ACCULINNA [6] provided the selection of the ^{8}He beam. The average intensity of the 26 A·MeV ^{8}He secondary beam was about 4000 pps. The energy spread was 5% (FWHM). A diameter of the beam spot on the cryogenic gaseous helium target was 10mm. The ^{8}He ions were monitored and discriminated from other beam components by a system of two plastic scintillators. A thin, was positioned up-stream in respect to the target, whereas a thick one downstream.

Two multiwire proportional chambers (MWPC) were installed in front of the target for tracking the beam ions. The effective thickness of the helium target cooled to T=35K and under a pressure of 11atm was $2.4 \cdot 10^{21} \cdot$ cm^{-2}. The detecting system consisted of two twin large-area, position sensitive ΔE(strip)-ΔE(strip)-E(CsJ) telescopes and a ΔE multi-wire proportional gas chamber (DEG). The telescopes measured the energies and positions of ^{8}He ions elastically scattered from the target. They were placed at a distance of 164 mm from the target with their geometric centers set symmetrically at 17.5^0 LAB. The forward-angle scattering data were obtained from inclusive measurements. To search for the backward scattering both products: the high-energy α-particles and low-energy ^{8}He had to be detected. DEG chamber was intended to measure the energy losses of the low-energy ^{8}He ions. It was installed at a distance of 60 mm from the target, so the beam passed through its central zone. This central 12-mm vertical zone was insensitive to the beam particles due to thick central anode wires

Results and Discussion

The integral ^{8}He beam fluxes on the target with the filled target and the empty one were respectively $4.3 \cdot 10^9$ and $2 \cdot 10^9$. The forward-angle elastic scattering was identified as a clear energy-angle for locus ^{8}He in the telescopes. The differential cross section for elastic scattering to forward angles was obtained by averaging statistic over 5^0 angular bins in CM frame and estimating the effective solid angle for each bin. However, only upper limits for the elastic-scattering cross section in the CM angular range of 130^0-165^0 was set because those coincident events, which were recorded, did not obey the condition of the total energy balance.

The forward angles elastic scattering was described terms of the phenomenological OM potentials. A rather unique 6-parameter OM potential of volume Wood-Saxon shape has been found: V=186.76, r_v=0.3904, a_v=0.9746, W=21.908, r_i=0.8933, a_i=0.9612. The values stand for the strength, radius and diffuseness of the real and imaginary WS parameters respectively with radius: R= $r_0 \cdot (8^{1/3} + 4^{1/3})$. The fit to the angular distribution data is shown in Fig.1 by a dot line.

The obtained potential yields the surprisingly large total reaction cross section for ^{8}He+^{4}He interaction at the energy of 26 A·MeV. The value of σ_R exceeds of 200mb, at least, that of ^{12}C+^{4}He at the same energy per nucleon [7]. In order to examine sensitivity of calculated σ_R values on uncertainty of the data normalization factor, which is estimated to be 20%, the angular distribution data points were multiplied by a factor of N = 1.2 and N = 0.8, and subjected to 6-parameter fitting. As expected, the N= 1.2 led to a reduction of σ_R = 944 mb and N = 0.8 produced σ_R= 1120 mb. In that way the total reaction cross section has been determined as σ_R = 1030+/-90 mb.

In order to evaluate the unobserved cross-section enhancement at backward scattering angles exchange processes were calculated by DWBA code FRESCO-18 [8]. We considered a one-step direct 4n transfer first, and a two-step sequential transfer of two pairs of neutrons through the intermediate ground and excited (2+) states of ^{6}He nuclei. For the one-step transfer the internal ^{8}He wave function was taken as that of 2-body: α+4n in 2S state with the binding energy equal to the ^{8}He 4-n separation energy (3.111 MeV). It means that 4n cluster is transferred with zero excitation energy what is obviously questionable. The spectroscopic amplitude (SA) for α+4n clustering was set equal to 1.0. The calculated potential scattering with coherent inclusion of the one-step transfer is shown in Fig. 1 as a dashed curve.

The contribution of the two-step transfers was estimated by taking into account all combinations of the both ^{6}He in their two states. The OM parameters for the intermediate channels were borrowed from the analysis ^{6}Li+^{6}Li scattering at the corresponding. The all SA relevant for the two-step transfers were taken to be equal 1.0. This approximation seems to be sufficient for the need of the present qualitative treatment. The analysis revealed that the intermediate system of the two excited ^{6}He nuclei yields the largest contribution to the backward-angle enhancement. The influence of the two-step processes is shown in Fig. 1. One can see that the direct transfer of a 4n cluster is much weaker than the sequential two-step process.

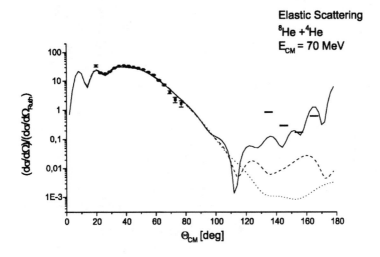

FIGURE 1. Experimental data for ^{8}He+^{4}He elastic scattering at 26 A·MeV. Symbols in the backward hemisphere indicate upper limits only. Curves are for the calculations: dot line for OM, dashed line for the inclusion of the one-step transfer, solid line for the coherent sum of potential scattering, one-step and two-step transfer processes.

d(^{8}He,4n)^{6}Li REACTION TO STUDY 4-n SYSTEM

The excitation energy spectrum of 4-n system could be measured by the missing mass method using the one-step α-particle transfer on a light target, e.g. deuterium.

The (d,^{6}Li) reactions have been used in the past for studies of α-particle clustering in light stable nuclei, e.g. see Ref.[9]. The α-particle transfers in d(^{8}He,4n)^{6}Li reaction populate several states in ^{6}Li residual nucleus. The ground (1^{+}) one and first excited 2.18MeV (3^{+}), the later is unbound, are known to be in a high degree of α+d clustering nature. The next state, which is particle stable and could contribute to the ^{6}Li energy spectrum, would is not populated in the one-step process due to its isospin. Others states of ^{6}Li are broad and their contributions could be discriminated. The α-particle within ^{8}He nucleus is rather a well defined object. Shell model calculations give for ^{8}He nucleus a large value of α spectroscopic factor, around 0.7 [10].

DWBA (FRESCO) predictions of d(^{8}He,4n)^{6}Li reaction are shown in Fig. 2. The two transitions, corresponding to the ground and the first excited states of ^{6}Li were calculated as the one-step, single L-transfer at LAB energy of 96 MeV. The excitation energy of 4n cluster was taken to be equal to +3.0 MeV above free 4-n threshold according to the suggestion in Ref. [2]. The calculations for such exotic system could be questionable. It concerns, first of all, Optical Model (OM) parameters applied. For the entrance channel a deuteron OM potential was taken from the (d,^{6}Li) reactions study [9] for stable nuclei with similar masses. The exit channel OM parameters were borrowed from the compilation for the α+^{6}Li interaction at 29.4 MeV [11]. The α-particle SA equal to 1.0 for ^{8}He and 1.06 for both states of ^{6}Li were taken.

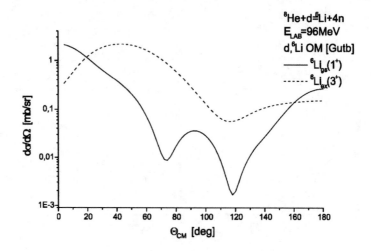

FIGURE 2. DWBA calculations for $d(^8He,4n)^6Li$ reactions leading to the two states ^6Li for the LAB beam energy of 96 MeV.

According to the calculations the cross section for the transition leading to the excited $^6Li(3^+)$ state is larger than that to the ground state.

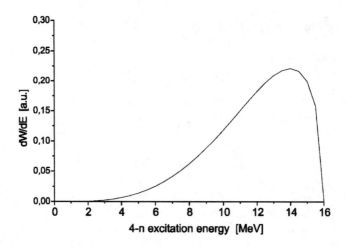

FIGURE 3. Excitation energy distribution of 4-neutron system for the 5 body (4n + ^6Li) phase-space with the total CM energy of 16 MeV.

Under an assumption that the reaction is that of 2-body kinematics, one can obtain the 4-n excitation energy spectrum from measurement of the relevant ^6Li energy spectrum. The expected structure in the spectrum, caused by the final state interaction

(FSI) among the 4 neutrons, has to be discriminated from spectra generated by other processes.

In the first approximation the other processes could be simulated by the relevant phase-space distribution which is not uniform. The phase-space distribution for 5 body system of 4 neutrons and ^{6}Li as a function of 4-n excitation energy is shown in Fig. 3. The total CM energy is equal to 16 MeV, 95.5 MeV in LAB system. The most probable 4-n excitation energy is 7/8 of the total CM energy. One can see that, up to 8 MeV of the excitation, the left shoulder of the distribution is smooth enough to make the observation of the structure originated by FSI effect possible.

ACKNOWLEDGMENTS

The partial support of the work by Russian Basic Research Fundation (grant No. 02-02-16550) is acknowledged.

REFERENCES

1. D. R. Tilley and H.R. Weller, *Nucl. Phys.* **A541,** 1 (1992).
2. Gutich I.F. et al., *Yad. Fiz,* **50,** 19 (1989) 19 (in Russian).
3. F. M. Marques et al, *Phys. Rev.* **C65,** 044006-1 (2002).
4. N. K. Timofeyuk, *nucl-th*/0203003, 2002.
5. M.V. Zhukov, A. A. Korsheninnikov and M. H. Smelberg, *Phys. Rev.* **C50,** R1 (1994).
6. A. M. Rodin et al., *Nucl. Inst. and Meth.* **B126,** 236 (1997).
7. Dao T. Khoa, G. R. Sachler and W. von Oertzen, *Phys. Rev.* **C56,** 954 (1996).
8. I. J. Thompson, *Computer Phys. Rep.* **7,** 167 (1988).
9. H. H. Gutbrod et al., *Nucl. Phys.* **A165,** 240 (1971).
10. S.D. Kurgalin and Yu.M. Tchuvil'sky, *"Spectroscopic Factors of Multineutrons in Nuclei"*, Proceedings of 2-d Intern. School on Nuclear Physics. 1992. INR (Kiev).
11. C.M. Perey and F.G. Perey, *Atomic Data and Nucl.Data Tables*, **17,** 1 (1976).

Structures of Continuum States in Halo Nuclei

Kiyoshi Katō*, Takayuki Myo* and Kiyomi Ikeda†

*Division of Physics, Graduate School of Science, Hokkaido University, Sapporo 060-0810, Japan
†RI-Beam Science Laboratory, RIKEN(The Institute of Physical and Chemical Research), Wako, Saitama 351-0198,Japan

Abstract. The complex scaling method has been studied as a very useful method to obtain many-body resonances, and we here show that not only resonant states but also rotated continuum states in the complex scaling method play an important role in analyses of the continuum structures. Using the extended completeness relation, it is shown that we can calculate strength functions of the Coulomb breakup reaction. This new formalism to calculate strength functions is applied to neutron halo nuclei [11]Be, [6]He and [11]Li and discuss structures of continuum states.

INTRODUCTION

Recently, much interest has been concentrated on the three-body resonances of two-neutron halo systems observed in neutron drip-line nuclei.[1] The so-called Borromean system[2] has a weakly-bound ground state which has a halo structure, but no bound states of any two-body sub-system. Most of excited states are unbound in such a system. Therefore, we need treat resonant and continuum states in addition to the weakly-bound states to study the Borromean systems with deep interest in mechanism of excitation from the halo structure.

Unbound states of a many-body system have various kinds of structure which are provided by characteristic interactions. Many thresholds for open channels appear even in three-body system. The structure of continuum states starting from each threshold is considered different from each other. Although those continuum states are obtained as solutions of the Schrödinger equation on each branch cut of the complex energy Riemann sheets, it is difficult to distinguish them only by seeing their energy positions in a usual method. This difficulty can be solved by the complex scaling method, because every cut degenerated on the real energy axis is rotated around the corresponding threshold energy by taking a finite scaling angle θ.[3]

In Fig. 1, possible thresholds and corresponding branch cuts are shown for a three-body system in a usual case (b) and the complex scaling method (c). We can distinguish the continuum states obtained on the rotated cuts in the complex scaling calculation with a finite θ value. Using this result in addition to resonant states, we can investigate the structure of the continuum states. For this purpose, we first studied matrix elements of an operator with the complex scaled wave functions.[4] The physical meanings of those matrix elements were discussed in association with a sum rule value.[5] From the discussions of the sum rule value, we proposed an idea of the extended completeness relation in the complex scaling method.[6]

Applying the extended completeness relation to calculations of Green's function, we

CP644, *Exotic Clustering: 4th Catania Relativistic Ion Studies*, edited by S. Costa, A. Insolia, and C. Tuvè
© 2002 American Institute of Physics 0-7354-0099-7/02/$19.00

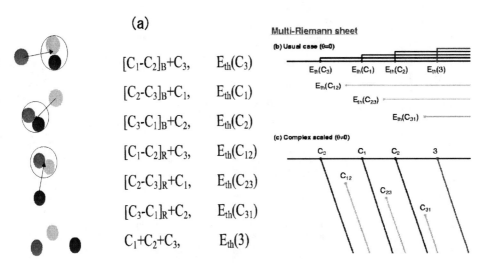

FIGURE 1. Various thresholds (a) of a three-body system and corresponding Riemann sheets in the usual case (b) and the complex scaling method (c).

study the strength functions of the Coulomb breakup reactions. To see reliability of the method, we investigate the Coulomb breakup reaction of the one-neutron halo nucleus ^{11}Be using the coupled channel model for ^{10}Be+n. We also apply this method to three-body systems of ^{4}He+n+n and ^{9}Li+n+n. Through analyses of the calculated strength functions, we discuss contribution from three-body resonances, two-body resonances and three-body continuum.

MATRIX ELEMENTS OF RESONANT STATES

Resonant states are generated by a potential in addition to continuum states being solutions of a kinetic energy operator. This common origin of resonant states with bound states suggests that they have very similar properties.

The wave function of a resonant state with a complex momentum $\kappa - i\gamma$ shows an exponentially divergent behavior, $e^{ikr} = e^{i(\kappa - i\gamma)r} = e^{\gamma r} \cdot e^{i\kappa r}$, at asymptotic distances, then it is considered not squared-integrable. However, by applying the complex scaling ($r \to re^{i\theta}$ for $\theta > 0$), this exponentially divergent behavior is transformed to a dumping behavior which is squared-integrable like bound states. It is possible to calculate matrix elements of any operator with normalized wave functions of resonant states. Such an integrable property is the inherent character of resonant states but is not necessary attributed to the complex scaling method.

The normalization of resonant states has been discussed by many people.[7] Gyarmati and Vertse[8] showed that the matrix elements of an operator $\hat{O}$ defined as

$$\langle \tilde{\Psi}_1^\theta | \hat{O}^\theta | \Psi_2^\theta \rangle = \int_{R^\theta} \Psi_1(\vec{r}e^{i\theta}) \hat{O}(\vec{r}e^{i\theta}) \Psi_2(\vec{r}e^{i\theta}) d(\vec{r}e^{i\theta}) \tag{1}$$

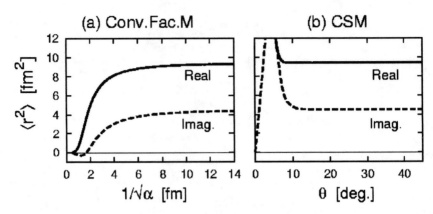

FIGURE 2. The convergences of the matrix elements $\langle r^2 \rangle$ with a resonance wave function in the factorization method (a) and the complex scaling method (b). Solid and dashed lines with Real and Imag represent the real and imaginary parts of the matrix element $\langle r^2 \rangle$, respectively.

are identical with those given by the factorization method:[7]

$$= \lim_{\alpha \to 0} \int_R \Psi_1(\vec{r}) \hat{O}(\vec{r}) \Psi_2(\vec{r}) e^{-\alpha r^2} d\vec{r} = \langle \tilde{\Psi}_1 | \hat{O} | \Psi_2 \rangle \tag{2}$$

Here, R^θ is the integration path which is given by a rotation $Re^{i\theta}$ from the real axis R. This means that this matrix element of resonant states is generally defined by an improper integral avoiding the singular point at an infinity, because the resonant state has the singularity only at an infinity.

In Fig. 2, we show the r^2 matrix elements calculated for θ and α in the complex scaling method and the factorization method, respectively. When θ goes over the angle of the resonance pole in the momentum plane, the matrix elements quickly reach to the converged real and imaginary values which coincide with ones calculated in the factorization method.

CONTINUUM STATES IN THE COMPLEX SCALING METHOD

The complex scaling method has usually been used for investigation of resonances. However, we discuss that the rotated continuum states are also very important in studies of structures of unbound states.[5, 6] Here, we define the continuum solutions obtained on the rotated cuts in the complex scaling method as the rotated continuum states.

The physical meaning of the rotated continuum states on the rotated cuts is expressed as continuum states which exclude the resonant states existing in a wedge region between a rotated cut and the real energy axis. This indicates that the continuum states on the real energy axis are divided into the resonances and the rotated continuum states.

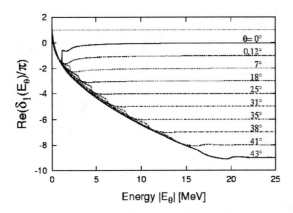

FIGURE 3. The θ dependence of the real part of the phase shift $\delta(E_\theta)$ in unit π.

In Fig. 3, we show a potential scattering phase shift $\delta(E_\theta)$ of the scattering states calculated on the -2θ rotated energy axis. When $\theta = 0$, we can see a sharp resonance behavior at $E_\theta = 1.2$ MeV. This resonance behavior disappears when θ increases beyond the angle $\theta_r = 0.119°$ of the resonance pole. Furthermore, the difference of asymptotic phase shifts $\delta(E_\theta)$ at large energy is seen to be π, when θ is chosen so as to correspond to angles between resonances. This result indicates that the relationship between the phase shift and numbers of bound and resonant states in the complex scaling method is expressed by an extended Levinson theorem.

The role of the rotated continuum states was also discussed in association with sum rule values of electric transitions.[5] The sum rule values calculated along the real energy axis are shown to be equal to the sum of values calculated for bound and resonant states and integrated over the rotated continuum states.

EXTENDED COMPLETENESS RELATION

The problems discussed above are deeply concerned in the completeness relation in the complex scaling method. Here, we consider the usual completeness relation, which corresponds to the case of $\theta = 0$. The usual completeness relation of bound states and continuum,

$$\sum_B |\Psi_B\rangle\langle\Psi_B| + \int_L dE |\Psi_k\rangle\langle\Psi_k| = 1, \tag{3}$$

is proven by taking Cauchy's integral along a closed semi-circle in the above half momentum plane.[9] This proof can easily be applied to the complex scaling method where the semi-circle is rotated as shown in Fig. 4. Since the rotated semi-circle includes the resonant poles whose number is N_θ, the resonance term is added as follows:

$$\sum_B |\Psi_B^\theta\rangle\langle\tilde{\Psi}_B^\theta| + \sum_R^{N_\theta} |\Psi_R^\theta\rangle\langle\tilde{\Psi}_R^\theta| + \int_{L_\theta} dE_\theta |\Psi_{k_\theta}\rangle\langle\Psi_{k_\theta^*}| = 1. \tag{4}$$

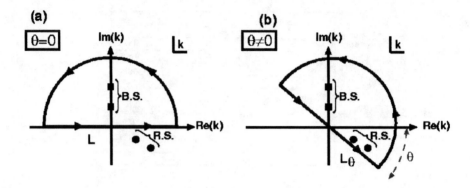

FIGURE 4. Semi-circle contours of Cauchy's integral in the usual case (a) and the complex scaling method (b).

This extended completeness relation is directly applied to calculation of strength functions.[6] The strength function for the transition operator $\hat{O}$ is expressed as

$$F(E) = \sum_{v} \langle \Psi_{gr} | \hat{O}^{\dagger} | \Psi_v \rangle \langle \tilde{\Psi}_v | \hat{O} | \Psi_{gr} \rangle \delta(E - E_v) \tag{5}$$

$$= -\frac{1}{\pi} \mathrm{Im} R(E). \tag{6}$$

Here $R(E)$ is the so-called response function defined by

$$R(E) = \int dr dr' \tilde{\Psi}_{gr}^{*}(r) \hat{O}^{\dagger} G(r,r') \hat{O} \Psi_{gr}(r'). \tag{7}$$

In the complex scaling method, this response function is calculated as

$$R(E) = \int dr dr' \tilde{\Psi}_{gr}^{*}(re^{-\theta}) \hat{O}^{\dagger}(re^{\theta}) G^{\theta}(r,r') \hat{O}(re^{\theta}) \Psi_{gr}(r'e^{\theta}), \tag{8}$$

where Green's function is calculated by using the extended completeness relation as

$$G^{\theta}(r,r') = \langle r | \frac{1}{E - H_\theta} | r' \rangle \tag{9}$$

$$= \sum_{B} \frac{\Psi_B(r,k_B) \tilde{\Psi}_B^{*}(r',k_B)}{E - E_B} + \sum_{R}^{N_\theta} \frac{\Psi_R(r,k_R) \tilde{\Psi}_R^{*}(r',k_R)}{E - E_R} \tag{10}$$

$$+ \int dE_\theta \frac{\Psi(r,k_\theta) \Psi^{*}(r',k_\theta^{*})}{E - E_\theta}. \tag{11}$$

Inserting this form of Green's function, we obtain the strength function separated into three-terms of bound-state, resonant-state and continuum-state contributions;

$$F(E) = F_B(E) + F_R^{\theta}(E) + F_k^{\theta}(E), \tag{12}$$

where the separation between resonances and continua depends on the scaling angle θ. However, the total strength $F(E)$ is independent on θ.

APPLICATIONS TO HALO NUCLEI; ^{11}BE, ^{6}HE AND^{11}LI

We apply this framework to the Coulomb breakup reaction of ^{11}Be which has a one-neutron halo structure. The result of a single channel ^{10}Be+n model was shown to be very successful,[6] but a more realistic calculation is necessary because the single channel model is too simple and the ^{10}Be-n interaction is artificially modified for even and odd parity states.

To understand deeply the halo structure and excited-state structures of ^{11}Be, we here perform a coupled channel calculation for the ^{10}Be+n orthogonality condition model.[10] In Fig. 5, we show the energy eigenvalues in the complex scaling method and the $E1$ strength distribution calculated with obtained wave functions. Although contributions from resonances are small, but we can see a good correspondence between calculation and experiment, and can confirm the reliability of the present method.

We also apply this method to the ^{4}He+n+n three-body system. The ^{4}He+n+n three-body model, where the Pauli principle is taken into account by the orthogonality condition model, is explained in detail in Ref. [12]. In the present calculation, we introduce a ^{4}He-n-n three-body force to describe effects from the internal degrees of freedom in the ^{4}He-core. This effective three-body force is expressed as $V_{\alpha nn}(r_1, r_2) = V_3 e^{-\nu(r_1^2+r_2^2)}$ with $V_3 = -0.218$ MeV and $\nu = (0.1/b)^2$ fm^{-2}, where $b = 1.4$ fm is a harmonic oscillator length parameter of the ^{4}He-core cluster wave function.[13] In Fig. 6 (a), we show the calculated energy spectrum of ^{6}He in comparison with the experimental data. It is seen that the 2_1^+-level observed experimentally as a genuine three-body resonance is well reproduced. Furthermore, the 2_2^+ resonance state is predicted. The detailed discussions of excited resonant states in ^{6}He are given in Ref. [12].

On the other hand, as shown in Fig. 6 (b), this ^{4}He+n+n model predicts no narrow

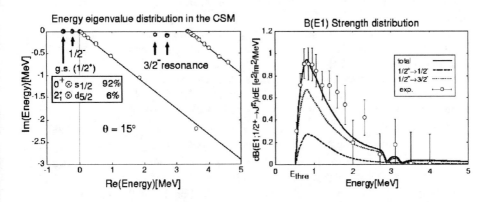

FIGURE 5. Comparison of the calculated $E1$ strength functions and experients[11].

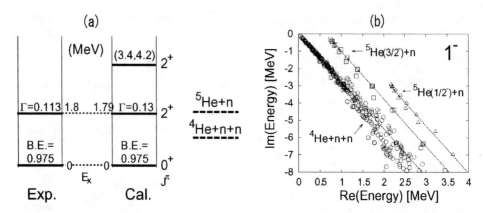

FIGURE 6. (a) The energy spectrum of ^{6}He. Values in parentheses show the resonance energy and the width of the second 2^+ state. (b) The 1^- energy eigenvalues of the complex scaled ^{4}He+n+n model Hamiltonian, where open circles show the 3-body continuum states, open squares and triangles the 2-body continuum states of ^{5}He($3/2^-$)+n and ^{5}He($1/2^-$)+n, respectively.

1^- resonances at the low energy region of ^{6}He. All eigenvalues of calculated 1^- states are distributed along the continuum lines of the ^{4}He+n+n three-body channel and of the two-body channels with ^{5}He($3/2^-$)+n and ^{5}He($1/2^-$)+n configurations. The strength distribution of the dipole transition from the 0^+ ground state to the 1^- continuum states provided us with interesting information on soft-dipole strengths and reaction mechanisms of the Coulomb breakup reaction. It is seen from the strength distribution into the ^{4}He+n+n three-body channel or ^{5}He($3/2^-$)+n and ^{5}He($1/2^-$)+n two-body channels whether ^{6}He dissociates directly into ^{4}He+n+n or by way of ^{5}He resonances.

In Fig. 7 (a), the calculated $E1$ strength distributions are shown. It is found that in the total strength, there is a low energy enhancement at around 1 MeV measured from the three-body threshold energy, which is just above the two body threshold of ^{5}He($3/2^-$)+n (0.74 MeV),[12] and the $E1$ strength decreases slowly with the excitation energy. The most interesting result is that the dominant transition strength comes from the two-body continuum component of ^{5}He($3/2^-$)+n with the low energy enhancement, and that the contributions from other components are relatively very small. This result indicates that the three-body dissociation strength of ^{6}He consists of dominantly the sequential breakup of a ^{6}He$\rightarrow^5$+n$\rightarrow^4$He+n+n process.

From this result, we can understand the structure of the ground state of ^{6}He as well. Although ^{5}He($3/2^-$) is a resonance, the wave function of ^{5}He($3/2^-$) in the ^{5}He($3/2^-$)+n channel has a large overlap with the ^{4}He+n($p_{3/2}$) configuration which is a dominant component in the ground state of ^{6}He. The result of the large transition strength into ^{5}He($3/2^-$)+n channel indicates that one of the neutrons of $(p_{3/2})^2$ in the ground state of ^{6}He is excited to a continuum state of s- or d-wave orbit by the external $E1$ transition field. Furthermore, we see that the three-body continuum states of ^{4}He+n+n does not contribute strongly. This result is also understood by considering that two $p_{3/2}$-neutrons in the ground state of ^{6}He, are hardly excited simultaneously in the breakup process. Another two-body continuum component of ^{5}He($1/2^-$)+n hardly contributes to the

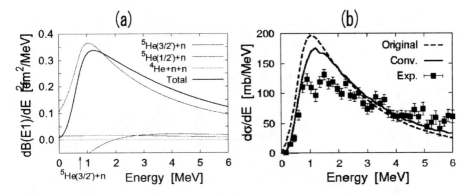

FIGURE 7. (a) E1 transition strength distributions, where dashed, dotted and dash-dotted lines are contributions from 2-body and 3-body continuum states. The thick solid line shows the total strength distribution. (b) Coulomb breakup cross sections of ^{6}He into ^{4}He+n+n with and without convolution. Experimental data are taken from Ref. [14]

strength because the probability of $(p_{1/2})^2$ is very small in the ground state, nearly a few percent. Furthermore, a large resonance width of ^{5}He($1/2^-$) (5.84 MeV) also reduces and broaden the strength, and so we can not see a distinguishable structure of the ^{5}He($1/2^-$)+n two-body component in the transition strength.

As shown in Fig. 7 (b), our calculated E1 strength distribution presents a good correspondence with the experimental data[14]. To compare with experimental data, we must take convolution with the experimental detector efficiency. We employed the convolution procedure which is the same as Aumann.[15]

As a typical two-neutron halo nucleus, ^{11}Li has extensively been studied from both experimental[1] and theoretical[2] sides. In theoretical studies, the ^{9}Li+n+n three-body model has extensively been studied by many people. However, the ^{9}Li+n+n three-body mode for ^{11}Li is less reliable in comparison with the ^{4}He+n+n model for ^{6}He. We discussed this problem from the viewpoint of the neutron $J^\pi = 0^+$ pairing correlation and the ^{9}Li-n interaction in connection with the spectroscopy of ^{10}Li in Refs. [16, 17, 18]

Recently, we proposed a new coupled-channel three-body model beyond the ^{9}Li+n+n model by throwing away the assumption of a single closed sub-shell configuration $(0s)^4(0p_{3/2})^1_\pi(0p_{3/2})^4_\nu$ for the ^{9}Li core. The wave function of ^{11}Li in the present new model is generally given as

$$\Psi(^{11}Li) = \sum_i \Phi_i(^9Li)\chi_i(\vec{r}_1, \vec{r}_2). \qquad (13)$$

Here, the suffix i denotes the channel specified by different configurations $\Phi_i(^9Li)$ for ^{9}Li and then $\chi_i(\vec{r}_1, \vec{r}_2)$ is the wave function of the active valence two neutrons.

Since paying much attention to the importance of the neutron $J^\pi = 0^+$ pairing in the binding of ^{11}Li, we describe the ^{9}Li core in ^{11}Li as

$$\Phi_{gr}(^9Li) = \alpha_0\Phi(C0) + \alpha_1\Phi(C1) + \alpha_2\Phi(C2) + \cdots, \qquad (14)$$

where

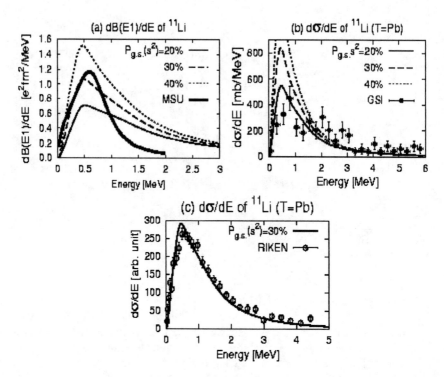

FIGURE 8. Coulomb breakup cross section of ^{11}Li into ^{9}Li+n+n. Experimental data are taken from (a) [21], (b) [22] and (c) [20], and we show the calculations of with different s-wave amptitudes in (a) and (b) and with the 40% fixed amplitude in (c).

$$(C0) \quad : \quad (0s)^4 (0p_{3/2})\pi(0p_{3/2})^4_\nu, \tag{15}$$

$$(C1) \quad : \quad (0s)^4 (0p_{3/2})\pi(0p_{3/2})^2_{\nu,J_1=0}(0p_{1/2})^2_{\nu,J_2=0}, \tag{16}$$

$$(C2) \quad : \quad (0s)^4 (0p_{3/2})\pi(0p_{3/2})^2_{\nu,J_1=0}(1s_{1/2})^2_{\nu,J_2=0}. \tag{17}$$

$$\cdots.$$

The coefficients $\{\alpha_i; \ i = 0, 1, 2, \cdots\}$ are determined by solving the Schrödinger equation $H_c(^9\mathrm{Li})\Phi_{gr}(^9\mathrm{Li}) = \mathscr{E}_{gr}\Phi_{gr}(^9\mathrm{Li})$ when the ^{9}Li core cluster is isolated from the valence neutrons. However, in ^{11}Li, they depend on the motion of valence neutrons in the ^{9}Li+neutrons system. They are described by the wave functions $\{\chi_i\}$ being solutions of the coupled-channel equation. We first applied this model to the ^{9}Li+n system, and successfully investigated ^{10}Li by solving the coupled-channel equation.[18]

In Ref. [19], this new model was applied to ^{11}Li, and it was shown that a long-standing problems of the small binding energy, the large r.m.s. radius and the large amplitude of $(s_{1/2})^2$ in ^{11}Li can be solved in consistent way with the spectroscopy of ^{10}Li, which could not give the consistent answer in the previous studies of the single channel case. Using the obtained wave functions of ^{11}Li, we calculate the Coulomb breakup cross sections and compare the results with experiments[20, 21, 22] as shown in Fig. 8. The calculated results depend largely on the s-wave amplitudes of halo neutrons in the ground state of ^{11}Li. In Figs. 8 (a) and (b), comparison between experiments of MSU[21] and GSI[22] and the results calculated with changing the s-wave amplitude. Furthermore, we can see that the data of RIKEN[20] is well reproduced by the calculation with the 40% s-wave amplitude.

CONCLUSION

In this paper, we discussed that the complex scaling method is very useful in not only calculating the resonance energies and widths but also studying the structure of continuum states of many-body systems. Due to the weak binding of drip-line nuclei, those nuclei breakup easily by adding an external field. The breakup spectrum depends on the halo structure of the ground state, and also is considered to show an influence from properties of continuum states. The present method using the complex scaling method is very promising for study on those Coulomb breakup reactions of halo nuclei.

REFERENCES

1. I. Tanihata, J. Phys. G: Nucl. Part. Phys. **22** (1996), 157.
2. M. V. Zhukov *et. al.*,Phys. Rep. **231** (1993), 151.
3. J. Aguilar and J.M. Combes, Commun. Math. Phys. **22** (1971), 269; E. Balslev and J.M. Combes, Commun. Math. Phys. **22** (1971), 280.
4. M. Homma, T. Myo and K. Katō, Prog. Theor. Phys. **97** (1997), 561.
5. T. Myo and K. Katō, Prog. Theor. Phys. **98** (1997), 1275
6. T. Myo, A. Ohnishi and K. Katō, Prog. Theor. Phys. **99** (1998), 801.
7. Ya. B. Zel'dovich, Zn. Eksp. I Theor. Fiz. **39** (1960), 776; Sov. Phys. JETP **12** (1961), 542.
   N. Hokkyo, Prog. Theor. Phys. **33** (1965), 1116
   T, Berggren, Nucl. Phys. **A109** (1968), 265.
   W.J. Romo, Nucl. Phys. **A116** (1968), 617.
8. G. Gyarmati and T. Vertse, Nucl. Phys. **A160** (1971), A160.
9. R. G. Newton, J. Math. Phys. **1** '1960), 319; B. G. Giraud and K. Katō, Preprint.
10. R. Suzuki *et.al.*, Proc. of Int. Symp. on Exotic Nuclear Structures, 15-20 May 2000, Debrecen, Ed. by Dombrádi and A. Krasznahorkay, pp.373-376.
11. T. Nakamura *et. al.*, Phys. Lett. **B331** (1994), 296.
12. S. Aoyama, S. Mukai, K. Katō and K. Ikeda, Prog. Theor. Phys. **93** (1995), 99; **94** (1995), 343.
13. T. Myo, K. Katō, S. Aoyama and K. Ikeda, Phys. Rev. **C63** (2001), 054313.
14. T. Aumann *et al.* Phys. Rev. **C59** (1999), 1252.
15. T. Aumann, private communication.
16. K. Katō and K. Ikeda, Prog. Theor. Phys. **89** (1993), 623.
17. S. Mukai, S. Aoyama, K. Katō and K. Ikeda, Prog. Theor. Phys. **99** (1998), 381.
18. K. Katō, T. Yamada and K. Ikeda, Prog. Theor. Phys. **101** (1999), 119.
19. T. Myo, S. Aoyama, K. Katō and K. Ikeda, Prog. Theor. Phys. **108** (2002), 133.
20. S. Shimoura *et. al.* Phys. Lett. **B348** (1995), 29.
21. D. Sackett *et. al.* Phys. Rev. **C48** (1992), 118.
22. M . Zinser *et. al.* Nucl. Phys. **A619** (1997), 151.

Shape effects, U(3) symmetry and heavy clusterization

A. Algora[1,2], J. Cseh[1] and P. O. Hess[3]

[1]*Institute of Nuclear Research of the Hungarian Academy of Sciences,*
Pf. 51, H-4001 Debrecen, Hungary
[2]*IFIC-Univ. Valencia, Apartado Oficial 22085, 46071 Valencia,*
Valencia, Spain
[3] *Instituto de Ciencias Nuclerares, UNAM, Circuito Exterior, C.U.,*
A.P. 70-543, 04510 México, D.F., Mexico

Abstract. The structural aspects of the fission channels of ^{252}Cf are studied from the microscopic viewpoint using the U(3) selection rule. Effective U(3) symmetry labels, which are consistent with deformation, are used to characterize the parent and daughter nuclei. Our study shows a non-uniform structural dependence of the relative preference for different fission channels.

INTRODUCTION

Recent experimental studies of the fission process using multi-detector arrays have revealed many interesting new features of this still puzzling phenomena. Among these new results it is worth mentioning the first direct identification of neutronless binary fission channels and the detection of exotic neutronless ternary processes in ^{252}Cf [1]. These new results and the more detailed spectroscopy in the second and third minima (of the energy surface) studied through fission [2] raises the question of to what extent microscopic effects may play an important role in the fission process.

In a simplified picture, fission can be viewed as a two step process: first, a cluster state is formed, and then the cluster state decays through the barrier. From this point of view a cluster basis can be considered as a natural basis for the study of the first stage of the process. To study this question an extension of the U(3) selection rule to heavy nuclei has been used to see if there is any microscopic preference for certain fission channels in the binary clusterizations of ^{252}Cf. In comparison with our previous study [3] the novel feature of the present work is that we have performed a systematic study over a large number of possible binary channels.

U(3) SELECTION RULE FOR CLUSTER CONFIGURATIONS

As it is well known in the cluster studies of light nuclei, the U(3) symmetry gives a connection between the Shell Model (SM) and the Cluster Model (CM) (see e.g. [4] and references therein). This connection gives rise to a simple U(3) selection rule, based on the microscopic nuclear structure [5]. In its original from it is based on a

CP644, *Exotic Clustering: 4th Catania Relativistic Ion Studies,* edited by S. Costa, A. Insolia, and C. Tuvè
© 2002 American Institute of Physics 0-7354-0099-7/02/$19.00

harmonic oscillator approximation [6], however, it turns out to be valid for more general interactions as well [7].

In practice the U(3) selection rule is used in the following way: let us characterize each nucleus with a single U(3) representation (the parent nucleus is labelled by the $[n_1, n_2, n_3]$ quantum numbers, and the daughter nuclei by the $[n_1^{C_1}, n_2^{C_1}, n_3^{C_1}]$, and $[n_1^{C_2}, n_2^{C_2}, n_3^{C_2}]$ Young patterns). Then the wavefunction of the cluster model carries the representation of:

$$[n_1^{C_1}, n_2^{C_1}, n_3^{C_1}] \otimes [n_1^{C_2}, n_2^{C_2}, n_3^{C_2}] \otimes [n, 0, 0] = \sum_k \oplus [n_{1,k}^c, n_{2,k}^c, n_{3,k}^c], \qquad (1)$$

where $[n, 0, 0]$ stands for the relative motion. The U(3) selection rule requires the matching between $[n_1, n_2, n_3]$ and the product representations of Eq. (1).

The question is how to determine the U(3) representations of the nuclei of interest. In the case of light nuclei this question can be easily addressed since it is well known that the U(3) symmetry is valid and the nuclear states can be characterized by a single irrep (leading irrep). For heavy nuclei the situation is not so simple since the U(3) symmetry is not valid and different irreps are mixed due to the spin-orbit and other interactions.

One possible solution to this problem is the use of what have been called "effective" representations [8]. An effective representation is a mixture of pure U(3) representations that behaves like an U(3) representation of its own. In ref. [9] a procedure for the determination of effective representations for strongly deformed prolate nuclei is outlined. The effective quantum numbers take into account the renormalization due to the coupling of higher shells. Another important feature of this work is that these effective representations can be endowed with microscopic content in terms of asymptotic Nilsson model states. Our approximation is based on the same concept, but compared with ref. [9] some further development was required. In a systematic study of the kind we present here not only strongly deformed prolate nuclei appear. The studied nuclei can have also oblate deformation, and the range of deformation may change from near spherical to well deformed. We have recently extended the procedure of [9] to fulfill these requirements [10].

The effective U(3) quantum numbers are determined from the occupation of the Nilsson-orbits [11], corresponding to the relevant deformations [12] using the relations of ref. [9, 10]. The number of quanta of the relative motion in (1) was determined using the Harvey prescription in order to take into account the Pauli principle [13]. To characterize quantitatively the "forbiddenness" of each fission channel the reciprocal forbiddenness defined in ref. [14] was used.

APPLICATION TO THE BINARY CLUSTERIZATIONS OF ^{252}CF: RESULTS.

We study the structure effects on the binary cluster configurations of ^{252}Cf:

$$^{252}_{98}Cf \rightarrow\ ^{A}_{Z}X + ^{252-A}_{98-Z}Y. \qquad (2)$$

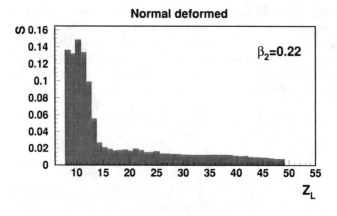

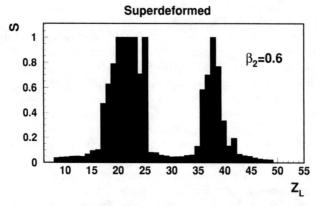

FIGURE 1. The reciprocal forbiddenness S versus the Z_{light} of the studied binary fission channels. According to (2) Z_{light} fully determines the fission channel in its first stage. The values of S correspond to mean values over channels that have the same Z_{light} and different A_{light}. In the upper part of the figure the parent nucleus has a deformation $\beta \sim 0.22$. In the lower part of the figure the parent nucleus has a 'hypothetical' superdeformation $\beta \sim 0.6$ The Z dependence shows a non-uniform behavior.

Once the fission channels are defined, the U(3) selection rule can be applied as it was outlined in the former section. From its application an insight can be obtained on the microscopic preference for different fission channels. The preference of different cluster configurations can be characterized by the reciprocal forbiddenness which takes the value of 1 for allowed clusterizations (generally $0 \leq S \leq 1$).

Using ground state deformations for the parent and daughter nuclei [12], all studied cluster configurations in question turn out to be forbidden (Fig. 1, upper part). A clear tendency towards cluster radioactivity (or very asymmetric fission) can be inferred from this figure. One aspect that seems to be of particular interest: the low-lying cluster configuration of two prolate nuclei is not the pole-to-pole one, rather both clusters are declined from the molecular axis. For example the lowest lying pole-pole configuration of ^{144}Ba + ^{108}Mo lies at higher excitation than the one with declined axes. This result shows that the pole-pole configurations, preferred by penetrability calculations are highly Pauli-forbidden. Therefore they can be only small components in the ground state wave function of ^{252}Cf. The "compact packing" is favored by structure.

There are many interesting questions that still can be addressed in this rather simple framework. One of particular interest is if there are allowed clusterizations in case we change the deformation of the parent nucleus (for example to superdeformation or hyperdeformation). For that we have studied the possible clusterizations of an 'hypothetical' superdeformed ^{252}Cf ($\beta_2 \sim 0.6$). The obtained results are presented in the lower part of Fig. 1. As in the former calculations the daughter nuclei are considered to have ground state deformations. It is interesting to see that in this case we have allowed clusterizations as well. The regions of allowed clusterizations correspond to two particular cases: a) both clusters having large prolate quadrupole deformation (region with $Z_{light} \sim 36$), b) one cluster with prolate quadrupole deformation and the other with oblate deformation (region with $Z_{light} \sim 22$).

In comparison with the penetrability calculations, our structure considerations reflects another aspect of the fission process. The penetrabilities are mainly related to the Coulomb forces, while the structural effects are governed by the Pauli principle.

ACKNOWLEDGMENTS

This work was supported by the MTA-CONACyT and the CSIC-MTA exchange programs and by the OTKA (grant No. 37502). A. A. acknowledges partial support of the EC HPMFCT-1999-00394 and FPA 2002-04181-C04-03 contracts.

REFERENCES

1. Hamilton J H *et al* 1994 *J. Phys. G* **20** L85, Ramaya A V *et al* 1998 *Phys. Rev. Lett.* **81** 947
2. Krasznahorkay A *et al.* 1998 *Phys. Rev. Lett.* **80** 2073; Krasznahorkay A *et al.* 2001 *Acta Physica Polonica B* **32** 657
3. Algora A, Cseh J and Hess P O 1998, *J. Phys. G* **24** 2111
4. Cseh J 1992*Phys. Lett. B* **281** 173; Cseh J and Levai G 1994 *Ann. Phys. (NY)* **230** 165
5. Cseh J and Scheid W 1992 *J. Phys. G* **18** 1419, Cseh J 1993 *J. Phys. G* **19** L97
6. Wildermuth K and Kanellopoulos Th 1958 *Nucl. Phys.* **7** 150
7. Cseh J, Levai G, Algora A, Hess P O, Intasorn A, Kato K 2000 *Acta Physica Hungarica New Series-Heavy Ion Physics* **12** 119
8. Rowe D J 1996 *Prog. Part. Nucl. Phys.* **37** 265
9. Jarrio M, Wood J L and Rowe D J 1991 *Nucl. Phys. A* **528** 409
10. Hess P O, Algora A, Hunyadi M, and Cseh J *Eur. Phys. J. A*, in press
11. Lederer C M and Shirley V S 1978 *Table of Isotopes* 7th edn (New York: Wiley) appendices p. 37 and references therein, Velazquez Aguilar V M and Hess P O, Nilsson code, unpublished
12. Möller P *et al* 1995 *At. Nucl. Dat. Tab.* **59** 185
13. Harvey M *Proc. 2nd Int. Conf. on Clustering Phenomena on Nuclei College Park, USDERA report* ORO–4856–26 p. 549
14. Algora A and Cseh J 1996 *J. Phys. G* **22** L39

Cluster Radioactivity

From the Discovery of Cluster Radioactivity to Future Perspectives

Alessandra Guglielmetti

Istituto di Fisica Generale Applicata, Università degli Studi di Milano
INFN, Sezione di Milano, Italy

Abstract. A brief historical overview on the experimental work done in recent years on cluster radioactivity is given. The experimental techniques are discussed and a few most representative cases are presented. A few open problems are discussed, as far as the experimental and the theoretical interpretations are concerned.

INTRODUCTION

Although the official date for the discovery of cluster radioactivity is always reported to be 1984, the year in which the spontaneous emission of ^{14}C clusters from ^{223}Ra was reported by Rose and Jones (1) and Alexandrov et al. (2), probably already in 1952 such a fascinating process was unknowlingly measured, when the signals from the particle decay of ^{232}U were attributed to spontaneous fission instead of the most probable ^{24}Ne emission, as was shown in recent years (3). Moreover, one should mention that already in 1980, 4 years before the experimental confirmation, was the pioneering theoretical paper of Sandulescu et al (4) published.

In the following years not only the Rose and Jones experiment was confirmed beyond any reasonable doubt, but many other cases were discovered and a kind of systematics was rapidly obtained. In 1984 ^{24}Ne emission was measured (5), followed in 1989 by ^{28}Mg (6), in 1991 by ^{20}O (7), in 1992 by ^{23}F (8). Other most important milestones are the first (and up to now, the only one) fine structure measurement in 1989 (9), the experiments on hindrance factors of odd-A emitters (1992-3)(10), the attempts to extend the measurements in the trans-tin region (1994-7)(11), the observation of the heaviest cluster, ^{34}Si (2000)(12).

Today twenty-three cases of cluster radioactivity are known from trans-lead nuclei, with partial half lives from 10^{-9} down to 10^{-17} (Fig. 1) (13); a dozen theoretical models have been proposed to explain this mechanism, which seems to have features in common both with alpha decay and (cold) spontaneous fission (for a review see other papers at this Conference). Comparison with theories has, in any case, helped to point out the main features of this decay mode, above all its strong dependence on the barrier penetration and consequently on the Q-value, which must be as high as possible in order to compensate for the small preformation factors typical of the heavy clusters being investigated. This explains why all measured cases led to the doubly magic ^{208}Pb residual nucleus, or its immediate neighbours.

CP644, *Exotic Clustering: 4th Catania Relativistic Ion Studies*, edited by S. Costa, A. Insolia, and C. Tuvè
© 2002 American Institute of Physics 0-7354-0099-7/02/$19.00

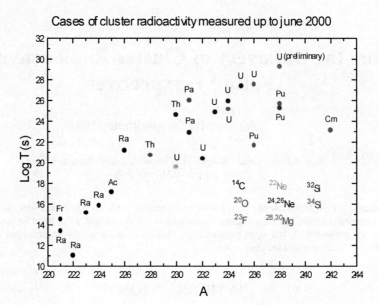

Fig. 1. Partial half lives vs. mass number of mother nucleus for all cases of cluster radioactivity measured up to now.

EXPERIMENTAL TECHNIQUES

Sources

Basically two different types of sources have been used in experiments on cluster radioactivity: chemical sources and ion-implanted ones. Chemical sources are obtained by chemical separation and generally by electrodeposition of the active material and allow to perform experiments "off-line". On the other hand, ion-implanted sources are most often used for experiments "on-line" which is the only possibility when the nuclide half-life is too short. A few examples are sources obtained at Isolde (10) and GSI (11). Common features of all sources are a rather strong activity (up to a few mCi), necessary for compensating the small branching ratios, a high purity or at least a very well known isotopic and elemental composition, to be able to disentangle contributions in cluster emission from competitive cases, and a relatively low thickness (typically less than 1 mg/cm^2), to enable clusters of 2-2.5 Mev/amu to escape from the source with enough energy to be detected and identified.

Detectors

Small decay constants and small branching ratios demand for a high efficiency and high selectivity detecting apparatus. High selectivity means high rejection power of the huge flux of alpha particles accompanying the rare events searched for: so far, the most used detection technique has been the one of solid state nuclear track detectors (SSNTD) (13). They are generally plastics or glasses which are able to record latent tracks of heavy ions whose reduced energy loss is above a given threshold, characteristic of the detector itself. A chemical etching is necessary for enlarging this damage up to dimensions visible under an optical microscope. What happens is that, while the non-irradiated surface is etched at a uniform velocity V_G, etching along the particle trajectory might proceed at a higher velocity V_T. In this situation, competition between the two velocities produces a conical track appearing as a dark spot in the bright field of an optical microscope. The physical basis underlying particle identification by means of SSNTDs is the dependence of V_T on the particle reduced energy loss which, in turn, is related by means of the Bethe formula to the particle charge and mass numbers and to its energy. In practice, identification is achieved by measuring the sensitivity $S=V_T/V_G$ at a given value of the particle range. This is essentially the same method as the ΔE-E one on which a silicon detector telescope is based.

The high flux of alpha particles can be rejected simply by choosing detectors which are not sensitive to alphas. Moreover these cheap plates can be tailored in whatever shape and a very high (100%) geometrical efficiency apparatus can be built. The main disadvantage of such a technique lies in the poor energy resolution (1-2 MeV for ions with E=30-80 MeV) which does not allow fine structure measurements in the decay spectra. The mass resolution (2-3 units for A=12-30) is also not enough to distinguish between adjacent isotopes; usually the mass attribution is achieved on the basis of the maximum Q-value criterion.

RECENT RESULTS

Uranium Isotopes Decays

The large systematics accumulated on uranium isotopes in recent years (A=232-236) (13) has pointed out the different behavior of the partial half-lives for spontaneous fission and cluster radioactivities with the mass number of the parent, the former being practically constant and the latter rapidly varying as a consequence of shell and other structure effects. In view of the additional feature that both spontaneous fission and cluster radioactivity have, for this particular isotopic series, partial half-lives of similar order of magnitude, it is interesting to attempt to extend the systematics both from the light and heavy sides. This is why both ^{230}U (14) and ^{238}U cluster decay have been recently studied (for the latter see the paper of S. P. Tretyakova at this Conference).

The well-known sensitivity of the cluster decay rate to the Q-value strongly restricts the number of possible clusters being emitted by a trans-lead nucleus. For ^{230}U, it is found that the maximum Q-value, 61.40 MeV, is reached for ^{22}Ne emission (by leading to ^{208}Pb); ^{24}Ne, the cluster measured for heavier (A=232-234) U isotopes (13) also looks like a good candidate (Q=61.36 MeV); other clusters like different isotopes of F or Mg give largely unfavorable Q-values. Several calculations have been performed on ^{22}Ne decay of ^{230}U giving branching ratios relative to α decay in the range 10^{-14}-10^{-15}. In order to be able to observe a few cluster decay events, such low branching ratios imply a strong ^{230}U radioactive source, at least in the mCi range. The source was produced by irradiating metallic thorium targets with intense proton beams: the ^{232}Th(p,3n) reaction gives ^{230}Pa which subsequently β decays to ^{230}U. After separation from ^{230}Pa, ^{230}U was electroplated on two platinum disks, as sources of 0.5 cm^2 area and 1.3 mCi and 0.6 mCi activity, respectively. The task of detecting a few Ne ions within an enormous background of α particles was accomplished by solid state nuclear track detectors. In comparison with similar experiments on cluster radioactivity, the present one was made difficult by the coexistence of a rather low branching ratio together with a relative light cluster (therefore having a relatively low ionizing rate). The solution to the problem of finding the track detector giving the optimum signal-to-noise ratio for this particular critical case was given by BP-1, a phosphate glass, etched under desensitizing conditions (HF, 50%, 25 °C). To maximize the detection efficiency while keeping the α integrated flux as low as possible, the BP-1 glasses were arranged in a 4π geometry by covering the inner surface of an aluminum sphere, Φ=17 cm. The task of scanning the large detector surface ($\approx$ 700 cm^2) was made easier by an automatic scanning device coupled to an image analyzer (8) . A total of 6 events was found and identification was made possible by comparing the sensitivity S=V_T/V_G measured for each track at several points along the particle trajectory with accelerator calibrations (see Fig. 2): all events were attributed to Ne ions. As far as the mass number of the Ne clusters, it is well known that track detectors do not have enough A resolution to allow an unambiguous identification.Therefore, in order to decide in favor of one of the two most probable Ne clusters, either ^{22}Ne or ^{24}Ne, we resorted to theoretical calculations finding that the former case is more favorable than the latter by two orders of magnitude. By taking into account the decayed number of ^{230}U atoms ($1.83*10^{14}$ in one hemisphere and $1.25*10^{14}$ in the other) and the detector coverage efficiency ($\approx$ 78% of the geometrical surface) we obtained: B_α =$(4.8\pm2.0)*10^{-14}$ for the branching ratio and $T_{1/2}$=$(3.7\pm 1.5)*10^{19}$ s for the partial half-life. When comparing the predictions given by a few theoretical models with our result, we find a satisfactory agreement, generally within one order of magnitude. Our result for ^{22}Ne is shown in Fig. 3 together with those existing in the literature (13) or recently obtained for the other U isotopes (15). It is interesting to see how the partial half-life continues to decrease with decreasing neutron number of the uranium mother nucleus: more specifically by comparing the lightest U isotopes (^{230,232}U) which both decay with very similar Q-values to the same bimagic ^{208}Pb daughter nucleus by emitting two different Ne isotopes (^{22,24}Ne) we can

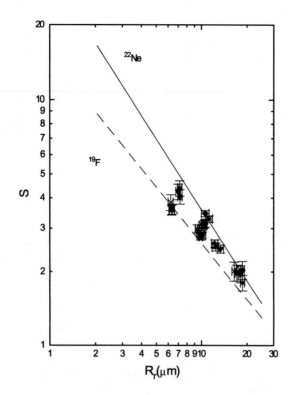

Fig 2- Comparison of the sensitivity S measured for the events detected from the ^{230}U decay at different points along the particle trajectory with accelerator calibration curves vs. R_r, the particle residual range.

trace mainly such an effect to the higher preformation probability of the lighter cluster (^{22}Ne) in respect to the one of the heavier (^{24}Ne). For what concerns the comparison between decay rate for spontaneous fission and cluster emission we note that while the former is practically constant for all uranium isotopes, being dependent on the almost stable Z^2/A fissility parameter, the latter is not. This is due to the presence of shell effects which critically determines a strong dependence on the Q-value.

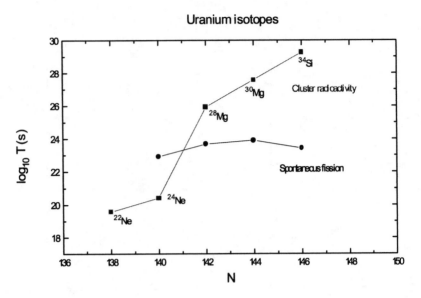

Fig. 3- Logarithm of the half life for cluster radioactivity (squares) and spontaneous fission (circles) vs. neutron number for different even-even uranium isotopes.

^{225}Ac Decay By ^{14}C Emission

One of the most important achievements in the yet ultra-decennial research on cluster radioactivity was the discovery of the sensitivity of its partial half-life to the microscopic properties of the mother-daughter nuclei. This resulted from many evidences like e.g. the fine structure in the energy spectrum of ^{14}C clusters in the well known ^{223}Ra decay, the hindered decay of odd-A emitters like ^{221}Fr, ^{221}Ra, ^{223}Ra, ^{231}Pa, ^{233}U and others, the anomalously high/low values of two clusters branching ratios such as ^{24}Ne/^{28}Mg, ^{23}F/^{24}Ne and others (13). Within such a framework, ^{225}Ac was a special case since while being an odd-Z nucleus its partial decay rate as well its branching ratio relative to α decay do not seem to exhibit any special hindrance like the one of other odd-A exotic emitters. By measuring the degree according to which and odd-A transition is slower in comparison with an even-even one having the same barrier penetrability, the so called hindrance factor, one finds typical values in the range 10-100 for ^{14}C emitters and a value close to 1 for ^{225}Ac$\rightarrow$^{14}C (10). The fact that such a case seems surprisingly to behave like an even-even one has been variously justified by using arguments based on the microscopic structure of ^{225}Ac. In order to have a firmer basis for any theoretical interpretation it was decided to repeat this experiment, already performed in 1993(10), under improved experimental conditions. A very strong ^{225}Ac source was prepared to be used in two independent experiments:

-1 an high efficiency but low energy resolution experiment to re-measure the integrated decay rate of the ^{225}Ac$\rightarrow$ ^{14}C+ ^{211}Bi decay.

-2 an high resolution experiment to study the energy spectrum of the emitted ^{14}C clusters.

The ^{225}Ac source was obtained by taking profit of the intense, mass separated 60 keV ^{225}(Fr+Ra) beam produced at ISOLDE, CERN. Here such nuclides are obtained by means of spallation reactions induced by 1 GeV proton beam on a thick ThC$_2$ target. In the course of 4.8 day irradiation time, about $2*10^{15}$ ^{225}(Ra+Fr) atoms were collected. The ^{225}Ac source was subsequently obtained by β decay. In the first experiment the source was placed in the center of an aluminum hemisphere, 19.5 cm diameter, covered inside with BP-1 glass, for 260 h. After etching and scanning by means of an automated system, 14 events were found in 86 cm^2 of detector surface. Figure 4 shows a comparison of sensitivity vs. residual range for these events and the ^{14}C calibration curve: the agreement is astonishing. From the known source intensity and irradiation time, the branching ratio for ^{14}C decay of ^{225}Ac came out to be $(4.5\pm1.4)10^{-12}$ and the corresponding partial half-life $T_{1/2}=(1.9\pm0.6)10^{17}$ s, in very good agreement with the previous measurement (10).

The goal of the second experiment was to perform a fine-structure measurement in order to find out which level of the ^{211}Bi residual nuclide is preferentially fed by the ^{14}C decay of ^{225}Ac. For this measurement it would have been impossible to make use of track detectors due to their poor energy resolution; the source was therefore placed inside the superconducting solenoidal spectrometer SOLENO of IPN-Orsay, whose selective features allow to focalize ^{14}C ions at the focal plane of the spectrometer while rejecting the enormous flux of α particles. By putting a Si detector at the focal point it is therefore possible in principle to measure ^{14}C energy spectrum with a good energy resolution. Unfortunately the 8.5 day long experiment turned out to be difficult due to pile-up and radiation damage of the Si detector because of the unusual high source activity and no unambigous result was found out.

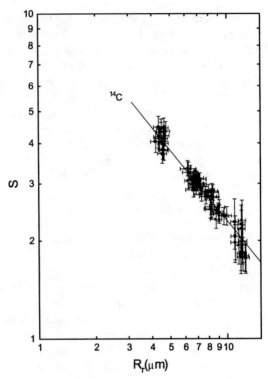

Fig.4- Comparison of the sensitivity S measured for the events detected from the ^{225}Ac decay with accelerator calibrations vs the particle residual range R_r.

OPEN PROBLEMS AND PROSPECTS

The results presented in the previous section and the systematics accumulated in the many years of intense investigation point out that cluster radioactivity is now a well established phenomenon which extended our knowledge on the ground state properties of nuclei and stimulated attempts toward finding an unified description of the different hadronic decay modes of nuclei, α and cluster decays and spontaneous fission (13). This does not imply, of course, that no space is left for further progress.

As a matter of fact, there are essentially three still open problems, namely (i) the possible existence of other islands of cluster radioactivity in which the residual nuclei are close to the other doubly-magic ones like ^{100}Sn and ^{132}Sn, (ii) the investigation of even-odd effects, both from the experimental but especially from the theoretical side, and (iii) the study of the emission of very heavy clusters in view of a possible solution of the α decay-like vs. fission-like model problem (12). As far as the first point is concerned, several attempts of measuring ^{12}C cluster radioactivity of ^{114}Ba have already been performed (11) leading to the conclusion that, unfortunately, the present experimental techniques do not guarantee a sensitivity high enough to allow detection of such cases involving nuclei very far away from the stability line.

The study of even-odd effects of which the already described ^{225}Ac case is one example could be extended, from the experimental side, by the ^{223}Ac decay by ^{14}C and ^{15}N emissions (see the paper of A. A. Ogloblin at this Conference) and ^{229}Th decay via Oxygen emission. However, the main progress is expected here from the theoretical side: apart from the attempt of Dumitrescu et al. (16) to microscopically reproduce the measured hindrance factors, no other calculations have been performed either by using the α decay or the fission models.

Finally, other heavy cluster emissions could be studied in principle, like Si from ^{241}Am and heavier (Cf) emitters.

It is desirable that from these and other future experiments an unambiguous and clear theoretical interpretation of this phenomenon could be obtained.

REFERENCES

1. H.J. Rose and G.A. Jones, *Nature* **307** (1984) p. 245
2. D.V. Alexandrov et al., *JETP Lett.,* 40 (1984) p. 909
3. R. Bonetti et al., *Phys. Lett.* **B 241** (1990), p. 179
4. A. Sandulescu, D.N. Poenaru and W. Greiner, *Sov. J. Part. Nucl.,* **11** (1980) p. 528
5. S.W. Barwick, P.B. Price and J.D. Stevenson, *Phys. Rev.* **C 31** (1985) p. 1984
6. S.P. Tretyakova et al., *Z. Phys.* **A 333** (1989) p. 349;
   S. Wang, D. Snowden-Ifft, P.B. Price et al., *Phys. Rev* **C 39** (1989) p. 1647
7. R. Bonetti et al., *Nucl. Phys.* **A 556** (1993) p. 115
8. P.B. Price et al., *Phys. Rev.* **C 46** (1992) p. 1939
9. L. Brillard et al., *C R Acad. Sci. Paris* **309** Ser II (1989) p. 1105
10. R. Bonetti et al., *Nucl. Phys.* **A 576** (1994) p. 21;
    R. Bonetti et al., *Nucl. Phys.* **A 562** (1993) p. 32
11. A. Guglielmetti et al., *Phys. Rev.* **C 56** (1997) p. R2912
12. A.A. Ogloblin et al., *Phys. Rev.* **C 61** (2000) p. 034301
13. R. Bonetti and A. Guglielmetti, "Measurements in Cluster Radioactivity- Present Experimental Status" in *Heavy Elements and Related New Phenomena*, vol II, edited by W. Greiner and R.K. Gupta (World Scientific, Singapore, 1999), p. 643
14. R. Bonetti et al., *Nucl. Phys* **A 686** (2001) p. 64
15. R. Bonetti et al, *Phys. Rev* **C 62** (2000) p. 047304
16. R. Bonetti, I. Bulboaca, F. Carstoiu, O. Dumitrescu, *Phys. Lett* **B 396** (1997) p. 15

Status of the Cluster Radioactivity

D. N. Poenaru*, R. A. Gherghescu*, Y. Nagame† and W. Greiner**

*Horia Hulubei National Institute of Physics and Nuclear Engineering,
PO Box MG-6, RO-76900 Bucharest-Magurele, Romania
†Advanced Science Research Center, Japan Atomic Energy Research Institute,
Tokai, Ibaraki 319-1195, Japan
**Institut für Theoretische Physik der Universität, Pf 111932,
D-60054 Frankfurt am Main, Germany

Abstract. Many heavy nuclei with atomic numbers in the range 87–96 were experimentally identified as cluster emitters since 1984, confirming earlier predictions. The measured partial half-lives against ^{14}C, ^{18,20}O, ^{23}F, $^{22,24-26}$Ne, ^{28,30}Mg and ^{32,34}Si radioactivities were found to be in good agreement with calculated values within analytical superasymmetric fission model. Experimental difficulties are mainly related to the low yield in the presence of a strong background of α-particles. Until now, only some of the most favourable cases were investigated, leading to magic or almost magic proton and neutron numbers of daughter nuclei. We present a systematics of experimental results compared to calculations, clearly showing other possible candidates for future experiments. Universal curves may be used to estimate the expected half-lives.

> **Motto:** "The goal of science ... is to discover
> simplicity in the midst of complexity"
> *L. Hartwell, Nobel Centennial, 2001.*

INTRODUCTION

Fission phenomena continue to surprise us even nowadays. After major discoveries, by experimentalists, of induced and spontaneous fission in 1939 and 1940, the series continued in 1946 with particle accompanied fission, with cold binary fission in 1981 and with α and ^{10}Be accompanied cold fission as well as the double and triple fine structure in a binary and ternary fission in 1998.

On the other hand cluster radioactivities had been predicted [1] in 1980, four years before the first (brilliant) experimental confirmation of ^{14}C radioactivity of ^{223}Ra was reported [2]. Predictions were initially [2, 3, 4] ignored, but shorter after claiming [5] patternity, it was subsequently acknowledged. At present the most cited two papers in the field are [1, 2], as starting points of theoretical and experimental works. Further studies were rapidly boosted (compare the reviews [6, 7, 8], the multiauthored book [9] and the references therein with [10, 11]). A charged particle heavier than ^{4}He but lighter than a fission fragment is spontaneously emitted in a cluster decay mode of an atomic nucleus. There is a whole family of such disintegration modes: ^{14}C radioactivity; ^{24}Ne radioactivity; ^{28}Mg radioactivity, and so on. The obtained until now data on half-lives and branching ratios relative to α-decay of ^{14}C, ^{18,20}O, ^{23}F, $^{22,24-26}$Ne, ^{28,30}Mg and ^{32,34}Si radioactivities are in good agreement with predicted values within the analytical superasymmetric fission (ASAF) model, as we will show below. A new "number" was

CP644, *Exotic Clustering: 4th Catania Relativistic Ion Studies*, edited by S. Costa, A. Insolia, and C. Tuvè
© 2002 American Institute of Physics 0-7354-0099-7/02/$19.00

introduced in Physics and Astronomy Classification Scheme (PACS): *23.70.+j Heavy-particle decay.*

Four theoretical models with predicting power were reviewed in 1980 [1]: fragmentation theory; penetrability calculations like in traditional theory of α-decay; numerical (NuSAF)- and analytical (ASAF) superasymmetric fission models. A new superasymmetric peak, experimentally confirmed as a "sholder", has been obtained in the ^{252}No fission fragment mass distribution calculation, based on the fragmentation theory and the two center shell model developed by the Frankfurt school. One of the eight decay modes by cluster emission, predicted in 1980 by calculating the penetrability, from sixteen even-even parents, has been ^{14}C decay of ^{222,224}Ra. Three variants of the numerical superasymmetric fission (NuSAF) models were developed since 1979 by adding to the macroscopic deformation energy of binary systems with different charge densities a phenomenological shell correction term, and by performing numerical calculations within WKB approximation. In this way a qood agreement with experimental half-lives was obtained for many even-even α-emitters over a range of 24 orders of magnitude.

A very large number of combinations parent – emitted cluster had to be considered in a systematic search for new decay modes. In order to check the metastability of more than 2000 nuclides with measured masses tabulated by Wapstra and Audi, against about 200 isotopes of the elements with $Z_2 = 2$–28, this number is of the order of 10^5. The numerical calculation of three-fold integrals involved in the models mentioned above were too time-consuming. The large amount of computations can be performed in a reasonable time by using an analytical relationship for the halflife. Since 1980 the ASAF model was developed to fulfil this requirement. The Myers-Swiatecki's liquid drop model potential barriers were adjusted with a phenomenological correction accounting for the known overestimation of the barrier height and for the shell and pairing effects in the spirit of Strutinsky's method.

Besides the half-life predictions in several papers since 1984, there are three large tables, two [12, 13] published and one unpublished [14]. In the last one cold fission is included too and the systematics is extended in the region of heavier emitted clusters (mass numbers $A_e > 24$), and of parent nuclei far from stability and superheavies.

The unified approach of cold binary fission, cluster radioactivity, and α-decay was extended to cold ternary fission [15] and to multicluster fission [16]. Microscopic approaches [17, 18] were published as well. In the region of highly excited states above the barrier, Moretto's model [19] describes in a unified way the light particle emission (evaporation) and fission. The purpose of the present work is to present a systematics of experimental results compared to calculations, from which one may suggest [20] other possible candidates for future experiments. Universal curves may be used to estimate the expected half-lives.

EXPERIMENTS

The main quantities experimentally determined are the partial half-life, T, and the kinetic energy of the emitted cluster, $E_k = QA_d/A$, where Q is the released energy, and A_d and A are the mass numbers of the daughter and parent nuclei. Usually the main experimental difficulties are caused by the very long half-lives and the small value of branching ratio relative to α-decay [21]. Sometimes the experimental sensitivity is not high enough to achieve a positive result, hence only an upper limit can be established. The longest upper limit determined up to now is $T \geq 10^{29.2}$ s for the ^{24,26}Ne radioactivity of ^{232}Th, and the smallest branching ratio $b_\alpha = 10^{-15.87}$ for ^{34}Si decay of ^{242}Cm. On the other side, the most favourable values are $T = 10^{11.01}$ s for ^{14}C radioactivity of ^{222}Ra, and $b_\alpha = 10^{-8.88}$ for ^{14}C decay of ^{223}Ra.

A $\Delta E \times E$ telescope of two silicon detectors directly viewing the source was used in the first experiment [2], which was running six months in order to obtain 11 events. The strong background of α particles produced multiple pile-ups of electric pulses and damaged the semiconductor detectors which were replaced with new ones, after irradiation with a large flux of α particles (over $10^9 \alpha/cm^2$).

The solution adopted in the elegant experiments performed at Orsay was to deflect the unwanted ^{4}He ions, simply or doubly ionized, by a strong magnetic field produced in the superconducting spectrometer SOLENO, and to select only ^{14}C clusters to reach the detector in the focal plane. A source about 300 times stronger than that used by Rose and Jones was employed so that a run of only five days was necessary to obtain 11 events. With this unique instrument it was possible to discover [22] the fine structure in cluster decay [23] and to perform the most accurate experiment [24] by using high quality implanted sources [25] of Ra isotopes made at the ISOLDE mass separator, CERN, Geneva. Similarly, an Enge split-pole magnetic spectrometer with a gas filled detector in its focal plane has been used at Argonne [26] to confirm the mass number of the emitted ^{14}C fragment from ^{223}Ra.

Another method extensively used [27] is based on solid state nuclear track detectors (SSNTD) [28, 29] which are not sensitive to alphas and other low-Z particles, because they need a certain threshold of ionization. Very frequently such detectors are made from polyethylene terephtalate or from a phosphate glass. They are cheap and handy, but like the photographic plates, do not deliver the information on-line; only after a suitable post-irradiation chemical etching are the tracks visible. The etching rate along the paths of the ions depends on the charge number of the ionizing particle. The plot of the etching rate versus the residual range yields the atomic number, Z, identification. Finally the track can be seen, located and measured by manual or automatic scanning with a microscope, and one may derive the characteristics of the incoming particle from its shape and dimensions. SSNTD are widely used in a large variety of nuclear physics experiments, particularly for rare events in spontaneous fission, cold fission, and spontaneously fissioning shape isomers.

The data obtained until now (see the reviews [27, 26, 8, 25, 29], the references therein, and the recently published papers [30, 31, 32, 33]) on half-lives and branching ratios of cluster radioactivities are in good agreement with predicted values from the ASAF model, as we shall show below.

A summary of experimental data may be seen in the Table 1. We selected some of the measured values without giving the error bars, which usually are considered to be relatively small. When the experimental method did not discriminate between two neighbouring isotopes of the emitted nucleus, both are mentioned with the same half-life. More than one cluster decay mode were detected for some isotopes of Pa, U, and Pu. The cluster emitters ^{221}Fr, $^{221-224,226}$Ra, ^{225}Ac, ^{228,230}Th, ^{231}Pa, $^{230,232-236}$U, ^{236,238}Pu, and

TABLE 1. Experimental Q-values, Half-lives and Branching ratios relative to α-decay of cluster emitters. Spontaneous fission half-lives (T_f) are also given for heavy nuclei.

Parent nucleus Z and A		Emitted cluster Z_e and A_e		Q MeV	$\log_{10} T(s)$	$\log_{10} T/T_\alpha$	$\log_{10} T_f(s)$
87	221	6	14	31.294	14.53	11.99	
88	221	6	14	32.402	13.39	11.47	
88	222	6	14	33.052	11.01	9.42	
88	223	6	14	31.839	15.15	8.88	
88	224	6	14	30.541	15.69	10.16	
88	226	6	14	28.198	21.22	10.49	
89	225	6	14	30.479	17.16	10.93	
90	226	8	18	45.731	> 16.76	> 13.37	
90	228	8	20	44.730	20.72	12.80	
90	230	10	24	57.765	24.61	12.10	> 25.80
90	232	10	24	54.491	> 29.20	> 11.44	> 28.50
90	232	10	26	55.973	> 29.20	> 11.44	> 28.50
91	231	10	24	60.413	22.88	11.76	> 24.80
91	231	9	23	51.854	> 26.02	> 14.90	> 24.80
92	230	10	22	61.390	19.57	13.14	> 18.10
92	232	10	24	62.312	20.42	10.90	21.40
92	233	10	25	60.736	24.84	12.06	> 24.93
92	233	10	24	60.490	24.84	12.06	> 24.93
92	233	12	28	74.235	> 27.59	> 14.81	> 24.93
92	234	12	28	74.118	25.74	12.72	23.68
92	234	10	24	58.831	> 25.92	> 12.90	23.68
92	234	10	26	59.473	> 25.92	> 12.90	23.68
92	235	10	24	57.358	27.42	10.85	26.50
92	235	10	25	57.717	27.42	10.85	26.50
92	235	12	28	72.162	> 28.09	> 11.52	26.50
92	236	12	28	70.558	27.58	12.59	23.90
92	236	12	30	72.280	27.58	12.59	23.90
93	237	12	30	74.791	> 27.57	> 13.41	25.50
94	236	12	28	79.674	21.67	13.57	16.82
94	238	14	32	91.198	25.27	15.68	18.18
94	238	12	28	75.919	25.70	16.11	18.18
94	238	12	30	76.801	25.70	16.11	18.18
94	240	14	34	91.038	> 25.52	> 14.07	18.56
95	241	14	34	93.931	> 25.26	> 15.06	21.50
96	242	14	34	96.519	23.15	15.87	14.35

^{242}Cm are either β-stable or not far from stability nuclei. The Green approximation for the line of β-stability crosses the following Z,N pairs: 87,133; 88,135; 89,136; 90,138; 91,140; 92,142; 93,144; 94,146; 95,148, and 96,150. The α-decay half-lives are taken from tables of experimental data [21] when available, or otherwise calculated with a semi-empirical formula as in [13].

The strong competition of α-decay may be seen from Table 1. While $10^{11} < T < 10^{30}$ and $10^{1.5} < T_\alpha < 10^{18}$, the branching ratio $10^{-16} < b_\alpha \equiv b_{ac} = 1/b_{ca} < 10^{-8}$. Spontaneous fission [34] starts to be important in the region of heavy cluster emitters with $10^{14} < T_f < 10^{29}$ (see Table 1). For Pa, U, Np, Am, and Pu isotopes the branching ratio $b_f = T_f/T \equiv b_{fc} = 1/b_{cf}$ is in the range $(10^{-7}, 10^2)$, but for ^{242}Cm it approaches 10^{-9}, making very difficult the measurement of ^{34}Si radioactivity [32, 33].

Data for the *fine structure* of ^{14}C radioactivity of ^{223}Ra (see the reviews [25, 10]) were not included because, surprisingly, the transition toward the first excited state of the daughter nucleus is stronger than that to the ground state. The physical explanation relies on the nuclear structure of the parent and daughter nuclei. If the uncoupled nucleon is left in the same state in both the parent and heavy fragment, the transition is favoured. Otherwise the difference in structure leads to a large hindrance $H = T^{exp}/T_{e-e}$, where T^{exp} is the measured partial half-life for a given transition, and T_{e-e} is the corresponding quantity for a hypothetical even-even equivalent, estimated either from a systematics or from a model. A transition is favoured if $H \simeq 1$, and it is hindered if $H > 5$.

Unlike in α-decay, where the initial and final states of the parent and daughter are not so different from one another, in cluster radioactivities of odd-mass nuclides, one has a unique possibility to study a transition from a well deformed parent nucleus with complex configuration mixing, to a spherical nucleus with a pure shell model wave function. It can be used as a spectroscopic tool to obtain direct information on spherical components of deformed states.

COMPARISON WITH ASAF PREDICTIONS. NEW CANDIDATES

The Q-value is one of the quantities which play a very important role in any spontaneous nuclear decay with emission of charged particles. A parent nucleus with atomic number Z and mass number A decays into an emitted cluster A_e, Z_e and a daughter A_d, Z_d. One has $^A Z \rightarrow {}^{A_e} Z_e + {}^{A_d} Z_d$, with conservation of hadron numbers (neutrons and protons): $N = N_e + N_d$ and $Z = Z_e + Z_d$, where $A = N + Z$, and similar relationships for A_e and A_d. The process is energetically allowed if and only if the released energy,

$$Q = M(A, Z) - [M_e(A_e, Z_e) + M_d(A_d, Z_d)] \tag{1}$$

is a positive quantity, $Q > 0$. The atomic masses M, M_e, and M_d in units of energy, are taken from tables of experimental values.

Shell effects can be seen in Figure 3 of [20] where we display Q-values for the following decay modes: ^{14}C, ^{18,20}O, ^{23}F, $^{22,24-26}$Ne, ^{28,30}Mg and ^{32,34}Si versus the neutron number of the daughter, N_d. The variation with daughter neutron number, N_d, is almost regular giving, as a rule, a maximum value at the magic number $N_d = 126$. Exceptions: ^{22,24}Ne; ^{28}Mg, and ^{32}Si decay modes with larger or equal values at $N_d = 125$ or $N_d = 124$. Pairing effects are enhanced leading to even-odd staggering for a cluster decay mode like ^{25}Ne. The variation with Z_d for $Z_d = 80 - 82$ is, as expected, increasing toward the magic value $Z_d = 82$, with an exception for ^{23}F radioactivity (almost the same values for $Z_d = 80$ and $Z_d = 81$). Only for ^{20}O and ^{26}Ne decay modes are Q-values decreasing when Z_d is increased past the magic number 82 ($Z_d = 83$ and $Z_d = 84$). For ^{23}F radioactivity $Z_d = 84$ gives Q-values smaller than $Z_d = 82$ but higher than $Z_d = 83$. They are very close to $Z_d = 82$ for ^{14}C, ^{18}O, ^{24,25}Ne, and ^{30}Mg decay modes, but continue to increase slightly for ^{22}Ne, ^{28}Mg, and ^{32,34}Si radioactivity.

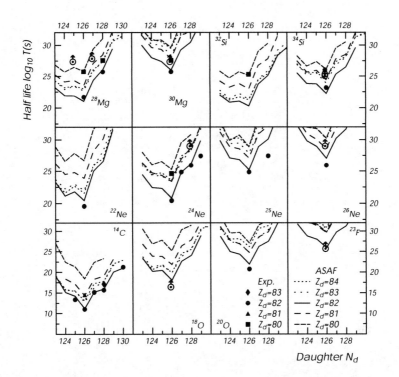

FIGURE 1. Predicted half-lives within ASAF model (lines) and measurements (points) for diferent kinds of cluster decay modes versus neutron number of the daughter nucleus. The calculations belonging to the same atomic number of the daughter are joined with a line of a style mentioned on the figure for $Z_d = 80 - 84$. The experimentally determined upper limits are marked with a vertical arrow. They correspond to: $Z_d = 82$ for ^{18}O and ^{23}F; $Z_d = 80$ for ^{24,26}Ne and ^{28}Mg; $Z_d = 81$ for ^{30}Mg radioactivity. The two limits for ^{34}Si radioactivity refer to $Z_d = 80$ (upper one) and to $Z_d = 81$ (lower one).

These remarkable properties of high Q-values were not enough exploited in practice until now, as can be seen in Figure 1, where a plot for the half-lives is shown, including the measured points. A strong shell effect can be seen in Figure 1: as a rule the shortest value of the half-life is obtained when the daughter nucleus has a magic number of neutrons ($N_d = 126$) and protons ($Z_d = 82$). There are few measurements for neighbouring daughters with $Z_d = 80$ or $Z_d = 81$, and only one for $Z_d = 83$. A striking exception is ^{32}Si decay for which the single measurement performed until now is far from other more favourable cases.

Possible candidates for future experiments, having reasonable half-lives and branching ratios relative to α-decay are given in Table 2. One should not forget about the competition of spontaneous fission disintegration in Pu, Am, and Cm isotopes.

TABLE 2. New candidates for eperimental searches of cluster radioactivities. Half-lives and Branching ratios relative to α-decay are predicted within ASAF model. Alpha decay life time is either measured (m) or estimated (es).

Parent		$lgT_i(s)$	Q_α MeV	$lgT_\alpha(s)$		Emitted		Q MeV	$lgT(s)$	lgT/T_α
Fr	220	1.43	6.800	1.64	m	C	14	30.716	17.13	15.49
Fr	222	2.94	5.830	5.49	es	C	14	30.084	18.33	12.84
Fr	223	3.12	5.434	7.30	es	C	14	29.006	19.07	11.77
Ac	223	2.12	6.783	2.47	m	C	14	33.068	12.79	10.32
Ac	224	4.02	6.327	5.61	m	C	14	32.007	16.30	10.69
Th	225	2.68	6.921	3.07	m	C	14	31.728	17.29	14.22
Th	229	11.36	5.167	11.61	m	O	20	43.412	26.20	14.59
Pa	229	5.08	5.835	8.12	m	Ne	22	58.958	23.90	15.78
						F	21	51.700	24.23	16.11
Pa	230	6.18	5.439	10.05	es	Ne	24	60.382	24.88	14.83
Pa	232	5.05	4.624	15.54	es	Ne	24	58.649	27.50	11.97
						Ne	25	59.043	27.77	12.23
U	231	5.56	5.577	9.82	m	Ne	24	62.218	22.73	12.91
Np	233	3.34	5.625	8.34	m	Ne	24	62.160	23.34	15.00
						Na	25	69.060	23.89	15.55
Pu	234	4.50	6.310	5.89	m	Mg	26	78.315	21.58	15.69
						Mg	28	79.156	22.06	16.17
Np	234	5.58	5.358	11.77	es	Mg	28	77.235	25.37	13.60
Np	235	7.53	5.191	12.66	m	Mg	28	77.099	24.09	11.43
Pu	235	3.18	5.951	7.70	m	Mg	28	79.664	22.90	15.20
Pu	237	6.59	5.749	11.17	m	Mg	28	77.734	25.35	14.18
Am	238	3.84	6.045	9.36	m	Si	32	94.767	25.26	15.90
Am	239	4.64	5.924	8.72	m	Si	32	94.510	24.11	15.39
Cm	239	4.03	6.586	7.03	m	Si	32	97.737	22.58	15.55
Cm	240	6.37	6.397	6.52	m	Si	32	97.558	21.29	14.77
Cm	241	6.45	6.185	8.61	m	Si	33	95.952	25.17	16.56
						Si	32	95.406	25.29	16.68

CLUSTER PREFORMATION AND UNIVERSAL CURVES

The (measurable) decay constant $\lambda = \ln 2/T$, can be expressed as a product of three (model dependent) quantities

$$\lambda = \nu S P_s \tag{2}$$

where ν is the frequency of assaults on the barrier per second, S is the preformation probability of the cluster at the nuclear surface, and P_s is the quantum penetrability of the external potential barrier. Not every quantity plays an equally important role. For α-decay and cluster radioactivities the penetrabilty dominates the half-life variation with A. The frequency ν remains practically constant, the preformation differs from one decay mode to another but it is not changed very much for a given radioactivity, while the general trend of penetrability follows closely that of the half-life. The external part of the barrier (for separated fragments), essentially of Coulomb nature, is much wider than the internal one (still overlapping fragments).

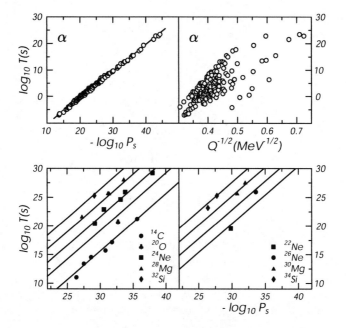

FIGURE 2. Universal curves for cluster radioactivities (bottom left and right) and α-decay (top left). In a typical "Geiger-Nuttal" plot for α-decay (top right) one sees considerable scattering of the data points.

Consequently, both fission-like and α-like models which take into consideration the external part of the barrier in the same manner, can provide a successful explanation for the measured half-lives. No wonder that majority of cluster theories (see the reviews [10, 17, 18, 35] and the references therein) are able to estimate the half-lives in good agreement with experiments. In fact the touching point is the scission configuration which proved to be also very important for spontaneous and induced fission phenomena [36, 37, 38].

According to [39], the preformation probability can be calculated within a fission model as a penetrabilty of the internal part of the barrier, which corresponds to still overlapping fragments

$$S = exp(-K_{ov}) \; ; \; K_{ov} = \frac{2}{\hbar} \int_{R_a}^{R_t} \sqrt{2B(R)E(R)}dR \qquad (3)$$

where R_a is the internal turning point ($E(R_a) = 0$), $R_t = R_1 + R_2$ is the separation distance of two fragments at the touching point configuration (scission), $B(R)$ is the nuclear inertia, and $E(R)$ is the deformation energy from which the Q-value was subtracted out.

By taking into account the above mentioned arguments, one may assume as a first approximation, that preformation probability only depends on the mass number of the emitted cluster, A_e, in the following manner:

$$\log S = \frac{(A_e - 1)}{3} \log S_\alpha \tag{4}$$

where the preformation probability of the α particle S_α may be determined by fit to experimental data. The next assumption is that $v(A_e, Z_e, A_d, Z_d) = \text{constant}$. From the fit one obtains: $S_\alpha = 0.0160694$ and $v = 10^{22.01}$ s^{-1}.

In this way one arrives at a single straight line *universal curve* on a double logarithmic scale

$$\log T = -\log P_s - 22.169 + 0.598(A_e - 1) \tag{5}$$

where

$$-\log P_s = c_{AZ}\left[\arccos\sqrt{r} - \sqrt{r(1-r)}\right] \tag{6}$$

with $c_{AZ} = 0.22873(\mu_A Z_d Z_e R_b)^{1/2}$, $r = R_t/R_b$, $R_t = 1.2249(A_d^{1/3} + A_e^{1/3})$, $R_b = 1.43998 Z_d Z_e/Q$, and $\mu_A = A_d A_e/A$. For all measurements performed until now the agreement is good, as can be seen in Figure 2. Usually a smooth behaviour can only be obtained for even-even nuclei. Nevertheless, with one exception (^{14}C radioactivity of ^{223}Ra for which the fine structure was observed), the measured parent nuclei with odd neutron or proton numbers were included on this plot, and they behave like the even-even ones.

Sometimes this universal curve is misinterpreted as being a Geiger-Nuttal plot. Geiger and Nuttal found in 1911 an empirical dependence of the α-decay partial half-life on the α-particle range in air. Nowadays this would correspond to a diagram of $\log T$ versus $Q^{-1/2}$. In this kind of systematics the experimental points are considerably scattered, as may be seen in Figure 2 (top right) for α-decay of even-even nuclei.

In **conclusion** up to now the ASAF model predictions have been confirmed. The strong shell effects of the daughter ^{208}Pb were not fully exploited. New searches for cluster decay modes can be made. Possible candidates are ^{220,222,223}Fr, ^{223,224}Ac, and ^{225}Th as ^{14}C emitters; ^{229}Th for ^{20}O radioactivity; ^{229}Pa for ^{22}Ne decay mode; ^{230,232}Pa, ^{231}U, and ^{233}Np for ^{24}Ne radioactivity; ^{234}Pu for ^{26}Mg decay mode; ^{234,235}Np and ^{235,237}Pu as ^{28}Mg emitters, as well as ^{238,239}Am and $^{239-241}$Cm for ^{32}Si radioactivity. Also ^{33}Si decay of ^{241}Cm could be observed. Universal curves provide a means to obtain rapidly estimates of the expected half-lives.

ACKNOWLEDGMENTS

This work was partly supported by the Japanese Society for the Promotion of Science (JSPS), Tokyo, by the Japan Atomic Energy Research Institute (JAERI), Tokai, by the Centre of Excellence IDRANAP under contract ICA1-CT-2000-70023 with European Commission, Brussels, by Bundesministerium für Bildung und Forschung (BMBF), Bonn, Gesellschaft für Schwerionenforschung (GSI), Darmstadt, and by the Ministry of Education and Research, Bucharest. One of us (DNP) would like to acknowledge the hospitality and financial support received during his research stage in Tokai and Frankfurt am Main, when the present work was completed.

REFERENCES

1. Săndulescu, A., Poenaru, D. N., and Greiner, W., *Sov. J. Part. Nucl.*, **11**, 528 (1980).
2. Rose, H. J., and Jones, G. A., *Nature*, **307**, 245(1984).
3. Aleksandrov, D. V. et al. *JETP Letters*, **40**, 909(1984).
4. Gales, S., Hourani, E., Hussonnois, M. et al. *Phys. Rev. Lett.*, **53**, 759(1984).
5. Săndulescu, A., Poenaru, D. N., Greiner, W., and Hamilton, J. H., *Phys. Rev. Lett.*, **54**, 490 (1985).
6. Poenaru, D. N., Ivaşcu, M., Săndulescu, A., and Greiner, W., Rep. E4-84-446, JINR, Dubna (1984).
7. Greiner, W., Ivaşcu, M., Poenaru, D. N., and Săndulescu, A., in *Treatise on Heavy Ion Science, Vol. 8*, Plenum, New York, 1989, p. 641.
8. Hourani, E., Hussonnois, M., and Poenaru, D. N., *Ann. Phys. (Paris)*, **14**, 311(1989).
9. Poenaru, D. N., Ivaşcu, M., and Greiner, W., in *Particle Emission from Nuclei, Vol. I–III* , CRC, Boca Raton, 1989.
10. Poenaru, D. N. and Greiner, W., editors, *Nuclear Decay Modes*, Institute of Physics, Bristol, 1996.
11. Poenaru, D. N., and Greiner, W., editors, *Handbook of Nuclear Properties*, Clarendon Press, Oxford, 1996; *Experimental Techniques in Nuclear Physics*, Walter de Gruyter, Berlin, 1997.
12. Poenaru, D. N., Greiner, W., Depta, K., Ivaşcu, M., Mazilu, D., and Săndulescu, A., *Atomic Data Nucl. Data Tables*, **34**, 423(1986).
13. Poenaru, D. N., Schnabel, D., Greiner, W., Mazilu, D., and Gherghescu, R., *Atomic Data Nucl. Data Tables*, **48**, 231(1991).
14. Poenaru, D. N., Ivaşcu, M., Mazilu, D., Gherghescu, R., Depta, K., and Greiner, W., Report NP-54, Central Inst. Phys., Bucharest (1986).
15. Poenaru, D. N., Dobrescu, B., Greiner, W., Hamilton, J. H., and Ramayya, A. V., *J. Phys. G: Nucl. Part. Phys.*, **26**, L97(2000).
16. Poenaru, D. N., Greiner, W., Hamilton, J. H., Ramayya, A. V., Hourany, E., and Gherghescu, R. A., *Phys. Rev. C*, **59**, 3457(1999).
17. Blendowske, R., Fliessbach, T., and Walliser, H., chap. 7 in [10] p. 337.
18. Lovas, R. G., Liotta, R. J., Insolia, A., Varga, K., and Delion, D. S., *Phys. Rep.*, **294**, 265(1998).
19. Moretto, L. G., *Nucl. Phys. A*, **247**, 211(1975).
20. Poenaru, D. N., Nagame, Y., Gherghescu, R. A., and Greiner, W., *Phys. Rev. C*, **65**, 054308(2002).
21. Rytz, A., *Atomic Data Nucl. Data Tables*, **47**, 205(1991).
22. Brillard, L. et al. *C. R. Acad. Sci. Paris*, **309**, 1105(1989).
23. Greiner, M., and Scheid, W., *J. Phys. G: Nucl. Phys.*, **12**, L229(1986).
24. Hourany, E. et al. *Phys. Rev., C*, **52**, 267(1995).
25. Hourany, E., chap. 8 in [10], p. 350.
26. Henning, W., and Kutschera, W., chap. 7 [9] p. 188.
27. Fleischer, R. L., Price, P. B., and Walker, R. M., *Nuclear Tracks in Solids. Principles and Applications*, Univ. of California, Berkeley, 1975; Price, P. B. and Barwick, S. W., Ch. 8 in [9], Vol. II, p. 205.
28. Tretyakova, S. P., *Sov. J. Part. Nucl.*, **23**, 156(1992).
29. Bonetti, R., and Guglielmetti, A., chap. 9 [10] p. 370.
30. Pan, Q., et al., *Phys. Rev., C*, **62**, 044612 (2000).
31. Bonetti, R., Carbonini, C., Guglielmetti, A., Hussonnois, M., Trubert, D., and Le Naour, C., *Nucl. Phys. A*, **686**, 64(2001).
32. Tretyakova, S. P., Bonetti, R., Golovchenko, A., et al., *Radiat. Meas.*, **34**, 241(2001).
33. Ogloblin, A. A., et al., *Phys. Rev. C*, **61**, 034301 (2000).
34. Hoffman, D. C., Hamilton, T. M., and Lane, M. R., chap. 10 [10] p. 393.
35. Royer, G., and Moustabchir, R., *Nucl. Phys. A*, **683**, 182(2001).
36. Nagame, Y. et al. *Phys. Lett. B*, **387**, 26(1996).
37. Zhao, Y. L., Nagame, Y., Nishinaka, I., Sueki, K., and Nakahara, H., *Phys. Rev., C*, **62**, 014612(9) (2000).
38. Zhao, Y. L. et al. *Phys. Rev. Lett.*, **82**, 3408(1999).
39. Poenaru, D. N., and Greiner, W., *Physica Scripta*, **44**, 427(1991).

Cluster Radioactivity Studies In Russia (I)

Alexey A. Ogloblin[1], Georgii A. Pik-Pichak[1], Svetlana P. Tretyakova[2]

1 – Kurchatov Institute, Moscow, Russia,
2 – JINR, Dubna, Russia

Abstract. A review of some cluster radioactivity studies in Russia is given. A short historical overlook is presented. Some emphasis is done on a semi-microscopic fission-like model by Pik-Pichak, which is applied to the calculations of full mass distributions of the emitted fragments both in the cluster radioactivity and cold fission regions. Some systematics of cold fission are predicted. The problem of the cluster radioactivity mechanisms is discussed. It is shown that study of the elastic scattering and deep fusion-fission of the daughter products provides independent information on the decay mechanism of the parent nucleus.

1. INTRODUCTION AND SHORT HISTORICAL REVIEW

The idea of the experiment on search of the decay of Radium isotopes with ^{14}C emission was proposed by a PhD student of one of us (AAO) B.G.Novatski as early as in 70-s. It was based on the consideration that if three successive α-decays are possible, the ^{12}C emission is also allowed. However, as the decay energy with ^{14}C emission is higher, this particular mode should be more probable. For some reasons the experiment on search of $^{223}Ra \rightarrow {}^{14}C$ decay was started only in the end of 1983. The first theoretical paper on the subject [1] did not recommend in direct way to study this particular decay. The effect (7 decays) was observed almost simultaneously with the publication of the pioneering work by Rose and Jones [2] using the similar experimental method (ΔE – E counters telescope). So the experiment by Kurchatov institute group [3] became the first independent confirmation of the existing of the new phenomenon.

The next important step was the observation of $^{231}Pa \rightarrow {}^{24}Ne$ decay by Tretyakova et al. [4] in Dubna using solid state track detectors (SSTD), the method which occurred to be the most effective for cluster radioactivity studies. This result demonstrated that the spontaneous emission of the light nuclei is not limited by a single case of ^{223}Ra and has a general character in accordance with the predictions [1]. This experiment was followed by a few others (see Table 1).

In the end of 80-s Kurchatov and Dubna groups combined their efforts and began joint experiments. The main aim of their program was to study the possibly heavier parent nuclei emitting the heaviest fragments. The first observation of the decays of ^{236}Pu, ^{236}U and ^{242}Cm (the latter together with Milano university) was achieved (Table1).

In the middle of 90-s the studies of the inverse processes, the elastic scattering and fusion-fission as a possible tool of investigation of the decay mechanism began both at

CP644, *Exotic Clustering: 4th Catania Relativistic Ion Studies*, edited by S. Costa, A. Insolia, and C. Tuvè
© 2002 American Institute of Physics 0-7354-0099-7/02/$19.00

Kurchatov institute [5] and Dubna [6]. Now both groups collaborate in this field as well.

TABLE 1. First observations of cluster decays

Decay	Year of observation	Method	Institute & Reference
$^{223}Ra \rightarrow {}^{14}C$	1984	Si, $\Delta E - E$	Oxford univ. [2] **Kurchatov Inst.** [3]
231Pa $\rightarrow {}^{24}Ne$	1984	SSTD	**JINR, Dubna** [4]
$^{222}Ra \rightarrow {}^{14}C$	1985	SSTD	LBL, Berkeley [7]
$^{224}Ra \rightarrow {}^{14}C$	1985	SSTD	LBL, Berkeley [7]
$^{226}Ra \rightarrow {}^{14}C$	1985	Magnet spectrom.	IPN,Orsay [8]
$^{232}U \rightarrow {}^{24}Ne$	1985	SSTD	LBL, Berkeley [9]
$^{233}U \rightarrow {}^{24}Ne$	1985	SSTD	**JINR, Dubna** [10]
$^{230Th} \rightarrow {}^{24}Ne$	1985	SSTD	**JINR, Dubna** [11]
$^{234}U \rightarrow {}^{24}Ne, {}^{28}Mg$	1987	SSTD	LBL, Berkeley [12]
$^{238}Pu \rightarrow {}^{28}Mg, {}^{32}Si$	1989	SSTD	LBL, Berkeley [13]
$^{236}Pu \rightarrow {}^{28}Mg$	1990	SSTD	**Kur. inst + JINR** [14]
$^{221}Fr \rightarrow {}^{14}C$	1993	SSTD	Milan Univ. [15]
$^{221}Ra \rightarrow {}^{14}C$	1993	SSTD	Milan Univ. [15]
$^{225}Ac \rightarrow {}^{14}C$	1993	SSTD	Milan Univ. [15]
$^{228Th} \rightarrow {}^{20}O$	1993	SSTD	Milan Univ. [16]
$^{236}U \rightarrow {}^{28}Mg$	1994	SSTD	**Kur. inst + JINR** [17]
$^{230}U \rightarrow {}^{22}Ne$	1999	SSTD	IAE, Peking [18]
$^{242}Cm \rightarrow {}^{34}Si$	2000	SSTD	**Kur. inst + JINR +** Milan Univ. [19]
$^{238}U \rightarrow {}^{34}Si$	2002	SSTD	**JINR** + Milan Univ. (preliminary)

The theoretical studies of cluster radioactivity in Russia resulted in development of two models. Pik-Pichak proposed a semi-microscopic fission-like model [20], which now is extended to cold fission [21]. Chuvilski – Furman – Kadmenski (CFK) developed microscopic α-decay-like model by [22].

Russian investigators published a few review articles at the early stage of cluster radioactivity studies [23-26]. Some talks on very recent Russian works were presented by the authors of this paper at different international conferences during last three years [21,27-31,39].

The whole material on recent activities of Russian groups was presented at CRIS Conference as two talks. The first one deals mostly with some theoretical problems and is dedicated to application of Pik-Pichak model. Besides, the mechanism of cluster radioactivity is discussed including the results obtained from the study of the inverse processes. The second talk describes current experimental situation with some emphasis on the experiments that are going on or planned. A separate talk by Chuvilski concerns some problems of CFK model.

2. DATA SYSTEMATICS AND COMPARISON WITH THEORY

The phenomenon called cluster radioactivity is a part of a wider class of cold decays. Their distinctive feature is the formation of the decay products in the ground or the lowest excited states. Evidently, alpha radioactivity is the most known member of this class of decays but it is out of scope of present discussion. Three regions of cold decays are discussed at the present time. They are: 1) "traditional" cluster radioactivity leading to the formation of the heavy daughters in the vicinity of double magic ^{208}Pb; 2) cold fission; 3) new region of cluster decays leading to double magic ^{100}Sn. Experimental systematics is presented in Fig.1.

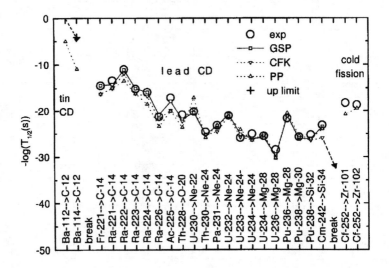

FIGURE 1. Experimental data on half-lives of cold decays (*). Comparison with three theoretical models is given (see the text)

In the region of "lead" radioactivity the values of logT change by ~ 20 orders of magnitude. All existing theoretical models reproduce the data with the accuracy 1-2 orders of magnitude. The comparison between the experiment and three theoretical models: most popular phenomenological fission-like model by Poeneru–Sandulescu–Greiner([32] and references there)   and those by Chuvilski–Furman– Kadmenski (CFK) and Pik-Pichak (α-decay-like and fission-like correspondingly), is shown in Fig.1.

Situation becomes different in the cold fission region where α-decay-like models underestimate the decay probabilities by many orders of magnitude. However, fission-like calculations reproduce the data quite satisfactory. As to the unexplored region of "tin" radioactivity the predictions of some models differ quite strongly. In particular, CFK model predicts much higher decay probabilities.

3. FULL MASS DISTRIBUTIONS OF THE FRAGMENTS

In order to establish the connection between different regions of cold decays and find the physical factors influencing the decay probabilities it is reasonable to calculate the complete emitted fragment mass distributions. This was done using Pik-

Pichak's model [20] because the latter well describes the whole set of existing data (including α-decay) and has a smaller amount of free parameters in comparison with the other universal models. Besides, the model takes into account the deformation both of parent and daughter nuclei. It was extended to the emission of fragments of any masses including the region of cold fission. Some results were presented in [21]. The adopted evolution of the geometrical shapes for some decays is shown in Fig.2.

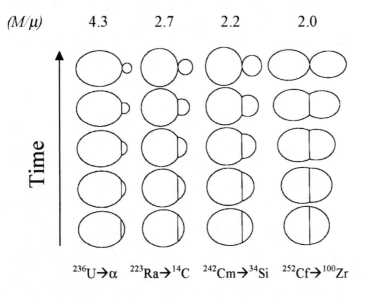

(M/μ) 4.3   2.7   2.2   2.0

$^{236}U \rightarrow \alpha$ $^{223}Ra \rightarrow {}^{14}C$ $^{242}Cm \rightarrow {}^{34}Si$ $^{252}Cf \rightarrow {}^{100}Zr$

FIGURE 2. The sequence of shapes of different decays. The upper row demonstrates the ratio of the effective mass to the reduced mass at the touching point..

An example of a full mass spectrum (^{234}U) is shown in Fig.3. One can define a few relatively intense groups of emitted fragments. Their emission probabilities enhancement depends on the nearness of the decay products to the magic nuclei and the presence of prolate deformations. $^{24}Ne - {}^{28}Mg$ peaks observed experimentally correspond to common "lead" radioactivity. A lower ^{48}Ca maximum reflects the double magic properties of this nuclide. A broad group $^{80}Ge - {}^{100}Zr$ originates from two effects: formation of the nuclei with proton and neutron numbers close to magic values Z=50 and N=82 values (e.g., $^{100}Zr + {}^{134}Te$) and nuclei with large static deformations (e.g., ^{152}Nd whose light partner is ^{82}Ge).

An interesting comparison with the experimental mass spectrum of induced cold fission of ^{234}U [33] can be done. We normalized the partial yields obtained in [33] to the known value of the $T_{1/2}$ of ^{234}U spontaneous fission. "Experimental" data obtained in such a way agree unexpectedly well with the calculated mass distribution.

Calculations of full mass spectra of different decaying nuclei allow predicting some features of cold fission. An interesting question is if the well-known dependence of logT on the fissility parameter Z^2/A is valid for cold fission as well (see Fig. 4). For the isotopes of Ra and heavier elements the linear dependence logT – Z^2/A is observed and occurs to be universal. So one can speak about some similarity between normal and cold fission.

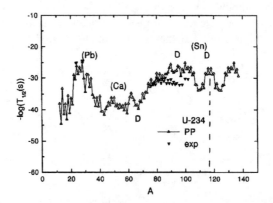

FIGURE 3. Mass distribution of cold decays of ^{234}U. Only the light fragments half of the spectrum is shown in full. Black triangles show experimental data. Letters in the brackets denote the "type" of radioactivity. Letter "D" indicates the regions of deformed daughter nuclei (see text).

However, it is not the case for the isotopes of lighter elements. Though the isotopes of each of them lye on the straight lines, the latter are not unique. This indicates that linear dependence here could be spurious and mask some other effects. Indeed, cold fission probabilities demonstrate rather universal dependence on the barrier height, what makes them closer to cluster decays. As to Z^2/A dependence of the calculated logT values of cold fission of the nuclei for which cluster radioactivity was observed (^{222}Ra, etc.) it demonstrates a clear linear behavior. Contrary to this, half-lives of cluster decays behave differently (see Figs.10, 11 in [21]).

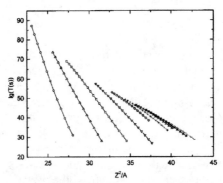

FIGURE 4. Dependence of symmetric cold fission on the fissility parameter Z^2/A for the isotopes (from left to right) of Ba, Hf, Hg, Ra, U, Cm, Cf, Fm. The selected cold fission groups include 30 mass units around the symmetric value each.

4. DECAY MECHANISM

All proposed models of cluster radioactivity reproduce the data in the region of "lead" decays quite well independently on the adopted mechanism. The reason lies in the fact that in all the cases the half-lives are determined by the penetrability of the Coulomb part of the barrier, and the role of its internal part, which is really sensitive to the decay mechanism, is small. The fission-like models, which consider the decay process to be adiabatic, are able even to explain the α-particle emission. However, there exist some evidence that emission of, at least, the lightest fragments is "sudden" in the sense that it can be described in the frame of R-matrix theory. Cluster decay mechanism for them is based on the following arguments:

- α-emission is excellently described by R-matrix theory not only in the particular case of α-radioactivity but in all other nuclear reactions.
- Cluster decays of some odd nuclei strongly depend on nuclear structure factors but not exclusively on the penetrability (fine structure of ^{223}Ra $\rightarrow$ ^{14}C spectrum is the best example).
- The heavy-ion (^{12}C, ^{16}O) scattering and fusion data are described by cluster-like but not fission-like shapes of internal parts of the barrier (see next section)

On the other hand, cold emission of heavier fragments definitely bears some features of the adiabatic, fission-like process. One can put the following arguments:

- The very fact of the existence of cold fission provides strong evidence against cluster mechanism of decays in this mass region because the reasonable spectroscopic factors are many orders of magnitude less than it is required for fitting the data.
- Fission-like models reproduce well the data in the large A region. Some predicted properties of cold fission (e.g., logT – Z^2/A dependence) are similar to those of normal fission.

So one has to expect some kind of transition between two mechanisms of cold decays. Of course, it should be noted that both α-decay-like and fission-like mechanisms, being some extremes, must not be taken too literally.

It is natural to expect that such transition should take place when the contribution of the internal part of the barrier to the overall penetrability becomes significant. The Coulomb and nuclear actions calculated for the potentials of the model [20,21] become equal at A-values of the emitted fragments ~ 40 (Fig.5). This finding correlates with calculations [34]: the penetrabilities of the internal part of the barrier for cluster decays, being treated as the preformation factors, begin to deviate from the theoretical values [35] at A ~ 30.

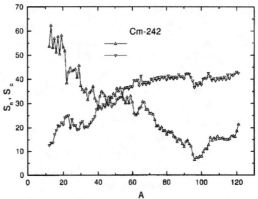

FIGURE 5. Coulomb ($\triangle$) and nuclear (∇) actions as function of A-values of the emitted fragments for ^{242}Cm cold decays.

Some indications of the transition under discussion were obtained [27] at the analysis of ^{242}Cm $\rightarrow$ ^{34}Si decay: a deviation from Geiger-Nuttal systematics (or, in the other words, from the universal curve [32]) was observed. If one plots the distances between the lines (equivalent to preformation factors) corresponding to the emission of different clusters as function of their A-values the ^{34}Si point goes down from the empirical law ΔlogT = 0.6 x ΔA. Though the deviation itself is not remarkable the change becomes evident if the cold fission data are included (Fig.6).

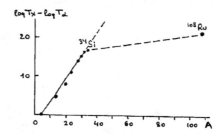

FIGURE 6. Differences between the lines in Geiger-Nuttal plot versus the emitted cluster masses. All the points, except for ^{108}Ru are extracted from experimental data. The preformation factor for ^{34}Si.deviates from the empirical law.

5. INVERSE PROCESSES

Different models of cluster radioactivity predict completely different shapes of the potential barriers (see, e.g. Figs. 6.2 and 6.18 in [32]). So the independent information about barriers, especially on their internal parts, is of great value for understanding the mechanism of cluster decays. As both decay products are formed in their ground states the study of their interaction, can, in principle, provide such information.

One of the processes directly connected with the interaction potential is elastic scattering. The problem is, however, that the internal part of the potential is normally obscured by strong absorption. Even in the scattering of the lightest projectiles of interest (carbon or oxygen isotopes) on lead the Fresnel diffraction mechanism dominates, which allows probing only the Coulomb part of the potential beyond the top of the barrier.

Nevertheless, it was shown [5,29,40] that the elastic scattering $^{12}C+^{208}Pb$ at the near barrier energies (75-76 MeV) becomes sensitive to the top of the barrier and even to its internal part if measured with high accuracy in the large angular interval up to the values 10^{-4} of the Rutherford cross-section.

The potential, which fits the data is shown in Fig.7 together with two theoretical potentials: ASAFM from fission-like model [32] and that of quasimolecular model by Buck and Merchant [36].

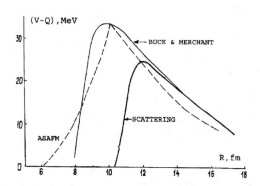

$^{12}C + ^{208}Pb$ POTENTIALS

FIGURE 7. $^{12}C+^{208}Pb$ potential obtained from fitting the elastic scattering data [29] and theoretical potentials for $^{220}Ra \rightarrow ^{12}C + ^{208}Pb$ decay for two models [32, 36].

The empirical potential is of cluster decay type (steep slope of the internal part). It is quite close to the folding model potentials. On the other hand it is evident that the main reason of inconsistency of both presented theoretical potentials with the scattering data is the too large width of the barriers independently from the physical grounds which they are deduced from: fission-like mechanism [32] or extreme cluster (quasimolecular) structure [36].

Of course, the crucial question is in what extent the obtained scattering potential is able to reproduce the decay probability. The corresponding data on ^{220}Ra are absent, but the predictions of the phenomenological model [32.37] can serve a good guide in this case. The calculations [37] predict $logT_{1/2} = 10.5$. Now, if we calculate the penetrability of the scattering barrier from Fig.7, take the microscopic spectroscopic factor from CFK model ($S = 1.4 \times 10^{-9}$) and a normal value for assault frequency $v = 1\times10^{21}$, we obtain for cluster decay of ^{220}Ra $logT_{1/2} = 10.3$ in excellent agreement with the "quasiempirical" value from [37]. So we got a new independent argument for α-decay-like mechanism for emission of the lightest fragments.

Another process, which can give independent information about the potentials is deep subbarrier fusion. Recent experiments on fusion-fission reactions [28], in which the cross-sections $\sim 10^{-32}$ cm^2 were reached demonstrated some deviation of the shape of the excitation function from the usual exponential fall-off. The preliminary analysis of the data [38] led to the potentials similar to the scattering one from Fig.7.

Basing on these results we plan to fulfill experiments, which would allow direct comparison of the inverse reactions data with the measured cluster decay half-lives of the compound nuclei. At the present time only two projectile – target combinations are available: ^{14}C + ^{208}Pb $\rightarrow$ ^{222}Ra and ^{22}Ne + ^{208}Pb $\rightarrow$ ^{230}U. If cluster decay of ^{223}Ac would be measured two other combinations become possible: ^{14}C + ^{209}Bi $\rightarrow$ ^{223}Ac and ^{15}N + ^{208}Pb $\rightarrow$ ^{223}Ac. All these experiments are under discussion or preparation by Kurchatov – Dubna collaboration.

ACKNOWLEDGMENTS

We are indebted to our numerous colleagues from Russian and foreign institutes for fruitful discussions. Especially we are grateful to R.Bonetti, Yu.M.Chuvilski, V.I.Furman, F.Goennenwein, S.A.Goncharov, W.Greiner, A.Guglielmetti, M.G.Itkis, S.G.Kadmenski, V.M. Mikheev, B.G.Novatski, Yu.Ts.Oganessian, D.N.Poenaru, A.Sandulescu,V.A.Shigin,W.Trzaska.

REFERENCES

1. Sandulescu, A., Poenaru, D.N., and Greiner, W., *Fiz.Elem.Chastits At. Yadra* **11**, 1334 (1980) [Sov.J.Part. Nuc. **11**, 528 (1980)]
2. Rose, H.J., and Jones, G.A., *Nature (London)* **307**, 245 (1984)
3. Alexandrov,D.V., Beliatsky, A.F., Glukhov, Yu.A., Nikol'sky, E.Yu., Novatsky, B.G., Ogloblin, A.A., Stepanov, D.N., *Pis'ma Zh. Eksp. Theor. Fiz* **40**, 152 (1984) [JETP Lett. **40,** 909 (1984)]
4. Sandulescu, A., Zamyatnin, Yu.S., Lebedev I.A., Myasoedov, V.F., Tretyakova, S.P., Hashegan, D., *JINR Rapid Commun.*,**5**, 84 (1984)
5. Dem'yanova, A.S., et al., in *Exotic Nuclei and Atomic Masses ENAM'95*, ed. by M. de Saint Simon and O.Sorlin, Editions Frontiers, pp. 401-402
6. Oganessian, Yu.Ts., Itkis, M.G., Kozulin, E.M., Pustylnik, B.I., Tretyakova, S.P., Calabretta, L., Guzel, T., *JINR Rapid Commun.,* **1[75]-96**, 123 (1996)
7. Price, P.B., Stevenson, J.D., Barwick, S.W., Ravn, H.L., *Phys.Rev Lett.* **54**, 297 (1985)
8. Hourani, E., Hussonnois, M., Stab, L., Brillard, L., Gales, S., Shapira, J.P., *Phys.Lett.* **B**, **160**, 375 (1985)
9. Barwick, S.V., Price, P.B., Stevenson, J.D., *Phys.Rev.* **C 31**, 1984 (1985)
10. Tretyakova, S.P., Sandulesku, A., Zamyatnin, Yu.,S., Kovotchin,Yu.S., Micheev, V.L., JINR Rapid Commun., **7**, 23 (1985)
11. Tretyakova, S.P., Sandulesku, A., Micheev, V.L., Hashegan, D., Lebedev, I.A., Zamyatnin, Yu.,S., Korotkin, Yu.S., *JINR Rapid Commun.,* **13**, 34 (1985)
12. Wang, S., Price, P.B., Barwick, S.W., Moody, K.J., Hulet, E.K., *Phys. Rev.* **C 36**, 2717 (1987)
13. Wang, S., Snowden, D., Price, P.B., Moody, K.J., Hulet, E.K., *Phys.Rev.,* **C 39**, R1647 (1989)
14. Ogloblin, A.A., Venikov, N.I.,Lisin, S.K., Pirozhkov, S.V., Pchelin, V.A., Rodionov, Yu.V., Semochkin, V.M., Shabrov, V.A., Shvetzov, I.K., Shubko, V.M., Tretyakova, S.P., Mikheev, V.L., *Phys. Lett.,* **B 235**, 35 (1990)
15. Bonetti, R.,Chiesa, C., Guglielmetti, A., Matheoud, R., Migliorino, C., Pasinetti, A., Ravn, H.L., *Nucl. Phys.,* **A 562**, 32 (1993)
16. Bonetti, R.,Chiesa, C., Guglielmetti, A., Migliorino, C., Cesana A., Terrani, M., *Nucl. Phys.,* **A 556**

115 (1993)

17. Tretyakova, S.P., Mikheev, V.L., Ogloblin, A.A., Shigin, V.A., Ponomarenko, V.A., *Pisma v JETF [JETP Lett.]*, **59**, 368 (1994)
18. Pan Qiang-yan et al., *Chinese Phys. Lett.*, **16**, 251 (1999)
19. Ogloblin, A.A., Bonetti, R., Denisov, V.A., Guglielmetti, A., Itkis, M.G., Mazzocchi, C.,Mikheev, V.L., Oganessian, Yu.Ts., Pik-Pichak, G.A., Poli, G., Pirozhkov, S.M., Semochkin, V.M., Shigin, V.A., Shvetsov, I.K., Tretyakova, S.P. *Phys.Rev.* **C 61**, 034301 (2000)
20. Pik-Pichak, G.A., *Sov.J. Nucl. Phys.*, **44**, 923 (1987)
21. Ogloblin, A.A., Pik-Pichak, G.A., Tretyakova, S.P., in *Fission Dynamics of Atomic Clusters and Nuclei*, ed. by Joao da Providencia, D.M.Brink, F. Karpeshine, F.B.Malik, Proc. of Int. Workshop Luso, Portugal, 15-19 May 2000, World Scientific, pp. 143-162
22. Kadmenski, S.G., Kurgalin, S.D., Furman, V.I., Chuvilski, Yu. M., *Sov. J. Nucl. Phys.*, **51**, 32 (1990)
23. Hasegan, D. and Tretyakova, S.P., in *Particle Emission from Nuclei*, ed. by Poenaru, D.N. and Ivasku, M.S., CRC Press, Inc. (1989), pp. 223-257.
24. Novatski, B.G. and Ogloblin, A.A., *Vestnik Acad. Nayk* [Proc. Sov. Acad. Science], **1**, 81 (1988)
25. Zamyatnin, Yu.S., Mikheev, V.L., Tretyakova, S.P., Furman, V.I., Kadmenski, S.G., Chuvilski, Yu. M., *Fiz.Elem.Chastiz At. Yadra*, **21**, 537-594 (1990)
26. Tretyakova, S.P., *Fiz. Elem. Chastiz At. Yadra*, **23**, 364-429 (1992)
27. Ogloblin, A.A., Bonetti, R., Denisov, V.A., Guglielmetti, A., Itkis, M.G., Mazzocchi, C.,Mikheev, V.L., Oganessian, Yu.Ts., Pik-Pichak, G.A., Poli, G., Pirozhkov, S.M., Semochkin, V.M., Shigin, V.A., Shvetsov, I.K., Tretyakova, S.P. in *Nuclear Shells – 50 Years*, Proc. Int. Conf. On Nucl. Phys., Dubna, Russia, 21-24 April 1999, ed. by Yu.Ts.Oganessian and R. Kalpakchieva, World Scientific, p.124
28. Tretyakova, S.P., Calabretta, L., Itkis, M.G., Kozulin, E.M., Kondratiev, N.A., Maiolino, C., Pokrovski, I.V., Prokhorova, E.V., Rusanov, A.Ya., Tretyakova, T.Yu., *ibid.*, p.151
29. Ogloblin, A.A., Artemov, K.P., Glukhov, Yu.A., Dem'yanova, A.S., Paramonov, V.V., Rozhkov, M.V., Rudakov, V.P., Goncharov, S.A., in *Nucleus-Nucleus Collisions*, Proc. of Conf. Bologna 2000, Bologna, Italy, 29 May – 3 June 2000, ed G. Bonsignori, World Scientific, p.409
30. Ogloblin, A.A. and Tretyakova, S.P., in *Physics of Isomers*, Proc of First Int Workshop, June 18-21, St.-Petersburg, 2000, ed. F.F.Karpeshin, p. 122
31. Tretyakova, S.P., Ogloblin, A.A., Pik-Pichak, G.A., *Progr. Theor. Phys. Suppl.* (to be publ.)
32. Poenaru, D.N. and Greiner, W. in *Nuclear Decay Modes* (Inst. of Phys., Bristol, 1996), Chap.6, pp. 275-336
33. Schwab, W., et el., *Nucl. Phys.* **A 577**, 674 (1994),
34. Poenaru, D.N. and Greiner, W., *Phys.Scr.* **44**, 427 (1991)
35. Blendowske, Fliessbach, T., and Walliser, H., in *Nuclear Decay Models* (Inst. of Phys., Bristol, 1996), Chap. 7, p.337
36. Buck, R., Merchant, A.C., and Perez, S.M., *Phys.Rev.Lett.*, **76**, 380 (1996)
37. Poenaru, D.N., et al., *Atomic Data and Nucl.Tables*, **48**,231 (1991)
38. Shilov, V.M., Proc. of 6th School-Seminar, Dubna, Russia, 22-27 Sept. 1997, ed by Oganessian, Yu.Ts. and R.Kalpakchieva, World Scientific, p.331
39. Ogloblin, A.A., Pik-Pichak, G.A., Tretyakova, S.P., *Izvestia Acad .Nauk* [Proc. of Russian Acad. Science], **65**, 11 (2001)
40. Rudakov, V.P., Artemov, K.P., Glukhov, Yu.A., Gonchrov, S.A., Dem'yanova, A.S., Ogloblin, A. A., Paramonov, V.V., Rozhkov, M.V. *ibid*, **65**, 56 (2001)

Cluster Radioactivity Studies in Russia (II)

Svetlana P. Tretyakova[1], Alexey A. Ogloblin[2]

1.Joint Institute for Nuclear Research, Dubna, Russia;
2. Kurchatov Institute, Moscow, Russsia

Abstract. We discuss the main results study of cluster radioactivity obtained up to the present time in Russsia and the status of the continued measurements and those under preparation. All these experiments use solid state detectors technique.

INTRODUCTION

Cluster radioactivity is a novel phenomenon occupying an intermediate position between alpha decay and spontaneous fission. At present time 19 nuclides from ^{221}Fr to ^{242}Cm emitting light nuclei from ^{14}C to ^{34}Si are known [1]. The all heavy residual daughter nuclei are grouped around the double magic ^{208}Pb.

These experiments have shown that the probability of ^{14}C emission is by about 10 orders of magnitude lower than of α decay. In the first experiments with aimed at search for ^{14}C clusters electronic methods of detection were employed. However, these methods have considerable limitations in the number of incident α particles (for semiconductor telescope) and in the target thickness (for a magnetic analyzers). As both methods use a small solid angle (~ 0.1sr) the detection efficiency is low, < 1% [2]. Clearly the electronic methods do not allow to reveal modes of decay with a relative probability of < 10^{-12} to α decay.

The most effective technique for studying CR occurred to be the usage of solid state nuclear track detectors (SSNTD), because of their unique capability of rejecting events due to low ionizing particles such as α particles and possibility to arrange geometrical efficiency approaching 2π. The sensitivity limit of CR partial half-lives practically reached is ~ 10^{-30} s. The SSNTD method has been developed in detail in [3, 4, 5]. Here we present only some feature of this method for it use in studies of cluster decay of nuclei.

For understanding of the mechanism of cluster decay it is could be useful to investigate the fusion reactions induced by the daughter products of the decay [6]. Fusion cross sections of such reactions are extremely small ($\leq$ 10^{-32} cm^{-2}), and this problem is directly associated with the sensitivity of the SSNTD method [7, 8].

CP644, *Exotic Clustering: 4th Catania Relativistic Ion Studies,* edited by S. Costa, A. Insolia, and C. Tuvè
© 2002 American Institute of Physics 0-7354-0099-7/02/$19.00

The Principal Peculiarities of SSNTD

For investigation of cluster decay we used polycarbonate (PC), polyethylene terephtalate (PET) and special phosphate glass types (PG) [4]. These detectors have different threshold sensitivity determined by ionization losses : PC (Rodyne, Lexan) records nuclei down to Z ~5 at the energies of ~2 MeV/u; PET (Mylar, Lavsan, Cronar) records nuclei down to Z ~ 8 and PG (various types) with charge thresholds in the range Z ~ 6 - 9. However, the cluster decay studies of nuclei with $Z \geq 90$ it is necessary to carry out identification in the conditions of high background of alpha-particles and fission fragments [4]. For cluster detection in presence of a high fluence of the other particles it is necessary to take into account two following effects. High fluence of α particles gives rise to the background etchable tracks of recoil nuclei produced by the interaction α particles with nuclei of the detecting material. They hamper the search of the events of interest and distort the shape of the tracks being searched for. Therefore, if the spectrometric measurements are being made, a limit exits on the α particle fluence density for polycarbonate, polyethylene terephtalate, and phosphate glass of the order of 10^{10}, 10^{12} and 10^{14} cm^{-2}, respectively; a limit on the fission fragment flux density for these detectors is not more than 10^{4}cm^{-2} [4].

For a reliable detection of clusters in conditions of a high background of α-particles and fission fragments with the fluence up to 10^{15} and 10^{6} cm^{-2}, respectively, a two-layer detector was developed at Dubna [9]. Its thin upper layer (6-10 μm) containing tracks of recoils is removed after etching. In order to eliminate the background of fission fragments an absorber, which stops the lightest fission fragments can be used. Here it is necessary to use an absorber without defects and uniform in composition and thickness. The second effect high fluence of particles is change of the detecting properties. To correct for this effect the special experiments were performed. The change in sensitivity of polyethylene terephtalate to ions with Z = 8 - 12 and phosphate glass (GOI – 104, Russia) to ions with Z = 14 are observed for fluence of α particles (Φ_{max}, α/ cm^2) larger than 10^{9} and 5×10^{14}, respectively [9,10].

At first SSNTD (polyethylene terephtalate) was used for study of [231]Pa [11]. The detectors were on the radioactive source and its exposure was performed in the air. The emitted fragments charge was determined using relations V_t/V_b (dE/dx) and V_t/V_b (R_{res}), where V_t and V_b are the etch rates along the ion track and in bulk material of detector, respectively; R_{res} is the residual range of the ion. Multiple etching makes it possible to obtain several values of V_t/V_b along the track and identify the track accurately enough. The error in charge identification reached $\Delta Z = \pm 0.15$ and that in the mass identification was $\Delta M/M = \pm 1$. The ion energy was determined from the total ion range with a relative accuracy $\Delta E/E = 3 - 5\%$. The mass of clusters with a known charge was found from the Q-value of the reaction. Normally the detection efficiency was 66% of 2π for one–layer detectors and several percent lower when the additional "antibackground" layer was used. This method registration and identification allows to apply very large area of thick radioactive sources.

For investigation of cluster decay of ^{236}Pu [12], ^{242}Cm [13] we used different techniques. Contrary to the first method the geometry of the detector exposure was carried out in vacuum and geometry of these experiments was arranged in such way that cluster entered the detector at the angles close to 90° to its surface. In this case clusters were identified using a replica made from special material. The measured parameters of the replica were transformed into a curve representing the dependence V_t/V_b (R_{res}). The detection efficiency was ~ 86% and ~73% of 2π solid angle, respectively for ^{236}Pu and ^{242}Cm. In this case of cluster registration the source area is limited.

Table 1 presents the types of SSNTD used for cluster decay investigation of various isotopes in Russia.

TABLE 1. Type of SSNTD used for cluster decay investigation of various isotopes.

Detector type	Φ_{max}, $\alpha/$ cm^2	Isotopes
Polycarbonate	$\leq 5 \times 10^{10}$	^{114}Ba
Polyethelene terephtalate	$\leq 5 \times 10^{12}$	^{230,232}Th, $^{230-236}$U, ^{237}Np, ^{231}Pa
Phosphate glass (GOI –104)	$\leq 5 \times 10^{14}$	^{236}Pu, ^{242}Cm, ^{231}Am

For registration of fission fragments in the fusion-fission reactions (^{12}C,^{16}O + ^{208}Pb [7, 8]) mica detectors were used [14]. The irradiated mica detectors were annealed 6 hours at 460°C for decreasing the background events coming from the interaction of the scattered ^{12}C and ^{16}O ions with the mica atoms. In this case the detection efficiency for fission fragments do not change and is ~ 100% . This method allows to investigate deep subbarrier reactions up to cross section level of 10^{-36} cm^{-2} and opens new possibilities of obtain additional information on the compound-states responsible for the cluster decay of nuclei.

EXPERIMENTAL INVESTIGATIONS OF CLUSTER DECAYS

Our program (JINR – Kurchatov Institute- Milan University collaboration) includes a series of experiments on search of the decays of the nuclides, which are important for understanding the nature of cluster radioactivity:

• ^{238}U, ^{242}Cm and ^{241}Am (odd nuclide), in which one could expect the observation of ^{34}Si emission.

• ^{223}Ac, being the isobar-alalogue of ^{223}Ra, for which the fine structure of the spectrum was observed. The emission of ^{14}C and ^{15}N is expected.

• ^{114}Ba,^{112}Ba with expected emission of ^{12}C and formation of double magic ^{100}Sn. Observation of these decays would demonstrate the existence of a new domain of cluster radioactivity.

The Obtained Experimental Data

The investigation of ^{242}Cm$\rightarrow^{34}$Si + ^{208}Pb decay was finished recently [13]. The search for the new region of cluster emitters near the double magic ^{100}Sn was undertaken, and the upper limit of ^{114}Ba $\rightarrow$ ^{12}C $+^{102}$Sn decay established (Table2). The analysis of results of ^{238}U $\rightarrow^{34}$Si + ^{204}Hg experiment is in progress.

^{242}**Cm.** We have already combined our efforts (Moscow, Dubna, Milan) in studying the decay ^{242}Cm $\rightarrow^{34}$Si + ^{208}Pb. The main difficulty encounteredin this study is that the spontaneous fission probability of ^{242}Cm is 10^9–10^{10} times higher than the expected probability of the cluster decay. ^{242}Cm was produced from ^{241}Am irradiated with thermal neutrons at the Kurchatov Institute reactor. After chemical separation two sources of ^{242}Cm about 0.23 and 0.17 mg with a thickness ~0.1 mg/cm^2 were prepared. The ^{242}Cm sources ~30 mm in diameter on Pt backing were located in the center of hemispheres ~190 mm in diameter. The interior surface of the hemispheres was inlaid with special phosphate glass track detectors. The detectors were covered with 20 μm thick polyimide film absorbers to prevent the action of the spontaneous fission fragments on the detectors. After penetrating 20μ polyimide ^{34}Si ions must have the residual energy ~34 MeV and range ~ 9 – 10μm. The setup was placed in a vacuum chamber. The total exposition was 292 days. The total number of ^{242}Cm nuclei that decayed during the exposition was 4.29 x 10^{17}. The duration of the exposition was fixed to ensure that the α - particle fluence onto detectors was $\leq 10^{14}$ α per cm^2. 15 tracks were found and identified as formed by particles with Z ~ 14. The mean kinetic energy value $(81.0 \pm 1.9) \pm 2.0$ MeV we measured is consistent with the expected one E_k = 82.97 MeV calculated from the ground state Q-value. The corresponding partial half-life is $(1.4 +0.5/-0.3) \times 10^{23}$ s. This value is in an agreement with the lower the limit obtained earlier [15]. The branching ratio relatively to alpha decay is 1.0×10^{-16} and the one relatively to spontaneous fission 1.6×10^{-9}. ^{242}Cm is the heavest nuclide for which cluster emission was studied.

The total experimental data on nuclear cluster decay obtained up to date in Russia are presented in Table 2. The some results were obtained with the participation of scientific groups of other countries (Italy, France, Romania) .

TABLE 2. Measured Values of Cluster Probabilities.

Initial nucleus and cluster emitted	$\lambda cl / \lambda\alpha$	$T_{1/2}$, yr
^{225}Ac $\rightarrow$ ^{14}C	(5.3 ± 1.4) x 10^{-12}	(5.0 ± 1.0) x 10^9
^{230}Th $\rightarrow$ ^{24}Ne	(5.6 ± 1.0) x 10^{-13}	(1.3 ± 0.3) x 10^{17}
^{232}Th $\rightarrow$ ^{26}Ne	≤ 3 x 10^{-12}	>5 x 10^{21}
^{231}Pa $\rightarrow$ ^{24}Ne	(3.8 ± 0.7) x 10^{-12}	(8.6 ± 1.6) x 10^{15}
^{233}U $\rightarrow$ ^{24,26}Ne	(7.5 ± 2.5) x 10^{-13}	(2.2 ± 0.8) x 10^{17}
^{234}U $\rightarrow$ ^{24,26}Ne	(3.9 ± 1.0) x 10^{-13}	(6.3 ± 2.1) x 10^{17}
^{234}U $\rightarrow$ ^{28}Mg	(2.3 ± 0.7) x 10^{-13}	(1.1 ± 0.4) x 10^{18}
^{235}U $\rightarrow$ $^{24-26}$Ne	<5 x 10^{-12}	>1.4 x 10^{20}
^{235}U $\rightarrow$ ^{28}Mg	<8 x 10^{-13}	>9 x 10^{20}
^{236}U $\rightarrow$ ^{24,26}Ne	<4 x 10^{-12}	>6 x 10^{18}
^{236}U $\rightarrow$ ^{30}Mg	~ 2 x 10^{-13}	>1.2 x 10^{20}
^{237}Np $\rightarrow$ ^{30}Mg	<4 x 10^{-12}	>5 x 10^{19}
^{236}Pu $\rightarrow$ ^{28}Mg	(2.7 ± 0.7) x 10^{-14}	1.1 x 10^{14}
^{241}Am $\rightarrow$ ^{34}Si	<5 x 10^{-15}	>9 x 10^{16}
^{242}Cm $\rightarrow$ ^{34}Si	1.0 x 10^{-16}	(0.44 ± 0.16) x 10^{16}
^{114}Ba $\rightarrow$ ^{12}C	$\lambda_{cl}/\lambda_{\beta}\leq 10^{-4}$	$\geq 10^{-4}$

The Current Experiments

^{238}U. The motivation to search the cluster decay of ^{238}U (with Milan University) is to look for the dependence of the decay probability on the number of neutrons in the parent nucleus. For uranium isotopes the data exist for ^{230,232,233,234,236}U (Table 2). It was shown [16] that cluster decay probability of ^{236}U falls out of the existing systematics and is more than one order of magnitude higher than it was expected. Our extrapolations based on the ^{236}U results show that the same kind of anomaly can take place also in the case of ^{238}U, and its time of life can reach the values ~3 x10^{20} years, what is much shorter than the predictions by Poenaru et al. [17] and Blendowsky R. and Walliser H. [18].

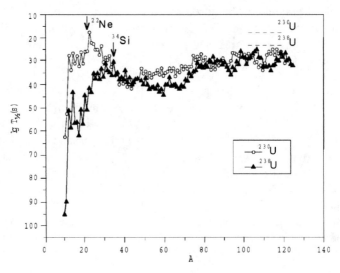

FIGURE 1. Mass distribution of cold decays of ^{230}U and ^{238}U. Only the light fragments half of the spectrum is shown in full. The life times for hot fission of ^{230}U and ^{238}U by the dashed lines are shown.

In Fig.1 the full mass for ^{230}U and ^{238}U calculated by Pik-Pichak model are shown. One can see that the distributions are similar on the region of cold fission but differ for emission of light fragments. An increase probability for the decay ^{230}U $\rightarrow$ ^{22}Ne is clearly seen.

For study of the cluster decay of ^{238}U (the Melinex track detectors and the metallic layers of ^{238}U with enrichment 99.9% (Good Fellow, England) total area 3000 cm^2 in 2-π geometry were used [19]. After two years exposition and scanning of 2500cm^2 area there was found 3 events of Si. This result is consistent with lgT$_{1/2}$ ~ 29.4 s (Fig.2).

This nuclide would become the 20th one for which cluster radioactivity was observed. Comparison with the existing models of cluster radioactivity is hampered by large uncertainties in the mass value of the daughter nuclide ^{204}Hg. At the moment of writing this paper the analysis is continued.

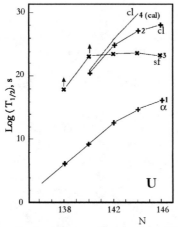

FIGURE 2. The partial half-lives for the emission of a α-particles (1), for the emission of clusters (2) and for the spontaneous fission (3) of even-even U isotopes versus the neutron number N in the original isotope. The dashed line (4) shows the estimates of the half-lives for the cluster decay according to Ref.17.

^{241}**Am** is a favorable odd-even nuclide to examine for possible ^{34}Si emission. ^{241}Am has an odd number of protons (Z=95) and, as well as ^{242}Cm is expected to emit the heaviest of observed till now fragments, ^{34}Si. It would be possible to conduct comparison of probabilities of decays of even-even and even-odd nuclei. The decay of the odd nuclei is considerably more responsive to nuclear structure than the decay of the even ones. The structure features of mother and daughter nuclei play important role in emission of clusters. The decay of ^{223}Ra could serve a good example [10]. At decay ^{241}Am there can be also considerable hindrance reducing the probability of the "fission –like" mechanism. The odd proton in Am-241 (I=5/2$^-$) should occupy an orbital $h_{9/2}$ according to Nilsson scheme. Thus transitions to all lower levels of ^{207}Tl are unfavorable, and their probabilities should be less than expected only from the barrier penetrability considerations. Moreover, we suppose to receive confirmation or refutation of a hypothesis that in the region of masses of clusters with A ~ 35 there is a transition between an "alpha – decay – like" and " fission-like" mechanism as it was observed for ^{242}Cm [19].

The different models predict branching ratios for emission of 80.6 MeV ^{34}Si ions ranging from 4 x 10^{-13} to 4 x 10^{-16} [21]which makes a determined search feasible. Its branching ratio for spontaneous fission relative to alpha decay is only ~ 4 x 10^{-12}. Several groups, including ours, using different types of detectors and irradiation geometry have already looked for ^{34}Si decay of ^{241}Am, with negative results [21]. The most low an upper limit of $\lambda_{cl}/\lambda_\alpha = 7.4$ x 10^{-16} was obtained in [21]. They used 8 mg of ^{241}Am and a six-month exposure.

We used 100 mg of ^{241}Am, vacuum hemisphere for detector irradiation, 1450 cm^2 of special phosphate glass detectors and exposition ~12 months. We expect to increase the lower limits of $T_{1/2}$ by two orders of magmitude in comparison with previous measurements [21]. The main difficulties of this experiment besides the low decay probability are caused by large background (10^{14} alpha particles and 10^5 fission fragments on 1 cluster). To make the observation of decay possible (~ 10 events) we use one year irradiation in the spherical geomerty of exposure. Under such conditions 1 cluster decay of ^{241}Am will correspond to logT = 27.3 s.

The Planned Experiments

^{223}**Ac.** We propose to search for the decays ^{223}Ac $\rightarrow$ ^{14}C + ^{209}Bi and ^{223}Ac $\rightarrow$ ^{15}N + ^{208}Pb. Among the possible new objects of cluster radioactivity studies, ^{223}Ac nuclide is of exceptional interest:

• One of the best evidences of the "α-decay-like" mechanism of cluster emission and of importance of nuclear structure effects was obtained from observation of the fine structure in ^{223}Ra $\rightarrow$ ^{14}C + ^{209}Pb decay spectrum. As ^{223}Ac is the isobar-analog of ^{223}Ra the study of its decay to ^{14}C + ^{209}Bi provides complimentary information in this most important case. In particular, if the transition hindrance is determined by the change of configurations of the odd nucleons one can expect that the decay ^{223}Ac $\rightarrow$ ^{14}C to the ground state of ^{209}Bi would be similar to the transition to the ^{209}Pb first excited state.

• Another important source of information on nuclear structure effects would be the measurement of ^{14}C/^{15}N emission probabilities ratio. It could be especially useful for testing the hypothesis of quasimolecular structure of light actinides. It should be emphasized that the daughter nuclei, ^{209}Bi and ^{208}Pb have well-known structure, what simplifies the theoretical analysis in comparison with the other cases.

• Cluster radioactivity models often predict quite different shapes of the potential barriers. Independent information on them, being of vital importance for the theory could be obtained from the study of the decay products interaction, elastic scattering or fusion-fission [6]. At present time direct comparison of the "empirical potentials" with cluster emission probabilities could be done in principle only for two systems: ^{222}Ra $\rightarrow$ ^{14}C + ^{208}Pb and ^{230}U $\rightarrow$ ^{22}Ne + ^{208}Pb, what is limited by the lack of available targets and (or) projectiles. Measurement of ^{223}Ac decay probabilities will add two more systems for such kind of investigation: ^{14}C + ^{209}Bi and ^{15}N + ^{208}Pb.

To summarize, the experiment with ^{223}Ac will make possible the overall comparison of the neighbor cluster decaying nuclei ^{222}Ra, ^{223}Ra and ^{223}Ac and provide new important information both on cluster radioactivity mechanism and structure of light actinides.

The main mode of ^{223}Ac decay is α radioactivity with $T_{1/2}$ =126s. For estimation of the effect we choose Ponaru predictions [17] being most popular and based on the phenomenological analysis of all exiting data.

The source will be produced by proton irradiation of Th target at a cyclotron :

$$^{232}Th(p.6n)^{227}Pa \rightarrow 38min, 86\% \rightarrow ^{223}Ac \rightarrow \alpha, ^{14}C, ^{15}N$$

The excitation function of the (p,6n) – reaction was measured [22]. The cross-section research its maximum 40 mb at the proton energy ~ 47 MeV. 10 mb level corresponds to the energies 41 and 62 MeV.

After irradiation protactinium source will be produced by chemical separation from the thorium target and reaction products (uranium, actinium and radium). The phosphate glasses allowing discrimination of carbon and nitrogen tracks will be used as SSNTD. The SSNTD exposition will be carried out in special vacuum chamber.

We propose to use the proton beam with E ~65 MeV and beam intensity ~ 25 μA, the times irradiation and subsequent SSNTD exposition times during 2 hours with the geometry efficiency ~ 30%. In this case we expect to collect about 250 events of ^{14}C and 10 tracks of ^{15}N.

As the result of this experiment we hope to observe for the first time the cluster decay of with emission of ^{14}C and ^{15}N fragments and measure the corresponding partial times of life. The observation of ^{223}Ac cluster radioactivity would allow to make comparisons with all existing models of cluster radioactivity and with the decay properties of the isobar analogue ^{223}Ra and ^{222}Ra. We expect to obtain information on structure of the light actinides.

^{112}**Ba.** Another experiment of principal importance, which is under discussion is aimed to search of new exotic nucleus ^{112}Ba and its cluster decay.

Observation of a new region of cluster radioactivity would be an extremely important step in understanding the nature of cold processes. First, it would demonstrate that "lead" radioactivity is not unique process. Secondly, the study of the decays in completely different regions of masses of parent nuclei could provide a critical test to existing theoretical models.

The only new domain of cluster radioactivity, which could be approached experimentally is connected with the formation of the daughter nuclei close to the double magic Z = N ^{100}Sn nuclide. There were several attempts [23 - 26] to observe the decay $^{114}Ba \rightarrow {}^{12}C + {}^{102}Sn$. However, only the lower limit of the partial half-life was established.

Not excluding the repetition of search of ^{114}Ba we are thinking over the experiment on search of even more exotic decay, that is $^{112}Ba \rightarrow {}^{12}C + {}^{100}Sn$. The very fact of synthesizing of ^{112}Ba is of great interest.

The heaviest known nucleus with N =Z is ^{100}Sn (N = Z = 50). The really exotic nucleus ^{112}Ba (N = Z = 56), which is twelve mass units heavier, was never searched, and any information on its properties (mass, decay modes, structure, etc) is important. ^{112}Ba can have many unusual features. Among them are:

- Possible exotic cluster structure consisted of the double-magic ^{100}Sn core plus either three α-particles or ^{12}C cluster.
- Predicted large static octupole deformation [27], the largest in this mass region.
- Possibility of different decay modes: α, β, p, 2p, ^{12}C.
- Possible extremely high ^{12}C decay probability in comparison with the other known cases of cluster radioactivity in trans-lead region

There are many uncertainties in planning this experiment. The ^{112}Ba mass is unknown, and different theoretical mass evaluations differ by 2 - 3 MeV. The probable Q-value for $^{112}Ba \rightarrow {}^{12}C$ decay lies in the limits 20 – 22 MeV. This makes the predictions of the decay probabilities (not only of cluster one but the competing α- or β-decays as well) uncertain to several orders of magnitude even in the frame of one particular model.

Most of the existing models of cluster radioactivity give rather pessimistic values of the cluster decay probability. However, one has to keep in mind that the parameters of most of these models were adjusted by comparison with the data in the quite different region of "lead" radioactivity. This attaches more confidence to the microscopic model of CFK, which as we have seen in the previous paper is most adequate to the description of the lightest clusters emission and is consistent with the scattering data due to use of folding potentials.

Depending on the Q-value the partial half-life for ^{12}C emission in CFK model could lie in the interval $\log T = 2$ s and $\log T = -2$ s. This makes the observation of cluster decay not hopeless because the probable competing β-decay is expected to have $\log T = -0.7$ s and $\log T = -1.4$ s, correspondingly.

Another important problem is the extremely low cross-section of the productive reaction. The only realistic reaction is ^{58}Ni + ^{58}Ni $\rightarrow$ 4n + ^{112}Ba. Different estimates give the cross-section integrated over the significant part of the excitation function the value $\sigma = 3 \times 10^{-35}$ cm^2. Under more or less realistic experimental conditions one can hope to collect about 100 ^{112}Ba nuclei, some part of which could decay by ^{12}C emission.

At present time our Kurchatov – JINR collaboration works on the proposal of this experiment.

ACKNOWLEDGMENTS

The authors are deeply thankful to Yu. Ts. Oganessian, M.G. Itkis, A.Sandulescu, for the great attention to the work, valuable advice and critical comments. Thanks are also due to V. L. Mikheev, Yu.S. Zamyatnin, G. A. Pik-Pichak, R. Bonetti, A. Guglielmetti, D. Poenary, Hasegan, V.M. Shilov, M. Hussonois V.A. Shigin for helpful discussions, K.I. Merkina, S. Burinova, V.A. Ponomarenko, for fulfilling the treatment of detectors and analysis obtained results.

This work was supported by the Russian Foundation of Fundamental Researches, grant 02-02-17297 and NATO under Reference PST.CLG. 976017.

REFERENCES

1. Tretyakova, S. P., Mikheev, V. L., Pyatkov, Yu. M., in *Advances In Nuclear Physics and Related Area*, ed. by D. M. Brink, M. E. Grypeous and S.E. Massen, Proc.of the European Conference 1997, Thessaloniki, Greece, 8 – 12 July 1997, Giahoudi – Giapouli Publishing, 1999, pp. 913 – 917.
2. Hourani, E., Hussonois M., and Poenaru, D.N., *Ann.Phys. Fr.* **14**, 311-396 (1989).
3. Fleisher, R.L., Price, P. B., and Walker, R. M., in *Nuclear Tracks in Solid*: Principle and Applications (University of California Press, Berkeley, 1975).
4. Tretyakova, S.P., *Fiz. Elem. Chastiz At. Yadra*, **23**, 364-429 (1992).
5. Zamyatnin,Yu.S., Mikheev,V.L., Tretyakova,S.P., Furman,V.I., Kadmenskii, S.G., Chuvil'skii, Yu.M., *Fiz. Elem. Chastits At. Yadra* **21**, 537-594 (1990).

6. Ogloblin, A.A. and Tretyakova, S.P., in *Physics of Isomers,* Proc of First Int Workshop, June 18-21, St.-Petersburg, 2000, ed. F.F.Karpeshin, p. 122.
7. Oganessian, Yu.Ts., Itkis, M.G., Kozulin, E.M., Pustylnik, B.I., Tretyakova, S.P., Calabretta, L., Guzel, T., *JINR Rapid Commun.,* **1[75]-96,** 123 (1996).
8. Tretyakova, S.P., Calabretta, L., Itkis, M. G., Kozulin, E. M., Kondratiev, N. A., Maiolino, C.,Pokrovski, I.V., Prokhorova, E.V., Rusanov, A.Ya., Tretyakova, T.Yu., in *Nuclear Shells – 50 Years,* Proc. Int. Conf. On Nucl. Phys., Dubna, Russia, 21-24 April 1999, ed. by Yu.Ts.Oganessian and R. Kalpakchieva, World Scientific, p.151.
9. Tretyakova, S.P. *Nucl. Tracks Radiat. Meas.,* **19,** 667-671 (1991).
10. Tretyakova, S.P., Mikheev, V. L., Kobzev A. P., Golovchenko, A. N., Ponomarenko, V. A., Timofeeva, O. V., Shigin, V. A.,and Hussonois, M., *Prib. Tekh. Eksp.* **4,** 53-57 (1998).
11. Sandulescu, A., Zamyatnin, Yu.S., Lebedev I.A., Myasoedov, V.F., Tretyakova, S.P., Hashegan, D., *JINR Rapid Commun.,* **5,** 84 (1984).
12. Hussonois, M., Le Du J. F., Trubert, D., Bonetti, R., Guglielmetti, A., Tretyakova, S.P., Mikheev, V. L., Golovchenko, A. N., and Ponomarenko, V. A., *Pis'ma Zh.Eksp. Teor. Fiz.* **62,** 685 – 689 (1995).
13. Ogloblin, A.A., Bonetti, R., Denisov, V.A., Guglielmetti, A., Itkis, M.G., Mazzocchi, C.,Mikheev, V.L., Oganessian, Yu.Ts., Pik-Pichak, G.A., Poli, G., Pirozhkov, S.M., Semochkin, V.M., Shigin, V.A., Shvetsov, I.K., and Tretyakova, S.P., *Phys.Rev.* **C 61,** 034301 (2000).
14. Tretyakova, S.P. *Nucl. Tracks Radiat. Meas.,* **19,** 665-666 (1991).
15. Mikheev, V. L., Tretyakova, S. P., Ogloblin, A. A., Denisov, V.A., Semochkin, V.M. , Shvetsov I. K., and D.N. Stepanov, *FLNR Scientific Report ,* **E7-93-57,** Dubna, JINR, (1993), p. 48.
16. Tretyakova, S.P., Mikheev, V. L., Ogloblin, A. A., Shigin, V. A., Golovchenko, A. N., and Ponomarenko, V. A., *JETP Let.,* **59,** 368 (1994).
17. Poenaru, D. N., Schnabel, D., Greiner, W., Mazilu, D. and Gherghescu R., *Atomic Data and Nucl. Tables,* **48,** 231 (1991).
18. Blendowsky, R. and Walliser, H., *Phys. Rev. Lett.* **61,** 1930 (1986).
19. Ogloblin, A.A., Pik-Pichak, G.A., Tretyakova, S.P., in *Fission Dynamics of Atomic Clusters and Nuclei,* ed. by Joao da Providencia, D.M.Brink, F. Karpeshine, F.B.Malik, Proc. of Int. Workshop Luso, Portugal, 15-19 May 2000, World Scientific, pp. 143-162
20. Hourani, E., Hussonnois, M., Stab, L., Brillard, L., Gales, S., Shapira, J.P., *Phys.Lett.* **B,** **160,** 375 (1985).
21. Moody, K. J., Hulet, E. K., Wang, S., Price, P. B., and Barwick, S. W., *Phys. Rev.* **C 36,** 2710 – 2712 (1987).
22. Sulk, H. C., Crawford, J. E., and Moore, R. B., *Nucl.Phys.* **A 218,** 418 - 428 [1974].23.
23. Oganessian, Yu. Ts., Lasarev Yu. A., Mikheev V. L., Muzychka, Yu, a., Shirokovskii, I. V., Tretyakova, S. P. And Utenkov, V. K., *Z. Phys.* **A 349,** 341 (1994).
24. Oganessian, Yu. Ts., Mikheev V. L., Tretyakova, S. P., Kharitnov, Yu. P., Yakushev, A. B., Timokhin, S.N., Ponomarenko, V. A., and Golovchenko, A. N., *Yad. Fizika,* **57,** 1178 –1182 (1994).
25. Guglielmetti, A., Bonetti, R., Poli, Price, P.B., Westphal, A. J., Janas , Z., Keller, H., Z., Kirchner, Klepper O., Piechaczek, A., Roeckl, E., Smidt K., Plochocki, A., Szeryp0, J., and Blank, B., Phys.Rev. C 52, 740 - 745 (1995); *Nucl. Phys.* **A583,** 867 – 872 91995).
26. Guglielmetti, A., Bonetti, R., Poli, G., Collatz, R., Hu, Z., Kirchner, R, Roeckl, E., Gunn, N., Price, P.B., Weaver, B. A., Westphal, A. and Szerypo, J., *Phys. Rev.* **C 56,** R2912 –R2916 (1997).
27. J.Skalski et al., *Phys. Lett.,* **B 238,** 6 (1990).

Microscopic approach to cluster radioactivity

V.P.Bugrov*, W.I.Furman[†], S.G.Kadmensky*, S.D.Kurgalin* and
Yu.M.Tchuvil'sky **

*Voronezsh State University; Voronezsh, Russia
[†]Joint Institute of Nuclear Research; Dubna, Russia
** Scobeltsyn Institute of Nuclear Physics, Moscow State University; Moscow, Russia

Abstract. Microscopic version of cluster decay theory is developed. The central point of the theory is the formalism of spectroscopic factors of heavy clusters workable all over the range of known examples of cluster decay. The approach is reasonably adequate to the experimental data. Obtained results confirm the assumption that the mechanism of the process is similar to α-decay one.

INTRODUCTION

Recently the set of experimental data, related to heavy cluster radioactivity, contains 22 measured examples and 14 low enough upper limits (or disputable results). In our view it is sufficiently large to be a subject of detailed theoretical conclusions. At the same time various theoretical approaches result by reasonable description of the data. So current state of the researches requires to extend the analysis to a wider area of nuclear processes including α-decay and nuclear fission. By now high-quality microscopic theory of α-decay is developed [1]. Classic subject of α-decay theory – $^{212}Po \rightarrow {}^{208}Pb + \alpha$ transfer is now well described by pure microscopic approach [1, 2]. Superfluid (BCS) model including very few simplifications of pure microscopic theory turns out to be capable to describe α-decay of even-even nuclei far from near-magic area with a very good (within a factor two) precision [3, 4]. Qualitative purposes of α-decay of odd nuclei are also clear enough in the framework of BCS approach [5].

The progress of α-decay theory makes the goal to describe using microscopic approach spontaneous emission of heavier clusters exciting. An approach of such type hold promise to make clear a mechanism of cluster decay and, possibly, to be a first step to microscopic theory of fission.

CLUSTER RADIOACTIVITY AND α-DECAY

The number of features of cluster decay demonstrates that this process and α-decay are closely related: α-decay readily fits to the systematics of cluster decay [6], kinetic energy of both alpha and cluster emission corresponds to the ground (or to the lower excited) states of a cluster and a daughter nucleus, in both cases well pronounced and irregular even-odd effect takes place. On the other hand these properties differentiate

CP644, *Exotic Clustering: 4th Catania Relativistic Ion Studies*, edited by S. Costa, A. Insolia, and C. Tuvè
© 2002 American Institute of Physics 0-7354-0099-7/02/$19.00

heavy cluster decay from fission. So as a first step it is reasonable to build the theory of the new type of radioactivity in analogy with α-decay.

Alpha-decay theory contains two basic points. First one is the problem of preformation of two-body – α-particle + low lying state of daughter nucleus – channel in the surface region of parent nucleus. Second one is the problem of matching of the wave function of this channel with asymptotic one. The first problem is the central point of α-decay theory. Widths of alpha-transitions of neutron resonances of mother nuclei to ground states of daughter one varying irregularly from resonance to resonance and being very small in average (in comparison with the results of two-body calculations) provide evidence that, first, the preformation factors should be introduced to the theory and, second, only a microscopic models may be capable to explain such behaviour of the preformation factors. Mentioned above direct and BCS calculations, solving the problem of preformation (spectroscopic) factors for the transfers between low lying states of even-even nuclei, require very much computer work. Naturally the difficulties essentially increase if one tries to describe not four- but many-nucleon spectroscopic factors which are necessary for the description of cluster decay. Strightforward BCS approach [3, 4, 9] turns out to be workable for clusters $X = 12, 14, 16$ but extension of the research to heavier clusters seems to be enormously hard task.

The second problem of alpha-decay theory is usually solved in terms of R-matrix scheme (see [3, 4, 5] for example). Taking into account that the amplitudes of asymptotic wave functions obtained in R-matrix approach are rather sensitive to the behavior of matched internal wave functions one should perform the calculations of the latter ones on an extremely large basis to achieve a correct description of A-nucleon wave functions in the region where asymptotic conditions are already fulfilled. Another computational difficulty is that the amplitude of asymptotic wave function in the subbarrier region varies by many orders of magnitude. It causes the problem of numerical stability. Usually the problem is solved in terms of quasiclassical approximation. To create analogous approach to cluster radioactivity one should rearrange α-decay theory taking into account that new troubles appear with the grows of cluster mass.

Coming now to the discussion of cluster radioactivity let us begin with the second problem. In the monograph [5] and preceding papers devoted to α-decay the simple method allowing one, first, to get rid of quasiclassical approximation and, second, to make the matching procedure more stable is presented. For this purpose matching conditions are imposed on the area of near-barrier maxima of both asymptotic and internal wave functions but not on logarithmic derivatives. Such conditions are really much more stable to the details of the wave functions and so allow one to make the basis of microscopic calculations smaller. A good description of alpha-decay widths is achieved in the frame of the method. In short this approach is as follows.

It is convenient to invert the picture and define two-body wave function of Gamov state $\Phi_{xc}(R)$ as the solution of Shroedinger equation with the potential term of the form (the S-wave example is discussed here for simplicity) $V_{x,A_f}^{nucl}(R) + V_{x,A_f}^{coul}(R)$, where the real part of the optical potential or the folding potential can be used as the first item, and the following asymptotics:

$$\Phi_{xc}(R) \to \sqrt{\frac{\hbar \Gamma_{xc} K_c}{Q_x}} G_c(R), \qquad (1)$$

where Γ_{xc} - experimental value of the decay width, x denotes the cluster, $c \ (\equiv J_f, \sigma_f, L)$ - decay channel, $K_c = \sqrt{\frac{2\mu_x Q_x}{\hbar}}$, μ_x - reduced mass, $G_0(R)$ - irregular Coulomb function describing Gamov state beginning from the point R_1, placed slightly to the left of outer turning point because imaginary part $iF_0(R)$ drops down very rapidly when calculations move from outer region to the center in the subbarrier region. A rapid grows of irregular Coulomb function provides the numerical stability of the calculations. This scheme is the second basic idea of the approach.

The phenomenological measure of clustering can be defined as:

$$W_{xc}^{cl} = \int_{R_{cl}}^{R_1} [\Phi_{xc}(R)]^2 dR, \qquad (2)$$

where $R_{cl} = 1,2(A_\alpha^{\frac{1}{3}} + A_f^{\frac{1}{3}})$ fm - conventional contact point. $\Phi_{xc}(R)$ possesses by the maximum to the right of R_{cl} so the value of W_{xc}^{cl} is loosely dependent on the choice of R_{cl}. The value W_{xc}^{cl} is used similarly to the reduced width $\gamma_{xc}^2(R)$. To predict this value theoretically one should build the microscopic channel wave function and calculate with it the integral analogous to presented in the exp. 2.

In case of alpha-decay the potential $V_{x,A_f}^{nucl}(R)$ is well-tested by the measurements of elastic and inelastic scattering and alpha-decay of neutron resonances so the shape of the potential is well-defined. Up to now the possibilities to build up the analogous optical potential for heavier clusters are poor because of dominating of the Coulomb scattering and inobservability of (n, X_0) reactions. What about the folding potentials for heavy cluster channels the results obtained by use of them depend strongly on N-N or N-nucleus input. Table 1 is an illustration of the problem under discussion. First three examples of potentials are built in the frame of single or double folding procedures with the different N-nucleus or N-N potentials. The forth one is the sole known optical potential where there are some bound cluster states [10]. One can see that the range of variation of W_{xc}^{cl} for ^{14}C is about four, and for ^{32}Si – about eight orders of magnitude. The fact is that only cluster decay research itself is the test on the potentials – no another process is capable of selection. So it is reasonable to choose one and only one potential and explore it for all examples of cluster decay. Such test showed that single folding potential proposed in [11] (see column 1 of the table 1) is preferable.

It is important to notice that the relative values – ratios of absolute values of the spectroscopic factors W_{xc}^{cl} of different transfers of one and the same cluster – are stable. This fact allows one to analyze even-odd effect, fine structure and other qualitative properties of the process.

TABLE 1. Spectroscopic factors $W_{xc}^{cl.}$, extracted using various potentials

Decay	Potential			
	1	2	3	4
$^{222}Ra \rightarrow^{14}C$	$2.0 \cdot 10^{-10}$	$1.1 \cdot 10^{-8}$	$1.5 \cdot 10^{-6}$	$1.2 \cdot 10^{-9}$
$^{223}Ra \rightarrow^{14}C$	$3.1 \cdot 10^{-12}$	$1.8 \cdot 10^{-10}$	$2.7 \cdot 10^{-8}$	$1.5 \cdot 10^{-11}$
$^{224}Ra \rightarrow^{14}C$	$2.2 \cdot 10^{-10}$	$1.5 \cdot 10^{-8}$	$2.3 \cdot 10^{-6}$	$8.9 \cdot 10^{-10}$
$^{225}Ac \rightarrow^{14}C$	$1.9 \cdot 10^{-10}$	$1.4 \cdot 10^{-8}$	$2.2 \cdot 10^{-6}$	$6.2 \cdot 10^{-10}$
$^{226}Ra \rightarrow^{14}C$	$0.8 \cdot 10^{-10}$	$6.6 \cdot 10^{-9}$	$1.1 \cdot 10^{-6}$	$2.2 \cdot 10^{-10}$
$^{230}Th \rightarrow^{24}Ne$	$2.2 \cdot 10^{-18}$	$1.2 \cdot 10^{-15}$	$2.4 \cdot 10^{-12}$	$2.7 \cdot 10^{-17}$
$^{231}Pa \rightarrow^{24}Ne$	$1.4 \cdot 10^{-19}$	$6.5 \cdot 10^{-17}$	$3.2 \cdot 10^{-13}$	$1.4 \cdot 10^{-18}$
$^{232}U \rightarrow^{24}Ne$	$3.3 \cdot 10^{-18}$	$1.5 \cdot 10^{-15}$	$3.6 \cdot 10^{-12}$	$3.3 \cdot 10^{-17}$
$^{233}U \rightarrow^{24}Ne$	$4.5 \cdot 10^{-20}$	$2.5 \cdot 10^{-17}$	$7.1 \cdot 10^{-14}$	$5.7 \cdot 10^{-19}$
$^{234}U \rightarrow^{24}Ne$	$1.6 \cdot 10^{-18}$	$1.1 \cdot 10^{-15}$	$3.1 \cdot 10^{-12}$	$1.5 \cdot 10^{-17}$
$^{234}U \rightarrow^{28}Mg$	$1.6 \cdot 10^{-22}$	$7.0 \cdot 10^{-19}$	$3.1 \cdot 10^{-12}$	$1.5 \cdot 10^{-17}$
$^{236}Pu \rightarrow^{28}Mg$	$2.0 \cdot 10^{-22}$	$1.9 \cdot 10^{-19}$	$1.2 \cdot 10^{-15}$	$7.2 \cdot 10^{-21}$
$^{238}Pu \rightarrow^{28}Mg$	$1.7 \cdot 10^{-21}$	$2.4 \cdot 10^{-18}$	$1.4 \cdot 10^{-14}$	$2.2 \cdot 10^{-20}$
$^{238}Pu \rightarrow^{32}Si$	$6.0 \cdot 10^{-24}$	$2.1 \cdot 10^{-17}$	$5.2 \cdot 10^{-16}$	$7.8 \cdot 10^{-21}$

MICROSCOPIC CALCULATIONS OF SPECTROSCOPIC FACTORS OF HEAVY CLUSTERS

To demonstrate the competence of the microscopic approach to cluster radioactivity one should build a formalism of the preformation (spectroscopic) factors and compare the results of their calculations with phenomenological values from exp.2. The basic goal is to create a method workable all over the range of known examples of cluster decay and slightly outside this range. The complexity of the calculations of these quantities increases drastically with the increasing of the cluster mass so a very compact formalism is required. The basic points of the formalism [12, 13] are as follows.

Let $\Psi_A(|A>)$, $\Psi_{A-x}(|A-x>)$, and $\Psi_x(|x>)$ be the internal wave functions of mother, daughter nuclei and cluster respectively. Formation amplitude (cluster formfactor) is defined as:

$$\Phi_{lm}(\rho) = <A \mid \hat{\mathscr{A}} \mid A-x, Y_{lm}(\theta\varphi)\delta(\rho-\rho')x>, \tag{3}$$

where $\hat{\mathscr{A}}$ is antisymmetrizer. Normalization constant

$$W_c = \int |\Phi_c(\rho)|^2 \rho^2 d\rho, \tag{4}$$

where c denotes channel index is referred to as the spectroscopic factor.

Let us use the definition of cluster formation amplitude in the channel spin representation. Using the expansion $\delta(\rho-\rho') = \sum_{nL}\phi_{nL}(\rho)\phi_{nL}(\rho')$ one can write down:

$$\Phi_L(\rho) = \sum_n < \hat{\mathscr{A}}\left\{U_L^{J_i\pi_i}\phi_{nL}(\rho')\right\} \mid \psi^{J_i\pi_iM_i} > \phi_{nL}(\rho), \tag{5}$$

where the first factor in the right hand side is the spectroscopic amplitude of disintegration $A \rightarrow (A-x) + x$. To calculate this value one should transform it from translationally-invariant to conventional shell model form.

Let us consider the auxiliary overlap containing oscillator wave functions $\varphi(\mathbf{R})$:

$$I \equiv \langle \varphi_{00}(\mathbf{R}_A), A \mid \varphi_{00}(\mathbf{R}_{A-x}), A - x\varphi_{n\lambda}(\mathbf{R}_x)x \rangle. \tag{6}$$

Talmi-Moshinsky transformation of the wave functions in the right hand side of exp. 6 associated with the transform $R_{A-X}, R_X \rightarrow R_A, \rho$, and integration over R_A bring the overlap to the form:

$$I = (-1)^n \left(\frac{A}{A-b} \right)^{-\frac{n}{2}} \langle A \mid A - x; \phi_{n\lambda}(\rho)x \rangle, \tag{7}$$

where the value of Talmi-Moshinsky coefficient of simple form is introduced. As the result one can obtain:

$$\Phi_L(\rho) = (-1)^L \binom{A}{x}^{\frac{1}{2}} \sum_n \left(\frac{A}{A-x} \right)^{\frac{n}{2}} \times \tag{8}$$

$$< \phi_{00}(R_A)\psi^{J_i \pi_i M_i} \mid \phi_{00}(R_{A-x}) \left\{ \psi^{J_f \pi_f}, \psi^{J_x \pi_x} \phi_{nL}(R_x) \right\}_{J_i \pi_i M_i} > \phi_{nL}(\rho).$$

Elliott-Skirme theorem connects Pauli-allowed wave functions of conventional and translationally-invariant shell models $\phi_{00}(\mathbf{R}_A) \mid A N^{min} \rangle = \Psi_{sh}^A$. It should be noticed that the theorem is a precise transformation in case of oscillator wave functions. Under a different conditions a superposition of shell model functions appears in the right-hand side of the theorem expression. Nevertheless for realistic wave functions in the exp. 8 Elliott-Skirme theorem seems to be a reasonable approximation. Indeed for $A \gg x$, first, the replacement $\rho \rightarrow R_x$ holds good and, second, the range of the sum over n Δn is not so wide for realistic shell model potential. So one can neglect by the multiplier $(A/(A-x))^{(n-n_{min})/2}$ and write:

$$\Phi_L(R_x) \simeq (-1)^L \left(\frac{A}{A-x} \right)^{n_{min}/2} \Phi_{sh.L}(R_x), \tag{9}$$

where:

$$\Phi_{sh.L}(R_x) \equiv < \psi_{sh.}^{J_i \pi_i M_i} \mid \hat{\mathscr{A}} \left\{ U_{sh.L}^{J_i \pi_i M_i} \delta(R_x - R_x') \right\} >, \tag{10}$$

$$U_{sh.L}^{J_i \pi_i M_i} = \left\{ \psi_{sh.}^{J_i \pi_i} \psi^{j_x \pi_x} Y_L(\Omega_{R_x}) \right\}_{J_i \pi_i M_i} \tag{11}$$

and the number n_{min} is determined by generalized Wildermuth prescription $n_{min} = N_A - N_{A-X} - N_X + q$ where q is the degree of the structural forbiddenness [14].

It should be stressed that the recoil effect involved to the formula of decay width through the factor $A/(A-x))^n$ play an important role in the cluster decay process. Being about 1.5 for alpha-decay the recoil effect turns out to be $\sim 10^2$ for $^{12,14}C$ and ^{16}O and becomes extremely large ($\sim 10^8 - 10^9$) for heavy clusters (^{30}Mg, ^{32}Si). For the oscillator wave functions $|A>$ and $|(A-X)>$ the formula 9 is precise one and no doubt that this approximation is much more well-grounded than neglecting of the recoil effect because the topological properties of many-nucleon wave function (numbers of nodes etc.) are the same in the oscillator and realistic model. Moreover for the realistic wave functions $|A>$ and $|(A-X)>$ presented approximation is lower limit of the recoil effect although underestimation is rather small.

To calculate the overlap in exp. 10 so-called multicluster representation (general form of Wildermuth-Kannelopulos [15] two-cluster representation of ^8Be wave function) of the heavy cluster wave function is used. Let us limit ourselves by considering on even-even cluster with $X/2 \geq Z_X$. Choosing multicluster disintegration of $|X>$ onto alpha-particle and bineutron wave functions one can write down:

$$| \psi^{J_x \pi_x M_x} > \equiv | xN = x - 4[f]L_x S_x J_x T_x >= \qquad (12)$$

$$a_{\{L_j\}} \mathscr{A} \left\{ \prod_{i=1}^{\beta} (\psi_{\alpha_i}) \prod_{k=1}^{\gamma} (\psi_{F_k})_{S_x} \prod_{j=1}^{\beta+\gamma-1} \left[\phi_{n_j l_j}(\rho_j) : \{L_j\} L_x \right] \right\}_{J_x \pi_x M_x},$$

where $\{L_j\}$ - the number of intermediate angular momenta, if L=0 than it is useful to put L_j=0, l_j=0. ψ_{α_i} - the internal wave function of alpha-particle and ψ_{F_k} - the same for bineutron, $\phi_{n_j l_j}(\rho_j)$ - the wave functions of Jacobi coordinates characterizing the system of β α-particles and γ bineutrons F_k. Here

$$\left(a_{\{L_j\}} \right)^{-1} = < xN = x - 4[f]L_x S_x \,|\, \hat{A} \,|\, \prod_{i=1}^{\beta} (\psi_{\alpha_i}) \prod_{k=1}^{\gamma} (\psi_{F_k})_{S_x} \qquad (13)$$

$$\prod^{\beta+\gamma-1} \left[\phi_{n_j l_j}(\rho_j) \right] L_x > .$$

is the normalization constant. The function in the right hand side of exp. 12 is not equal to zero identically if the following conditions are fulfilled: $n_j = F_k$ ($\gamma > 1$), if β=1, n_1=4, $n_j = F_k$ ($\gamma \geq 1$), if β=2 etc. The exp. 12 is valid until the quantum numbers of right hand side are determined the function unambiguously. That is true for the neutron number of a cluster $N_Z \leq 10$. In opposite case (for heavier clusters) analogous SU(3) scheme (angular momenta must be replaced to the Elliott symbols $(\lambda \mu)$) provides one-to-one correspondence. In that case one should rearrange the SU(3) wave function to the superposition of functions with definite L_j, l_j. Clebsh-Gordan expansion of SU(3) group is used for this purpose.

Let us come back to the exp. 11. It is obvious that left hand side of it should be expand onto the product of the wave function of (A-X) nucleons and the proper number of functions of biproton and bineutron pairs. It can be done by multicluster fractional parentage expansion [16]

$$| p_i >= \sum_{\Delta, \varepsilon} < p_i \,|\, p_f(\Delta), p(\varepsilon) > | p_f(\Delta) > | p(\varepsilon) > \qquad (14)$$

147

where $| p_i > . | p_f(\Delta) >$ - wave functions of proton and neutron components of parent and daughter nuclei respectively, Δ and ε - the proper quantum numbers, and

$$| p(\varepsilon) >= \prod_r | p_r^{(2)}(\varepsilon_r) >, \tag{15}$$

where $| p_r^2(\varepsilon_r) >$ - two-proton wave functions. As the result one can write down:

$$\Phi_{sh.L}(R_x) = \left[\frac{A!}{(A-x)!(4!)^\beta (2!)^\gamma} \right]^{\frac{1}{2}} a_0 \times \tag{16}$$

$$\sum_{\{\Delta_r\}\{\varepsilon_r\}\{\Delta'_r\}\{\varepsilon'_r\}} < \prod_{i=1}^\beta \tilde{\Phi}_{\alpha_i}^{\varepsilon_i \varepsilon_{i'}}(R_{\alpha_i})$$

$$\prod_{k=1}^\gamma \tilde{\Phi}_{2n_k}^{\varepsilon'_{\beta+k}}(R_{2n_k}) \mid \prod_{j=1}^{\beta+\gamma-1} \left[\phi_{n_j l_j} = 0(\rho_j) : L_x = 0 \right] Y_L(\Omega_{R_x})\delta(R_x - R'_x) >$$

$$\prod_r < p_{f_{r-1}}(\Delta_{r-1}) \mid p_{f_r}(\Delta_r) p_r^{(2)}(\varepsilon_r) >$$

$$\prod_r < n_{f_{r'-1}}(\Delta'_{r'-1}) \mid n_{f_{r'}}(\Delta'_{r'}) n_{r'}^{(2)}(\varepsilon'_{r'}) >,$$

where the first multiplier appeared due to the acting of antisymmetrizer. Formfactors $\tilde{\Phi}_{\alpha_i}^{\varepsilon_i \varepsilon_{i'}}(R_{\alpha_i})$ and $\tilde{\Phi}_{2n_k}^{\varepsilon'_{\beta+k}}(R_{2n_k})$ are three-dimensional formation amplitudes of α-particles and bineutrons for the configurations $\{\varepsilon_i, \varepsilon'_i\}$. $\varepsilon_{\beta+k}$ respectively. Then one can sum over $\{\varepsilon_r\}\{\varepsilon'_r\}$ and obtain:

$$\Phi_{sh.L}(R_x) = a_0 \eta_\alpha^\beta \eta_{2n}^\gamma \sum_{\{\Delta_r\}\{\Delta'_r\}} \prod_{i-1}^\beta \left(W_{sh.\alpha_i}^\Delta \right)^{\frac{1}{2}} \prod_{k=\beta+1}^{\beta+\gamma} \left(W_{sh.2n_k}^\Delta \right)^{\frac{1}{2}} \times \tag{17}$$

$$\prod_{i=1}^\beta \Phi_{sh.\alpha_i}^{'\Delta_i \Delta_{i-1} \Delta'_i \Delta'_{i-1}}(R_{\alpha_i}) \prod_{k=1}^\beta \Phi_{sh.2n_k}^{'\Delta'_{k+\beta} \Delta'_{k+\beta-1}}(R_{2n_k}) \mid \times$$

$$\mid \prod_{j-1}^{\beta+\gamma-1} \left[\phi_{n_j l_j = 0}(\rho_j) : L_x = 0 \right] Y_L(\Omega_{R_x})\delta(R_x - R'_x) >,$$

where $\Phi_{sh.i}^{'\Delta}(R_i)$ is normalized three-dimensional formfactor of the light cluster and the normalization constant is the spectroscopic factor of corresponding cluster by definition. The values η_α and η_{2n}^γ are introduced to take into account the difference in the oscillator parameters of light constituents of heavy clusters and real light particles. These values are similar for clusters $x \geq 12$, and for $\hbar\omega = \hbar\omega_{16O}$ they are $\eta_\alpha^2 \simeq 2.7$ with η_{2n}^2 close to unity.

The procedure of calculation of the presented value looks very complicated. However it may be essentially simplified by introducing of the following approximation. The matter is that the sum in the formula 17 contains the dominating component if BCS approach is used. Indeed, the large enhancement of the spectroscopic factors W_α^0 and W_{2n}^0 of the transfers between 0^+ states arises in that case. Alpha-decay investigations demonstrate that magnitude of the effect is about 10^3. So any term of the sum over Δ_r (Δ_r') containing some nonzero values of angular momenta is suppressed by the factor 10^{-3} in comparison to dominating one. This fact make it possible to neglect such terms. Furthermore in BCS approach alpha-particle formation amplitudes are very similar for different nuclei both in shape and in absolute value. In addition the deformation sensitivity of the amplitudes are rather weak [1, 4]. So it is reasonable to use the unified form for them. This form is obtained by BCS calculations of ^{218}Rn alpha-decay with the constant strength parameter of paring forces. The value of energy gap was used to control relation between this parameter and the basis size. Under these assumptions the spectroscopic factor of a heavy cluster takes the form:

$$W_{xc} = \left(\frac{A}{A-x}\right)^{n_{min}} a_0^2 \eta_\alpha^{2\beta} \eta_{2n}^{2\gamma} \prod_{i=1}^{\beta}(W_{sh.\alpha_i}^0) \prod_{k=1}^{\gamma}(W_{sh.2n_k}^0) \mathcal{J}_x, \tag{18}$$

$$\mathcal{J}_x = \int J_x^2(R_x)R_x^2 dR_x, J_x(R_x) = <\prod_{i=1}^{\beta}\Phi'^{\Delta_{0i}\Delta_{0i-1}\Delta'_{0i}\Delta'_{0i-1}}(R_{\alpha_i}) \times \tag{19}$$

$$\prod_{k=1}^{\gamma}\Phi'^{\Delta'_{0k+\beta}\Delta'_{0k+\beta-1}}(R_{2n_k}) \mid \prod_{j=1}^{\beta+\gamma-1}\left[\phi_{n_j l_j=0}(\rho_j) : L_x = 0\right] Y_L(\Omega_{R_x})\delta(R_x - R_x') > .$$

As it was mentioned above to calculate the formation amplitude of a cluster with neutron number $N_X > 10$ SU(3) Clebsh-Gordan expansion is used.

Presented expressions constitute complete formalism of spectroscopic factors of heavy even-even clusters from $1p$- and $(2s-1d)$-shells. What about clusters of odd mass it is not necessary to rearrange the formalism. Notation F_k may denote not only bineutron but odd nucleon also.

Table 2 demonstrates the quality of microscopic approach. Values W_{xc}^{cl} are extracted from the experimental data using the potential of [11]. Taking into account the variation of the calculated value W_{xc} in the range of many orders of magnitude it is possible to consider the agreement between the experiment and the theory as satisfactory.

Coming now to the discussion of general properties of the microscopic approach to cluster radioactivity it should be noticed that the most important feature of the mechanism is a huge-scale superfluid-enhancement effect. As it was discussed above the factor of enhancement is about $k_x \simeq 30$ per nucleon pair in going from the shell model which neglects BCS correlations in the expressions of binucleon spectroscopic factors. It is known from alpha-decay theory [3, 4, 5] that the subsequent transformations of the BCS and shell model wave functions corresponding to the transformations of one-nucleon coordinates to Jacobi coordinates of intracluster motion do not involve additional enhancement or suppression. So the factor of superfluid enhancement for cluster decay is roughly estimating about $k_x^{x/2} \sim 10^{3x/4}$. That's why conventional shell model is invalid to describe the effect under discussion.

149

TABLE 2. Comparison of calculated W_{xc} and extracted W_{xc}^{cl} spectroscopic factors of heavy clusters.

x	W_{xc}	W_{xc}^{cl} mother nucleus	x	W_{xc}	W_{xc}^{cl} mother nucleus
^{8}Be	$6.6 \cdot 10^{-7}$		^{28}Mg	$1.5 \cdot 10^{-21}$	$6.1 \cdot 10^{-22}$ ^{234}U
^{12}C	$1.4 \cdot 10^{-9}$		^{30}Mg	$5.7 \cdot 10^{-24}$	$1.8 \cdot 10^{-22}$ ^{236}U
^{14}C	$5.9 \cdot 10^{-11}$	$2.0 \cdot 10^{-10}$ ^{222}Ra	^{32}Si	$5.1 \cdot 10^{-25}$	$6.0 \cdot 10^{-24}$ ^{238}Pu
^{16}O	$3.2 \cdot 10^{-12}$		^{34}Si	$5.0 \cdot 10^{-27}$	
^{24}Ne	$7.0 \cdot 10^{-19}$	$2.2 \cdot 10^{-18}$ ^{230}Th	^{40}Ca	$6.0 \cdot 10^{-33}$	
^{26}Ne	$2.8 \cdot 10^{-21}$		^{48}Ca	$\sim 10^{-42}$	

Alpha-decay of odd nuclei displays pronounced superfluid effect. If spins of mother and daughter nuclei are equal than diagonal mechanism of α-decay allows one to shape α-cluster from two superfluid nucleon pairs and thus enhancement effect remains as large as the that in case of even-even nucleus. Such transitions are called favored. First-order forbidden transitions are dominating when the spins of initial and final states are different and so correspond to the break up of one superfluid . As it is clear form just mentioned properties of BCS model the forbiddenness factor of about 30 appears. This qualitative property is well pronounce in the α-decay widths. The same is true for heavy cluster decay also. However there are two ^{14}C-transitions (decay of ^{225}Ac and dominating line of the fine structure of ^{223}Ra) possessing by the experimental values of the spectroscopic factors typical for transitions in even-even nuclei although the initial and final spins are different. On one hand it is clear that the origin of the property is associated with the fact that cluster decay unlike alpha-one is followed by a strong rearrange of nuclear shape from deformed to spherical. On the other hand the description of even cluster transfer in odd nucleus is rather complicated problem because many mechanisms of shaping of the final states are in competition. This problem remains unsolved recently. The fact that there is no satisfactory microscopic theory of a decay in case of strong channel coupling is, in the end, the origin of such indefinite situation.

CONCLUSION

So the microscopic approach based on reasonably well-justified assumptions turns out to be adequate for the description of quantitative data and qualitative features of cluster radioactivity. Later measurements confirm it's predicting power.

As a consequence the assumption that mechanism of the process is similar to α-decay one is supported. This point is, in our opinion, essential because in opposite to α-decay heavy cluster decay of neutron resonances is not observed and the sole reliable experimental confirmation of the mechanism is specific form of fine structure of ^{14}C spectrum of ^{223}Ra where low energy line is much more intensive than ground one.

The sole phenomenological component of the model is the choice of the fixed potential from the large number of proposed in literature. Proposed formalism of spectroscopic factors do not correspond Fliessbasch prescription [17] in analogy with BCS formalism of α-decay [3, 4, 5] where this abandonment results to much better description of the experimental data. Mathematics of the approach demonstrates the number of dramatic features of the process less pronounced in α-decay such as recoil effect and superfluid enhancement.

At the same time the theory is not complete, some assumptions are not suitable for precise calculation, and some features of the discussed process still remain incomprehensible. For example the problem of the lack of hindrance in heavy cluster decay of some odd nuclei are not solved yet. Both improvements of the approach in some points and search for new examples to test it are of importance.

It is clear however that the list of perspective emitters of heavy clusters looks rather poor. Theoretical search for these emitters in near-magic regions seems to be finished. The sole region, for which certain expectations to find some interesting examples remain, is, in our opinion, the area of proton drip line.

Creation of a general (multichannel) theory of the composite particle interaction is of great importance for solving the discussed problems. Cluster radioactivity seems to be a good laboratory for testing of such theory.

ACKNOWLEDGMENTS

Work partially supported by RFBR, grant No. 00-02-16683.

REFERENCES

1. Lovas R.G., Liotta R.J., Insolia A. et al *Phys. Rep.* **294**, 265-362 (1998).
2. Varga K., Lovas R.G., Liotta R.J. *Nucl. Phys.* **A550**, 421-434 (1992).
3. Delion D.S., Insolia A., Liotta R.J. *Nucl. Phys.* **A549**, 407-419 (1992).
4. Delion D.S., Insolia A., Liotta R.J. *Phys. Rev.* **C49**, 3024-3028 (1994).
5. Kadmensky S.G., Furman W.I. *Alpha Deecay and Related Nuclear Reactions*, Energoatomizdat, Moscow, 1985.
6. Novatsky B.G., Ogloblin A.A. *Vestnik AN SSSR* **1**, 81-91(1988).
7. Delion D.S., Insolia A., Liotta R.J. *J. Phys.* **G19**, L189-L192 (1993).
8. Delion D.S., Insolia A., Liotta R.J. *J. Phys.* **G20**, 1483-1498 (1994).
9. Florescu A., Insolia A., Liotta R.J. *Phys. Rev.* **C52**, 726-734 (1994).
10. Christensen P.R., Winter A. *Phys. Lett.* **B65**, 19-22 (1976).
11. Gareev F.A., Ivanova S.P., Kalinkin B.N. *Izv. AN SSSR Ser. Fiz.* **32**, 1690-1694 (1968).
12. Kadmensky S.G., Furman W.I., Tchuvil'sky Yu.M. *Izv. AN SSSR Ser. Fiz.* **50**, 1786-1795 (1986).
13. Zamyatnin Yu.S., Mikheev V.L., Tret'yakova S.P. et al *Sov. J. Part. Nucl.* **21**(2), 231-256 (1990) .
14. Smirnov Yu.F., Tchuvil'sky Yu.M. *Phys. Lett.* **B134**, 25-28 (1984).
15. Wildermuth K., Kannelopulos T. *Nucl. Phys.* **9**, 449-456 (1958).
16. Glozman L.Ya., Tchuvil'sky Yu.M. *J. Phys.* **G9**, 1034-145 (1983).
17. Blendowske R., Fliessbach T., Walliser H. *Nucl. Phys.* **A464**, 75-86 (1987).

Dynamics of Cluster Formation

Liquid-vapor phase transitions in finite nuclei

L. G. Moretto*, J. B. Elliott*, L. Phair* and G. J. Wozniak*

*Nuclear Science Division, Lawrence Berkeley National Laboratory,
University of California, Berkeley, California 94720

Abstract. The leptodermous approximation is applied to nuclear systems for $T > 0$. The introduction of surface corrections leads to anomalous caloric curves and to negative heat capacities in the liquid-gas coexistence region. Clusterization in the vapor is described by associating surface energy to clusters according to Fisher's formula. The three-dimensional Ising model obeys rigorously Fisher's scaling up to the critical point. Multifragmentation data from several experiments including the ISiS and EOS Collaborations provide strong evidence for liquid-vapor coexistence. The phase diagram is obtained for the finite system and an extrapolation is made to infinite nuclear matter.

Nuclei are leptodermous, mesoscopic clusters. Their thin skin leads naturally to an expansion of their energy in powers of $A^{-1/3}$. This is the basis of the liquid drop model, which manages to reproduce the binding energies of nuclei to within 1%. A similar leptodermous treatment of nuclear systems at $T > 0$ should lead to an equivalently good reproduction of nuclear thermodynamical properties.

The appearance of a vapor phase at $T > 0$ opens two complementary perspectives for the characterization of phase coexistence: the liquid perspective and the vapor perspective. From the liquid perspective, one can determine the caloric curve in terms of vaporization enthalpy. From the vapor perspective one considers the extent to which nucleons are aggregated into clusters, as an indicator of incipient liquid condensation.

In the first part of this presentation we take the liquid perspective and derive analytically the caloric curve and the (negative) heat capacity for a drop undergoing an isobaric phase transition. In the second part we take the vapor perspective and show that the clusterization in the three-dimensional Ising model can be accounted for in terms of the leptodermous expansion.

THE ISOBARIC PHASE TRANSITION OF A LIQUID DROP

The thermodynamical equilibrium properties of first order phase transitions are completely describable in terms of the thermodynamic state variables associated with the individual separate phases.

Renewed attention to phase transitions has been generated by studies of models with either short range interactions (e.g. the Ising model [1, 2, 3, 4, 5] or the lattice gas model [6]). Features expected to disappear in the thermodynamic limit were noticed and were claimed to be essential, characteristic indicators of phase transitions in mesoscopic systems [7]. For instance, first order phase transitions were associated with anomalous convex intruders in the entropy versus energy curves, resulting in back-bendings in the

CP644, *Exotic Clustering: 4th Catania Relativistic Ion Studies*, edited by S. Costa, A. Insolia, and C. Tuvè
© 2002 American Institute of Physics 0-7354-0099-7/02/$19.00

caloric curve, and in negative heat capacities [7]. These anomalies have been attributed to a variety of causes, the foremost of which are surface effects.

In the context of nuclear physics, microcanonical models of nuclear multifragmentation have associated the anomalies of a convex intruder with the onset of multifragmentation [7]. Recently, the claim has been made of an empirical observation of these anomalies, such as negative heat capacities in nuclear systems [8]. It would be highly desirable to ground any evidence for these anomalies, theoretical or otherwise, on thermodynamics itself, minimally modified to allow for the possible role of surface effects related to the finiteness of the system.

In this section we investigate the subject of caloric curves and heat capacities of finite systems in the coexistence region and the underlying role of varying potential energies ("ground states") with system size on the basis of simple and general thermodynamical concepts. Our study applies to leptodermous (thin skinned) van der Waals-like fluids and to models such as Ising, Potts, and lattice gas.

Let us consider a macroscopic drop of a van der Waals fluid with A constituents in equilibrium with its vapor. The vapor pressure p at temperature T is given by the Clapeyron equation

$$\frac{dp}{dT} = \frac{\Delta H_m}{T \Delta V_m} \tag{1}$$

where ΔH_m is the molar vaporization enthalpy and ΔV_m is the molar change in volume. It gives a direct connection between the "ground state" properties of the system and the saturation pressure along the coexistence line. In fact, we can write $\Delta H_m = \Delta E_m + P\Delta V_m \sim \Delta E_m + T$. For $T \ll \Delta E$, $\Delta H_m \approx a_v$ can be identified approximately with the liquid-drop volume coefficient in the absence of other terms like surface and Coulomb, etc. Assuming $\Delta V_m \approx V_m^{\text{vapor}}$, that the vapor is an ideal gas and that ΔH_m is constant with T, Eq. (1) can be integrated to give

$$p \simeq p_o \exp\left(-\frac{\Delta H_m}{T}\right). \tag{2}$$

Equation (2) represents the p-T univariant line in the phase diagram of the system if ΔH_m is assigned its bulk value ΔH_m^0. It was observed long ago that for a drop of finite size ΔH_m must be corrected for the surface energy of the drop [9]

$$\Delta H_m = \Delta H_m^0 + \Delta H_m^s = \Delta H_m^0 - \gamma S_m = \Delta H_m^0 - a_s \frac{A^{2/3}}{A} \tag{3}$$

where γ is the surface tension, S_m is the molar surface of the drop of radius r, and a_s is the surface energy coefficient. Substitution in Eq. (2) leads to

$$p = p_0 \exp\left(-\frac{\Delta H_m^0}{T} + \frac{a_s}{A^{1/3} T}\right) = p_{\text{bulk}} \exp\left(\frac{a_s}{A^{1/3} T}\right). \tag{4}$$

At constant T the vapor pressure increases with decreasing size of the drop (see Fig. 1).

Let us now consider the case of isobaric evaporation of a drop starting from a drop with A_0 constituents and evaporating into a drop with $A < A_0$ constituents. Let us now

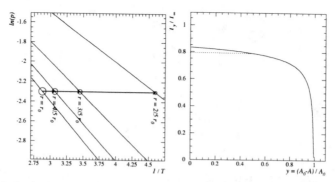

FIGURE 1. Left: the log of the saturated vapor pressure as a function of inverse temperature for different droplet radii. The size of the open circles is proportional to the droplet radius. Arrows illustrate the path of evaporation at constant pressure. Right: The temperature as a function of droplet size for a drop evaporating at constant pressure (open boundary conditions). The solid line shows the case of a spherical drop, while the dotted line shows the case of a finite cubic lattice evolving as in Fig. 2 top.

define the drop size parameter $y = \frac{A_0 - A}{A_0}$. At constant pressure

$$p_0 \exp\left(-\frac{\Delta H_m^0}{T}\right) = p_0 \exp\left(-\frac{\Delta H_m(y)}{T_y}\right), \tag{5}$$

from which follows

$$\frac{T_y}{T_\infty} \simeq \frac{\Delta H_m(y)}{\Delta H_m^0} \simeq 1 - \frac{1}{A^{1/3}} \simeq 1 - \frac{1}{A_0^{1/3}(1-y)^{1/3}}. \tag{6}$$

Thus, a slight <u>decrease</u> in temperature is predicted as the drop evaporates isobarically, thus leading to a negative isobaric heat capacity in the coexistence region as illustrated in Fig. 1. As the drop is evaporating at constant pressure, the drop moves from one coexistence curve to another according to its decrease in radius, and thus to progressively lower temperatures. This slight effect is due <u>not</u> to an increase in surface as the drop evaporates, since the drop surface of course <u>diminishes</u> as $A^{2/3}$, but to the slight increase of molar surface which <u>does</u> increase as $A^{-1/3}$ (see Fig. 2). Also, the formation of bubbles in the body of the drop is thermodynamically disfavored by the factor $f = \exp(-\gamma \Delta S / T)$ where ΔS is the surface of the bubble.

Let us now move to the amply studied cases of lattice gas, Ising, and Potts models. We consider first an evaporating finite system in three dimensions of size $A_0 = L^3$, with open boundary conditions. This case is essentially identical to the case of a drop discussed above (see Fig. 1). For maximal density at $T = 0$ (the ground state) $y = 0$ and the entire cubic lattice is filled. For decreasing densities, always at $T = 0$ a single cluster of minimum surface is present, which evolves from a cube to a sphere. The associated change in surface is shown in Fig. 2. The caloric curve from $y = 0$ to $y = 1/2$ is essentially <u>flat</u> like in the infinite system, and the heat capacity is trivially infinite.

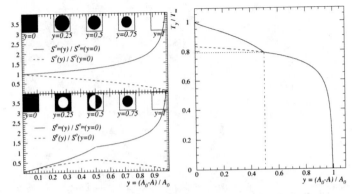

FIGURE 2. Left Top: The surface S^o (dashed) and molar surface S_m^o (solid) area of a drop for open boundary conditions normalized to their values at $y = 0$. Left Bottom: The surface S^p (dashed) and molar surface S_m^p (solid) area of a drop for periodic boundary conditions normalized to their values at $y = 0$. In-sets show the configurations at various values of y. Right: the temperature as a function of droplet size for a drop evaporating at constant pressure in a system with periodic boundary conditions. The solid line shows the case of a finite cubic lattice with periodic boundary conditions evolving as in Fig. 2 bottom, while the dotted line and the dashed line are the same as in Fig. 1 and the vertical dash-dotted line indicates the case of 50% lattice occupation.

The introduction of periodic boundary conditions rids the system of "dangling bonds," as it were, by repeating a cubic lattice of side L periodically along the three coordinates. These conditions, lead to peculiar consequences.

At $y = 0$, the lattice is filled with particles so that $\Delta H_m(0) = \Delta H_m^0$ characteristic of the infinite system. As y increases at fixed lattice size, a <u>bubble</u> develops in the cube and <u>surface</u> is rapidly created (see Fig. 2). The bubble develops since the periodic boundary conditions prevent evaporation from the surface. The bubble grows with increasing y until it touches the sides of the lattice. This occurs for $y \approx 1/2$. At nearly $y = 1/2$ and beyond, the "stable" configuration is a drop that eventually vanishes at $y = 1$. The change in surface associated with the range $0 \leq y \leq 1$ as well as the molar surface are shown in the bottom left panel of Fig. 2.

The evaporation enthalpy thus becomes

$$\Delta H_m(y) \simeq a_v \left(1 - \frac{y^{2/3}}{A_0^{1/3} (1-y)} \right) \tag{7}$$

from $y = 0$ to $y = 1/2$, and

$$\Delta H_m(y) \simeq a_v \left(1 - \frac{1}{A_0^{1/3} (1-y)^{1/3}} \right) \tag{8}$$

from $y = 1/2$ to $y = 1$. As a consequence, for periodic boundary conditions

$$\frac{T_y}{T_\infty} \simeq 1 - \frac{y^{2/3}}{A_0^{1/3} (1-y)} \tag{9}$$

from $y = 0$ to $y = 1/2$, while from $y = 1/2$ to $y = 1$ Eq. (6) holds.

The dramatic effect of periodic boundary conditions can be seen in Fig. 2. The temperature decreases substantially with increasing y, due to the fact that the molar enthalpy at $y = 0$ assumes its bulk value ΔH_m^0 and must meet the previous case of open boundary conditions for $y = 1/2$. This may well explain the calculated negative heat capacities reported in literature, as due to the unnatural choice of boundary conditions.

With the lessons learned above we can evaluate the heat capacities for nuclei. It is apparent from the above arguments that the key quantity is ΔH_m and its dependence on the drop size, irrespective of the physical causes that determine its magnitude and dependence. In the case of nuclei the quantity ΔH_m is determined by all the terms in the liquid drop model, such as the Coulomb and symmetry energy all of which contribute to the mean binding energy per nucleon. One can immediately infer that when the binding energy per nucleon <u>decreases</u> with A, the heat capacity should be positive, and vice-versa. Thus, since the maximum binding energy per nucleon occurs at $A \sim 60$, negative heat capacities should be possible only for $A < 60$. Let us proceed more precisely. We can rely again on the Clapeyron equations to calculate the heat capacity as follows

$$C_p = \left.\frac{dH}{dT}\right|_p = -\left.\frac{dH}{dA}\right|_p \left.\frac{dA}{dT}\right|_p = -\Delta H_m(A)\left.\frac{dA}{dT}\right|_p \tag{10}$$

but

$$\left.\frac{dT}{dA}\right|_p = \left.\frac{dp}{dA}\right|_T \frac{dT}{dp}. \tag{11}$$

From the integrated form of the Clapeyron equation we have

$$\left.\frac{dp}{dA}\right|_T = -\frac{1}{T}\frac{d\Delta H_m}{dA}p \tag{12}$$

so

$$\left.\frac{dT}{dA}\right|_p = -\frac{1}{T}\frac{d\Delta H_m}{dA}p\frac{TV_m}{\Delta H_m} = -\frac{T}{\Delta H_m}\frac{d\Delta H_m}{dA}. \tag{13}$$

Finally

$$C_p = \frac{\frac{(\Delta H_m(A))^2}{T}}{\frac{d\Delta H_m}{dA}}. \tag{14}$$

The derivative in the denominator can be evaluated approximately from the dependence on the binding energy per nucleon B upon the mass number

$$\frac{d\Delta H_m}{dA} = \frac{dB}{dA}. \tag{15}$$

The liquid drop model allows us to estimate such a derivative. Without the Coulomb term, of course, we recover the results presented above for a drop. Thus negative heat capacities should be expected and the caloric curves should look like that shown in Fig. 1. The Coulomb and symmetry terms, however, become important at large values of A, say, along the line of β-stability. From Fig. 3 it is apparent that the binding energy

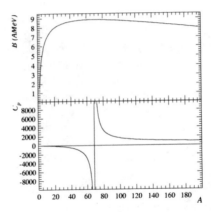

FIGURE 3. The binding energy of atomic nuclei (top) and the associated heat capacity (bottom).

decreases with A for $A >\sim 60$. Consequently in all this region of A, positive specific heats should be expected. Only for $A <\sim 60$, negative specific heats are predicted.

This straightforward result based on elementary thermodynamics and ground state binding energies raises serious questions as to the meaning of the negative heat capacities that have been claimed for large nuclear systems.

3D ISING MODEL: PARADIGM FOR MULTIFRAGMENTATION

Two features associated with the fragment multiplicities are found to be quite pervasive in all multifragmentation reactions. They have been named "reducibility" and "thermal scaling" [10, 11].

Reducibility is the property that the probability of observing n-fragments of a given size is expressible in terms of an elementary one-fragment probability. This property can occur only if fragments are independent of one another and it coincides with stochasticity. Both binomial, and its limiting form, Poissonian reducibilities have been extensively documented experimentally for nuclear multifragmentation [10, 11].

Thermal scaling is the linear dependence of the logarithm of the one-fragment probability with $1/T$ (an Arrhenius plot). It indicates that the emission probability for fragment type i has a Boltzmann dependence

$$q_i = q_0 e^{-B_i/T} = q_0 e^{-c_0 A^\sigma/T} \tag{16}$$

where B_i is a "barrier" corresponding to the production process.

The combination of these two empirical features attests to a statistical mechanism of multifragmentation in general, and to liquid-vapor coexistence specifically [14].

The three-dimensional Ising model satisfies both the criteria of simplicity in its Hamiltonian and lends itself to a thermal treatment with nontrivial results [15]. In this light we analyze the Ising results in the same way as has been done with nuclear multifragmentation data [10, 11]. Fig. 4 shows the multiplicity distributions for a sample of fragment

160

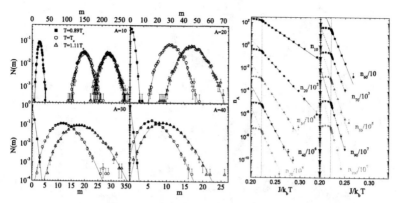

FIGURE 4. Left: The probability distributions for obtaining m fragments of size A at the three temperatures indicated. The solid lines are Poisson distributions with means given by the Monte Carlo data. Right: Arrhenius plots of the cluster distributions. The lines are fits of the form given in Eq. (16). The critical temperature is indicated by the dashed line.

sizes and temperatures. The solid lines represent Poisson distributions calculated from the corresponding mean multiplicities. The distributions are remarkably close to Poissonian for all masses and all temperatures below, at and above T_c.

If the fragment distributions exhibit thermal scaling, of the form given in Eq. (16), its Arrhenius plot (a semi-log graph of the number of clusters of size A (n_A) vs. $1/T$), should be linear. As shown in Fig. 4, this is the case over a wide range of temperatures ($0 < T < T_c$) and fragment sizes.

The features of reducibility and thermal scaling discussed above can be found united in Fisher's formula [12, 13] for the cluster abundance in a vapor as a function of cluster size and temperature. The formula is

$$n_A(T) = q_0 A^{-\tau} \exp(\frac{A\Delta\mu}{T}) \exp(\frac{c_0 A^\sigma}{T_c}) \exp(-\frac{c_0 A^\sigma}{T}) \qquad (17)$$

where q_0 is a normalization constant, τ is a topological critical exponent, $\Delta\mu$ is the difference in chemical potential of the system and the liquid and c_0 is the surface energy coefficient at zero temperature.

For the present work, the external magnetic field is zero and thus the chemical potentials of the liquid and gas phases are equal ($\Delta\mu = 0$), and Eq. (17) becomes:

$$n_A(T)A^\tau/q_0 = \exp(-c_0 A^\sigma \varepsilon/T) \qquad (18)$$

with $\varepsilon = (T_c - T)/T_c$. Therefore, a graph of the scaled cluster distributions ($n_A(T)A^\tau/q_0$) as a function of $\varepsilon A^\sigma/T$ should collapse the distributions of all cluster sizes onto a single curve. This scaling behavior can clearly be seen in Fig. 5.

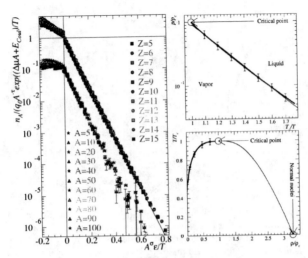

FIGURE 5. Left: the scaled yield distribution versus the scaled temperature for the ISiS data (upper) and $d = 3$ Ising model calculation (lower). For the Ising model, the quantity $(n_A/q_0A^{-\tau})/10$ is plotted against the quantity $A^\sigma \varepsilon / 1.435T$. Data for $T > T_c$ is scaled only as $n_A/q_0A^{-\tau}$. Upper right: The reduced pressure-temperature phase diagram: the thick line shows the calculated coexistence line, the points show selected calculated errors and the thin line shows a fit to the Clausius-Clapeyron equation. Lower right: The reduced density-temperature phase diagram: the thick line shows the calculated low density branch of the coexistence curve, the points show selected calculated errors and the thin lines show a fit to and reflection of Guggenheim's equation.

EXPERIMENTAL PROPERTIES OF THE PHASE DIAGRAM

The ISiS data sets

The ISiS charge yields from AGS experiment of 8 GeV/c π + Au fragmentation data were fit to the following modified form of Eq. (17) which incorporates the Coulomb energy release when a particle moves from the liquid to the vapor:

$$n_A = q_0 A^{-\tau} \exp\left(\frac{A\Delta\mu - E_{Coul}}{T} - \frac{c_0 \varepsilon A^\sigma}{T}\right). \tag{19}$$

The scaled data shown in Fig. 5 collapse to a single line over six orders of magnitude. This line is the liquid-vapor coexistence line, as shown below, and provides direct evidence for the liquid to vapor phase transition in excited nuclei.

Using the values of the parameters determined above for the ISiS experiment and Eq. (19), the coexistence curve observed in the scaled fragment yields in Fig. 5 can be cast into a more familiar form. Fisher's model assumes that the non-ideal vapor can be approximated by an ideal gas of clusters. Accordingly, the total pressure is the sum of their partial pressures: $p/T = \sum n_A$. The reduced pressure is given by:

$$\frac{p}{p_c} = \frac{T \sum n_A(T)}{T_c \sum n_A(T_c)}. \tag{20}$$

162

The coexistence line for finite neutral nuclear matter is obtained by using the $n_A(T, \Delta\mu = 0, E_{Coul} = 0)$ from Eq. (19) in Eq. (20) (see Fig. 5). Recalling the Clausius-Clapeyron equation: $dp/dT = \Delta H/T\Delta V$ one obtains: $p/p_c = \exp(\Delta H/T_c(1 - T_c/T))$ which describes many fluids up to T_c [16]. Fitting the coexistence line and using the above value of T_c gives $\Delta H = 26 \pm 1$ MeV, the enthalpy of evaporation of a cluster from the liquid. This value, after a correction $pV = T$ and allowing for the average size of the fragments, leads to $\Delta E \approx 15$ AMeV, remarkably close to the nuclear bulk energy coefficient.

The system's density can be found from $\rho = \sum An_A$, and the reduced density from

$$\frac{\rho}{\rho_c} = \frac{\sum An_A(T)}{\sum An_A(T_c)} \tag{21}$$

which is the low density branch of the coexistence curve of finite neutral nuclear matter, shown in Fig. 5. Following Guggenheim it is possible to determine the high density branch as well; empirically, the ρ/ρ_c-T/T_c coexistence curves of several fluids can be fit with the function [17]:

$$\frac{\rho_{l,v}}{\rho_c} = 1 + b_1(1 - \frac{T}{T_c}) \pm b_2(1 - \frac{T}{T_c})^{1/3}. \tag{22}$$

Fitting the coexistence curve from the ISiS E900a data with Eq. (22) one obtains an estimate of the full ρ_v branch of the coexistence curve and changing the sign of b_2 gives the full ρ_l branch of the coexistence curve of finite neutral nuclear matter. If normal nuclei exist at the $T = 0$ point of the coexistence curve and the parameterization of the coexistence curve in Eq. (22) is used, then the critical density is found to be $\rho_c \sim 0.3\rho_0$.

The EOS data sets

The EOS Collaboration has collected data for the reverse kinematics reactions 1.0 AGeV Au+C, 1.0 AGeV La+C and 1.0 AGeV Kr+C [18, 19].

Fig. 6 shows the Fisher plot of fragment mass yield distribution scaled by the power law pre-factor, the chemical potential and Coulomb terms. The scaled data for all three systems collapse onto a single line over several orders of magnitude. This collapse provides direct evidence for a liquid to vapor phase transition in excited nuclei.

The p-T and T-ρ coexistence curves can be determined from this analysis by transforming the information in Fig. 6 into the phase diagrams in Fig. 7. From these it is possible to make an estimate of the bulk binding energy of nuclear matter and the $\Delta E/A \approx 14$ MeV, close to the nuclear bulk energy coefficient of 15.5 MeV.

FINITE SIZE EFFECTS AND THE EXTRAPOLATION TO INFINITE NUCLEAR MATTER

Finite size effects are paramount in nuclei. For instance, the binding energy per nucleon decreases from the ~ 15.5 AMeV calculated for nuclear matter to about 8 AMeV for typical nuclei. This lowering is due to the surface and Coulomb energy.

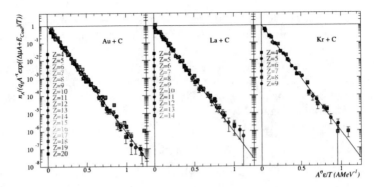

FIGURE 6. The scaled yield distribution versus the scaled temperature for the gold, lanthanum and krypton systems. The solid line has a slope of c_0.

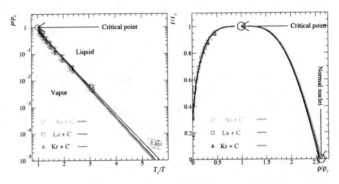

FIGURE 7. EOS data results. Left: The reduced pressure-temperature phase diagram: the points show calculations performed at the excitation energies below the critical point and the lines show fits to the Clausius-Clapeyron equation. Right: The points are calculations performed at the excitation energies below the critical point and the lines are a fit to and reflection of Guggenheim's equation.

We can expect that such a drastic reduction affects the critical temperature as well. The Ising model can be used again as a simple testing ground. Like in nuclei, we have a volume energy: if a finite system is considered (no periodic boundary conditions) a surface is generated with the attendant surface energy. This allows us to write a "liquid drop" formula for the Ising model: $E = a_V A + a_S A^{2/3}$.

We now determine the critical temperature for various sizes (lattices) and check its dependence on the lattice size (see Fig. 8). We now naively guess that, for a finite system all the quantities expressed in energy scale with the binding energy per site, but corrected for the surface energy. We can write

$$\frac{T_c^{A_0}}{T_c^\infty} = \frac{a_V A_0 + a_S A_0^{2/3}}{a_V A_0} = 1 - \frac{1}{A_0^{1/3}} = 1 - \frac{1}{L} \tag{23}$$

where A_0 is the number of sites in the lattice and L is the linear lattice side. As we can see in Fig. 8, this scaling works quite well.

164

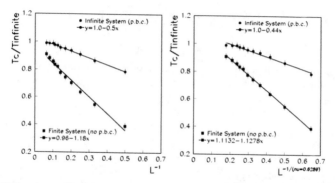

FIGURE 8. Finite size scaling of the critical temperature of the three-dimensional Ising model. Left: the naive estimate of finite size scaling. Right: a more sophisticated estimate of finite size scaling. The data points and fits on the top of both figures show the results for lattices with periodic boundary conditions (p.b.c.) which more closely represent an infinite system. The datapoints and fits on the bottome of both figures show the results for lattices with open boundary condistion (no p.b.c.) and more closely represent the case of finite systems like nuclei.

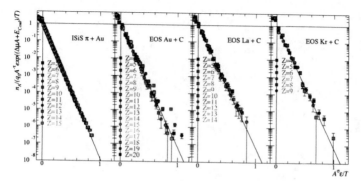

FIGURE 9. Fisher scaling and finite size scaling analysis of the ISiS and EOS data sets.

The result of this exercise is to show that the critical temperature of infinite nuclear matter should be approximately equal to that determined above for finite nuclei time the ratio of the binding energy of infinite nuclear matter to the binding energy of the finite nucleus. In fact, we can do better than that. In each of the three EOS reactions, remnants of different sizes (and thus of different critical temperatures) are characterized. In this way a good range of A_0 values is accessible. Now we can try to fit the entire universe of EOS data as done above, but using the scaling give by Eq. (23).

We do this by performing the Fisher scaling fit as above, but in ε in Eq. (19) use $T_c(A_0)$ from Eq. (23) and $T_c(\infty)$ left as a fit parameter. The preliminary results are shown in Fig. 9 when this is done for both the EOS and ISiS data sets individually. The extracted values for the critical temperature of infinite nuclear matter are ~ 13.6 MeV from the ISiS data and ~ 12.9 MeV from the EOS data. These values agree well with various theoretical estimates of the critical temperature of bulk nuclear matter.

165

ACKNOWLEDGMENTS

We thank Prof. C.M. Mader, Prof. R. Ghetti and Prof. J. Helgesson for their input and Ising model calculations and we acknowledge the experimental efforts of the ISiS and EOS collaborations. This work was supported by the US Department of Energy.

REFERENCES

1. A. Coniglio and W. Klein, J. Phys. A **13**, 2775 (1980).
2. J. L. Cambier and M. Nauenberg, Phys. Rev. **B 34**, 8071 (1986).
3. J. S. Wang, Physica A **161**, 249 (1989).
4. J. S. Wang and R. H. Swendsen, Physica A **167**, 565 (1990).
5. A. M. Ferrenberg and D. P. Landau, Phys. Rev. **B 44**, 5081 (1991). A. Coniglio, Nucl. Phys. A **681**, 451c (2001).
6. F. Gulminelli and Ph. Chomaz, Phys. Rev. Lett. **82** 1402 (1999).
7. D. H. E. Gross, Phys. Rep. **279**, 119 (1997).
8. M. D'Agostino *et al.*, Phys. Lett. B, **473**, 219 (2000).
9. L. Rayleigh, Phil. Mag. **34**, 94 (1917).
10. L. G. Moretto, *et al.*, Phys. Rep. **287**, 249 (1997).
11. L. Beaulieu *et al.*, Phys. Rev. Lett. **81**, 770 (1998).
12. M. E. Fisher, Physics **3**, 255 (1967).
13. M. E. Fisher, Rep. Prog. Phys. **30**, 615 (1969).
14. J. B. Elliott *et al.*, Phys. Rev. Lett. **85**, 1194 (2000).
15. C.M. Mader *et al*, LBNL-47575, nucl-th/0103030 (2001).
16. E.A. Guggenheim, "Thermodynamics", 4th ed. (North-Holland, 1993).
17. E.A. Guggenheim, J. Chem. Phys., **13**, 253 (1945).
18. J. A. Hauger *et al.*, Phys. Rev. C **57**, 764 (1998)
19. J. A. Hauger *et al.*, Phys. Rev. C **62**, 024626 (2000).

Dynamics and thermodynamics of constrained systems

M.J.Ison*, P.Balenzuela[†], A.Bonasera** and C.O.Dorso[†]

*Departamento de Física, Facultad de Ciencias Exactas y Naturales, Universidad de Buenos Aires, Pabellón 1, Ciudad Universitaria, 1428 Buenos Aires, Argentina
[†]Departamento de Física-Facultad de Ciencias Exactas y Naturales, Universidad de Buenos Aires, Pabellón 1 Ciudad Universitaria, 1428 Buenos Aires, Argentina
**INFN-Laboratorio Nazionale del Sud, Via S.Sofia 44, I-95123 Catania, Italy

Abstract. In this work we study the effects of imposing equilibrium to a system composed by 147 particles interacting via a Lennard-Jones (LJ) potential. The consequences of constraining the system in a spherical volume are studied by means of its dynamical and thermodynamical properties. We found the presence of a vapor branch in the caloric curve showing strong evidence of a phase transition, which is dependent on the density of the system.

INTRODUCTION

The fragmentation process of highly excited drops in heavy ion collisions has been analyzed in terms of a possible phase transition since the mid eighties [1]. Several models have been proposed to analyze the phenomena which can be briefly divided into two main groups: statistical and dynamical models.

Statistical models [2] asume that the fragmenting system reaches equilibirum at a given volume (freeze-out volume). These models predict the presence of a vapor branch in the caloric curve (*CC*) (functional relationship between the temperature and the excitation energy of the system). On the other hand, dynamical models, which can be further subdivided into transport models and microscopic models, do not make any asumption on the degree of equilibration of the system. Molecular dynamics (*MD*) simulations of exploding drops [3] do not show a vapor branch in the *CC*.

Since these two views of the process of fragmentation differ in the degree of equilibration od the system, it is of primary importance to perform *MD* simulations in an confining volume, thus allowing the system to reach equilibrium. Some studies have been performed in this direction, see for example [2, 4, 5, 6].

In this paper we focus on the analysis of drops formed by 147 Lennard Jones particles enclosed in different volumes performing a study in terms of the amount of energy added to the system and its density. We pay special attention to the dynamical characterization of such a system through the analysis of the Maximum Lyapunov Exponent (*MLE*), which measures the rate of exponential divergence of initially close trajectories in phase space.

CP644, *Exotic Clustering: 4th Catania Relativistic Ion Studies*, edited by S. Costa, A. Insolia, and C. Tuvè
© 2002 American Institute of Physics 0-7354-0099-7/02/$19.00

The Model

In this work, we perform simulations of classical systems interacting via a Lennard Jones potential, which reads:

$$V(r) = \begin{cases} 4\varepsilon \left[\left(\frac{\sigma}{r}\right)^{12} - \left(\frac{\sigma}{r}\right)^{6} - \left(\frac{\sigma}{r_c}\right)^{12} + \left(\frac{\sigma}{r_c}\right)^{6} \right] & r < r_c \\ 0 & r \geq r_c \end{cases} \tag{1}$$

We fix the cut-off radius as $r_c = 3\sigma$. Energy and distance are measured in units of the potential well (ε) and the distance at which the potential changes sign (σ), respectively. The unit of time used is: $t_0 = \sqrt{\sigma^2 m / 48\varepsilon}$.

Although our system is purely classical and no direct connection with nuclear systems can be established, one has to take into account that the main features of the nuclear interaction (strongly repulsive at very short range and attractive at a longer range) are present in this interaction potential. Then it is quite plausible that the main features of L.J. systems should appear in nuclear systems. In this respect, and of great importance for this work, both systems present an equation of state of the same type.

For the constrained case, we have defined the walls of our container using a very strongly short ranged repulsive potential (cut and shifted) defined as

$$V_w = \begin{cases} \exp(1/(r-R)) - \exp(1/(r_c - R)) & \Leftrightarrow \quad r_c \leq r \leq R \\ 0 & \Leftrightarrow \quad r < r_c \end{cases} \tag{2}$$

Where r_c defines the skin of the constraining volume.

Caloric curves of free expanding systems

When there are no walls present, the system is free to expand and fragment. In order to properly characterize the fragmentation process, different algorithms have been developed. Two of the most used are the MST (Minimum Spanning Tree), which detects fragments in configurational space, and the ECRA (Early Cluster Recognition Algorithm), which recognize clusters in phase space.

MST clusters are defined as: given a set of particles $i, j, k, \ldots$, they belong to a cluster C if,

$$\forall\ i \in C\ , \exists\ j \in C\ /\ \left| \mathbf{r}_i - \mathbf{r}_j \right| \leq R_{cl} \tag{3}$$

where $\mathbf{r}_i$ and $\mathbf{r}_j$ denote the positions of the particles and R_{cl} is a parameter usually referred to as the clusterization radius, and is usually related to the range of the interaction potential. In our calculations we took $R_{cl} = 3\sigma$.

On the other hand, the early cluster fragment algorithm (ECRA) [7], is based on the next definition: ECRA clusters are those that define the most bound partition of the system, i.e. the partition (defined by the set of clusters $\{C_i\}$) that minimizes the sum of the energies of each fragment according to:

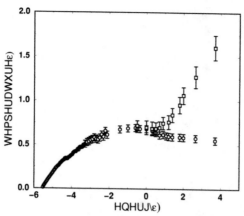

FIGURE 1. Caloric curve calculated for the expanding system (circles). The empty squares correspond to an estimation of a "fake temperature" that does not take into account in a proper way the collective motion and is simply calculated as a fraction of the total kinetic energy

$$E_{\{C_i\}} = \sum_i \left[\sum_{j \in C_i} K_j^{cm} + \sum_{j,k \in C_i} V_{j,k} \right] \tag{4}$$

where the first sum is over the clusters of the partition, and K_j^{cm} is the kinetic energy of particle j measured in the center of mass frame of the cluster which contains particle j.

A thermodynamic quantity that remains useful in small systems is the CC. In a free expanding system, it is defined as the temperature of the system at fragmentation time (τ_{ff}), defined as the time at which ECRA clusters become stable in time [3].

In fig.1) we show the CC for the unconstrained system. Two main features are to be noticed; first the CC develops a maximum, and second the CC has no "vapor branch" but develops a rather constant behavior.

CALORIC CURVES OF CONSTRAINED SYSTEMS

In constrained systems, the presence of the walls impose equilibrium into it, allowing us to calculate the temperature with the standard prescription in the microcanonical ensemble, i.e. the kinetic energy (K) is related to the temperature as:

$$T = \frac{2}{3(N-1)} < K > \tag{5}$$

where $N = 147$ is the number of particles, and <> stands for temporal average.

In fig.2) we show the CC for different densities. We can see that for the less dense case ($R = 15\sigma$) a loop in the CC is obtained. But as the density is increased this loop

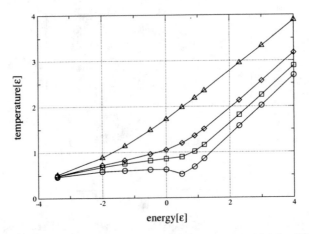

FIGURE 2. In this figure we show the CC for 147 Lennard Jones particles for different sizes of the constraining volumes. Circles denote the CC corresponding to a constraining volume of radius $R = 15\sigma$, squares for $R = 8$⟩

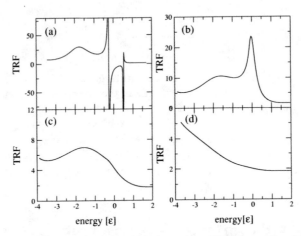

FIGURE 3. TRF as a function of the energy for $a)R = 15\sigma$, $b)R = 8\sigma$, $c)R = 6\sigma$, $d)R = 4\sigma$. In $a)$ the TRF becomes negative in the region where the corresponding CC displays a loop. As the density is increased the TRF shows a maximum (b), and becomes featureless at very high densities (c and d).

disappears and is replaced by a change in slope. At even higher densities the CC looks essentially straight.

We now analyze the Thermal Response Function (TRF) of such a system, defined as:

$$TRF = \left(\frac{dT}{dE}\right)^{-1} \tag{6}$$

In fig.3) we show the behavior of the TRF as a function of the energy. At low densities, two poles and negative values are attained by this quantity. This has been signaled as an

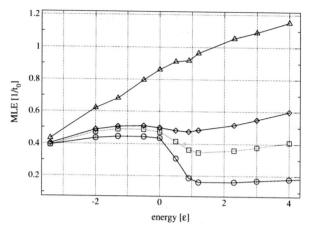

FIGURE 4. In this figure we plot the *MLE* as a function of the energy deposited in the system for the four values of the constraining volume considered in this work. Symbols have the same meaning as in fig.2). It can be seen that the *MLE* clearly signals the transition from a liquid-like regime to a vapor-like regime.

evidence for a first order phase transition [2]. It is due to the fact that surfaces appear in the system. As the density is increased above a certain threshold, the *CC* only displays a change in the slope and the two poles merge in a single finite maximum (finite-size effect).

MAXIMUM LYAPUNOV EXPONENT

One of the main tools to study a chaotic system is the maximum Lyapunov exponent [8], which is a measure of the sensitivity of the system to initial conditions. Given two very close initial conditions in phase space the *MLE*, λ, is given by the following relation :

$$\lambda = \lim_{t \to \infty} \lim_{d(0) \to 0} \left[\frac{1}{t} ln \frac{d(t)}{d(0)} \right].$$
(7)

where $d(t)$ is the distance in phase space between two trajectories which initially differ in a very small quantity d_0.

We calculate the *MLE* following a method given in Ref.[9]. In this method, after a time step $\tau \ll \tau_{sat}$, the distance $d(\tau) = d_1$ is re-scaled to d_0 in the maximum growing direction and the quantity $ln[d_1/d_0]$ is saved. Repeating the procedure at every time step τ, the logarithmic increments $ln[\frac{d_i}{d_{i-1}}]$ are collected. The *MLE* is defined as:

$$\lambda = \lim_{n \to \infty} \frac{1}{n\tau} \sum_{i=1}^{n} ln \left| \frac{d_i}{d_{i-1}} \right|$$
(8)

In fig.4) we show the *MLE* for several densities (the same densities at which the Caloric Curves are shown in fig.2)). The following features are relevant: For energies below zero ($E < 0$), the *MLE* is an increasing function of the energy, following the behavior of the *CC*. As energy is increased the behavior of the *MLEs* changes abruptly. In the range $0 \leq E \leq 1$ we find that, for the low density case the *MLE* displays a very pronounced loop which is in correspondence with the loop displayed by the Caloric Curve. On the other hand for the next two densities a clear loop is present in the *MLE* while the corresponding *CC* only show a change in the slope in this region. Finally for the highest density considered in this work the *CC* is featureless in this region while the *MLE* shows a valley.

CONCLUSIONS

We have presented a study of some important quantities to characterize the fragmentation process of constrained systems.

We found that the *CC* is strongly dependent on the density of our system. As the density is increased the behavior of this function goes from displaying a loop (which we refer to as first order like behavior) to one that only presents a change in the slope (in this case we talk about a second order like behavior). In the first case the Thermal Response function displays two poles, which converge into a single maximum as the density surpasses a given threshold ($R \sim 8.7\sigma \Rightarrow \rho \sim 0.05\sigma^{-3}$). Three regions can then be recognized, The liquid like, the transition region and the vapor like region.

We also studied the *MLE*, and found that it is very sensitive to the transition from liquid like to vapor like states of the system, even in the most dense case, when the *CC* is almost featureless and no clear signal of a phase transition can be found.

ACKNOWLEDGMENTS

This work was partially supported by the University of Buenos Aires (UBA) via grant No tw98, and CONICET via grant No 4436/96.

C.O.Dorso is a member of the carrera del investigador (CONICET), P.Balenzuela is a fellow of the Conicet, M.J.Ison is a fellow of UBA.

REFERENCES

1. A.Bonasera, M.Bruno. C.O.Dorso and P.F.Mastinu , Rivista del Nuovo Cimento **23**, (2000) 2.
2. D. H. E. Gross, *Microcanonical Thermodynamics*, (World Scientific, 2001).
3. A.Strachan and C.O.Dorso, Phys. Rev. C **59**, (1999) 285.
4. A. Chernomoretz, M.Ison, S.Ortiz and C.O.Dorso, Phys. Rev. C **64**, (2001) 024606.
5. F. Gulminelli, Ph. Chomaz, V. Duflot, EuroPhys. Lett. **50**, (2000) 434.
6. X. Campi and H.Krivine, arXiv:cond-mat/0005348
7. C.O.Dorso and J. Randrup Phys.Lett.B **301**, (1993) 328.
8. A.Lichtenberg and M.Lieberman, *Regular and Stochastic Motion* (Springer-Verlag, 1983).
9. G.Benettin,L.Galgani and J.M.Strelcyn, Phys.Rev.A **14**, (1976) 2338.

Cluster Characterization by EXAFS Spectroscopy

G. Faraci

Dipartimento di Fisica e Astronomia, Università di Catania, Via S. Sofia 64, 95123 Catania, Italy
Istituto Nazionale di Fisica della Materia, Unità di Catania

Abstract. This presentation includes a short synthesis of some researches on cluster characterization by Extended X-ray Absorption Fine Structure (EXAFS) spectroscopy, concerning the determination of structural and vibrational properties of atomic and molecular clusters.

INTRODUCTION

I thank the organizers of the CRIS-2002 meeting for inviting this contribution in the field of atomic and molecular clusters, certainly out of the main interests of this Conference; they give me the opportunity to emphasize the importance of this topic, both from the fundamental research point of view and from the technological one. In fact, quantum effects due to the reduced dimensionality are opening innovative frontiers in pure and applied research. A few examples in a very wide list can be sufficient: I mention the geometrical, electronic, optical and magnetic properties of artificial "materials" such as doped semiconductor nanostructures, transition metal and rare gas clusters.

This presentation however, will be limited to a short synthesis of some researches on cluster characterization by Extended X-ray Absorption Fine Structure (EXAFS) spectroscopy, concerning the determination of structural and vibrational properties of atomic and molecular clusters. Since a more detailed paper will soon appear on this subject by the same author [1], I give here a general presentation of the argument, referring the interested reader to a wide literature where many experimental and theoretical details can be found [1-10].

EXAFS SPECTROSCOPY

The determination of the structural configuration of small aggregates cannot be performed by the usual x-ray diffraction, because of the absence of long range order typical of a crystal. However, the x-ray absorption represents an excellent tool for investigating the short range space around a specific element present in a cluster. In fact, the usual absorption law for a radiation beam of intensity I_0, partially transmitted through a thin layer of material is:

$$I = I_0 \exp(- \mu \, d)$$

CP644, *Exotic Clustering: 4th Catania Relativistic Ion Studies*, edited by S. Costa, A. Insolia, and C. Tuvè
© 2002 American Institute of Physics 0-7354-0099-7/02/$19.00

where μ is the absorption coefficient, d the thickness of the material crossed by the x-ray beam, and I the intensity of the beam transmitted from the layer. The coefficient $\mu(E)$ is experimentally observed energy dependent, since the photoelectric cross section, involving electron transitions, does depend on the photon energy. In a convenient photon energy range, the absorption spectrum, e.g., of a thin metal foil, or of a layer containing metal clusters (see for example Fig. 1), is represented by a typical curve showing an absorption edge, corresponding to a core electron excitation from an internal shell, followed by a structured curve. In this curve, some oscillations, which can persist up to about 1000 eV after the threshold, are clearly visible. These oscillations, absent in a gas, are caused by an interference effect of the photoelectron wave emitted by the absorbing atom, with the surrounding atoms in the condensed state. This effect is the Extended X-ray Absorption Fine Structure (EXAFS).

The extraction of this signal, after a proper background subtraction, and its conversion from the photon energy to a function of the photoelectron wave vector is a curve containing important information. This can be obtained after a detailed analysis, involving Fourier transforms, accurate reference systems, and careful fitting procedures. The experimental data are described by (and should be compared with) a theoretical expression where contributions from a few coordination shells around the absorber are present. I do not give here any detail limiting myself to indicate the parameters obtainable by this spectroscopy; in short, for each coordination shell around the absorber, three parameters can be extracted: i) the radial distance from the absorber to the corresponding atomic shell; ii) the coordination number of that shell; iii) the so called Debye-Waller factor, taking into account the relative mean square thermal vibrational displacement of the pair absorber- backscatterer, for that shell.

This technique is quite accurate (accuracy of the order of 0.001 nm for the relative distance), non destructive and sensitive to a specific element; in addition, it does not require long range order and therefore it is often the unique tool for obtaining structural information for clusters as well as for diluted solutions, catalysts, biological systems, disordered alloys etc. Of course, the sample preparation with a usually low concentration of small sized clusters requires an accurate estimation of the acquisition time for an optimal statistical signal-to-noise ratio; this implies the use of high flux synchrotron radiation which has the advantage to be tunable in the required photon energy range by means of suitable monocromators.

CLUSTER CHARACTERIZATION

I mention now some relevant results recently obtained in my group, referring the reader to the literature for other important contributions in this subject. Most of the recent experiments were performed at 77 K, by EXAFS spectroscopy at the GILDA beamline of the European Synchrotron Research Facility (ESRF) Grenoble, France.

i) Co clusters [11-13]. As well known, clusters of magnetic materials are opening a number of very exciting researches. We prepared Co clusters diluted in a silver matrix by several methods; silver evaporation on a silicon substrate was performed concurrently with Co ion implantation in the growing silver layer so as to obtain a layer of 400

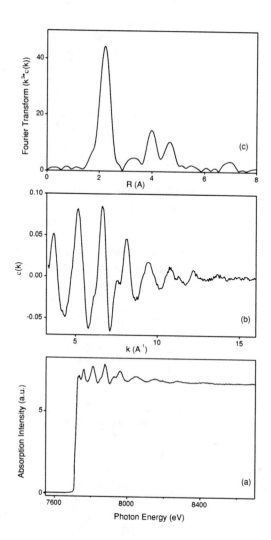

FIGURE 1. a) Lower curve: Raw fluorescence spectrum at 77 K for Co/Ag sample at 6 at. % after annealing. In the silver matrix, Co clusters with 136 atoms/cluster (average value) were detected. Clearly visible the EXAFS oscillations after the Co K-threshold. b)Middle curve: EXAFS oscillations (χ(k)) as a function of the electron wave vector k, extracted after the threshold, by subtracting the continuous contribution in the raw spectra. c) Upper curve: Fourier Transform of the EXAFS spectrum χ(k) weighted by k^3. The peaks correspond to the coordination shells around the cobalt absorber; they give the radial position of each shell after phase shifts corrections.

nm of Ag, with a Co concentration of 1.9 at. %; a second method was the simple Co implantation at 50 keV in Ag layers, for Co dilution between 0.1 and 0.7 at. %; a third method was the molecular beam epitaxial co-evaporation of Ag and Co, at rates chosen so as to obtain a Co concentration of 6 at. %. At the lowest dilutions, evidence of Co dimers dispersed in the Ag matrix was found; these dimers were demonstrated as distributed orthogonal to each other in a chainlike configuration, inside the face centered cubic lattice of the Ag crystal. These results were achieved analysing the distance and coordination number of three Co-Co and Co-Ag coordination shells. Similarly, at higher concentration, we obtained Co clusters with 8 and 20 atoms/cluster in the as prepared samples, whereas, after annealing at 350 C the size of the clusters grew up to 136 and 80 atoms/cluster, respectively. It is worth pointing out a structural phase transition from fcc to hcp, detected when the size of the cluster is larger than about 100 atoms.

ii) $InBO_3$, SnO_2 clusters [14,15]. These materials are high gap semiconducting oxides quite important for technological applications. Our aim was the confinement of size selected quantum dots (QD) of these molecular compounds. In this case a simple melting procedure was adopted, choosing boron oxide as a matrix, because of its low glass transition range and melting point. The analysis of EXAFS spectra gave clear indication of QD of $InBO_3$ with radius of 3 nm in the dilution range 0.3-2.0 %; for SnO_2 the size of the clusters was dependent on the dilution: the average radius of these agglomerates embedded in a B_2O_3 substrate was evaluated in 0.5, 1.0, and 1.4 nm for 0.1, 0.2, and 0.3 % dilution of the semiconductor in the glass, respectively. In both cases, it is interesting to emphasize the nucleation of molecular aggregates such as InO_3, InO_6, SnO_4 in the boron oxide matrix at very low concentration; this phase can be the early stage of cluster nucleation.

iii) Rare gas clusters [16-19]. Rare gases are a class of very important elements for fundamental physics with a very low binding energy of Van der Waals type. Crystals of inert gases are transparent insulators. When introduced into a solid matrix, these inert gases tend to form clusters or bubbles. Ion implantation represents the most simple method for obtaining an uniform ensemble of rare gas agglomerates into a host layer few hundreds nm thick, using the proper ion beam, ion fluence and implantation energy. Small clusters of a rare gas can be confined in solid phase at room temperature, using the large pressure exerted by the host lattice against the foreign atoms, pushed into nanocavities within the matrix. Of course the possibility of obtaining solid noble gas clusters at room temperature, without resorting to more complicated systems (as the diamond anvil cells for getting high pressure), stimulated many studies for fundamental and applied reasons. In implanted metals, Ar, Kr, and Xe form 1-2 nm radius solid clusters with fcc lattice, already at concentration 3-5 at %. The nanocrystal inclusions at RT are pressurized by the host matrix with a large pressure of the order of 1 GPa, as confirmed by the pressure-volume curves. For rare gas implanted in Si, the amorphization of the matrix can cause a more complicated situation: in fact, while Ar was clearly detected in as implanted samples as crystalline agglomerates, Kr and Xe were not observed in the early agglomeration stage after simple implantation at high fluence; in the annealed samples, on the contrary, the rare gas was detected in compressed solid or fluid phase. We point out here that for fluid bubbles the atoms of the clusters can be in quite a disordered configuration similar to liquid or amorphous systems. For such cases, the standard EXAFS expressions should be corrected, if the absorber-backscatterer pair distance has a non-Gaussian distribution and/or an anharmonic vibrational behaviour. EXAFS characterization of fcc Ar clusters in Al and Si at 300 K gave, in as implanted samples, a nearest-neighbours contraction of 0.031 nm for Ar/Al, whereas a higher value of 0.042 nm was deduced for Ar/Si. The overpressure obtained by this contraction was evaluated in 2.5 and 4.4 GPa respectively. An increase of the Debye temperature Θ_D was due to the overpressure: the compressed nanocrystals reach a harder value of 212 K in Al and 300 K in Si, whereas crystalline bulk Ar at low temperature has a Θ_D=93 K. Actually, as-implanted Xe in Si showed, at room temperature, compressed fluid bubbles with 1 nm radius and a first coordination number of only 6 Xe, with a mutual distance of 0.422 nm, reduced with respect to liquid Xe. The reduction of the coordination when the temperature increases beyond the transition temperature between the solid and the liquid phase, could be explained by a shell model with a solid small nucleus surrounded by a wide shell in liquid or amorphous phase.

REFERENCES

1. G.Faraci: Cluster Characterization by EXAFS Spectroscopy, Encyclopedia of Nanoscience and Nanotechnology in press.
2. P.A. Lee, P.H. Citrin, P. Eisenberger, and B.M. Kincaid: Reviews of Modern Physics , 53, 769, 1981.
3. Boon K. Teo , EXAFS: Basic principles and Data Analysis, Springer Verlag, Berlin 1986.
4. Edward A. Stern and Steve M. Heald, Handbook on Synchrotron Radiation, vol. 1, ed. E.E.Koch, North Holland Publ. Amsterdam, p. 955, 1983.
5. A. Filipponi and A. Di Cicco: Phys. Rev. B, 51, 12322, 1995.
6. J. J. Rehr and R.C. Albers: Rev. Mod. Phys., vol. 72, 621, 2000.
7. Adriano Filipponi, J. Phys.: Condens. Matter, 13, R23, 2001.
8. A. Pinto, A. R. Pennisi, G. Faraci, S. Mobilio, and F. Boscherini: Phys. Rev. B 51, 5315, 1995.
9. P. Decoster, B. Swinnen, K. Milants, M. Rots, S. La Rosa, A. R. Pennisi, and G. Faraci: Phys. Rev. B 50, 9752, 1994.
10. J. Mustre de Leon, J. J. Rehr, S. I. Zabinsky, and R. C. Albers: Phys. Rev. B 44, 4146 (1991); J. J. Rehr, R. C. Albers, and S. I. Zabinsky: Phys. Rev. Lett. 69, 3397 (1992); J. J. Rehr, S. I. Zabinsky, A. Ankudinov, and R. C. Albers: Physica B 208-209, 23, 1995; E. A. Stern, M. Newville, B. Ravel, Y. Yacoby: ibid. p.117.
11. Giuseppe Faraci, Agata R. Pennisi, Antonella Balerna, Hugo Pattyn, Gerhard Koops, and Guilin Zhang: Physics of Low Dimensional Systems, ed. J.L. Moran-Lopez, Kluwer Academic-Plenum Publishers, p. 33, 2001.
12. Giuseppe Faraci, Agata R. Pennisi, Antonella Balerna, Hugo Pattyn, Gerhard E. J. Koops, and Guilin Zhang: Phys. Rev. Lett. 86, 3566, 2001.
13. G. Faraci, A. R. Pennisi, A. Balerna, G. Koops, H. Pattyn, and G. Zhang: "Evidence of Co dimers in Ag" ESRF Highlights - 2001 p. 84, 2002.
14. G. Faraci, A. R. Pennisi, R. Puglisi, A. Balerna, and I. Pollini: Phys Rev. B 65, 24101, 2002.
15. G. Faraci, A. R. Pennisi, A. Balerna, submitted to Eur. Phys. J. B
16. F. Zontone, F. D'Acapito, G. Faraci and A. R. Pennisi: Eur. Phys. J. B 19, 501, 2001.
17. Giuseppe Faraci, Agata R. Pennisi, and Jean Louis Hazemann : Phys. Rev. B 56, 12553, 1997.
18. G. Faraci, A. R. Pennisi, A. Terrasi, and S. Mobilio: Phys. Rev. B 38, 13468, 1988.
19. G. Faraci, S. La Rosa, A. R. Pennisi and S. Mobilio and G. Tourillon: Phys. Rev. B 43, 9962, 1991

Dynamical Approaches to Cluster Formation and Nuclear Multifragmentation

Jürn W. P. Schmelzer

Department of Physics, University of Rostock, 18051 Rostock, Germany

Abstract. Multifragmentation in heavy-ion collisions is interpreted as a consequence of a first-order liquid-gas phase transition in expanding nuclear matter. In contrast to widely found attempts, it is shown that Fisher's model is, in general, not applicable for the analysis of experimental results concerning the shape of intermediate mass fragment (IMF) distributions. It may be valid only provided the fragmentation proceeds in the immediate vicinity of the liquid-gas critical point of nuclear matter but not in the thermodynamically metastable or unstable regions beyond it. But even in the vicinity of the critical point its applicability is questionable by different reasons one of them being the static nature of the model. Beyond the liquid-gas critical point, IMF size distributions of the shape $N(A) \propto A^{-\tau}$ may be explained straightforwardly based on a dynamic model of clustering in expanding nucleonic matter. This model leads to qualitatively similar results widely independent on the initial excitation energy of the nucleonic system. In dependence of the initial conditions the parameter τ may vary hereby in the range from $2 \leq \tau \leq 6$. Moreover, the model allows also to explain straightforwardly deviations from such a simple picture and the origin for them.

INTRODUCTION

In the variety of attempts to describe the dynamics of cluster formation in first-order phase transformations, three major theoretical tools have been developed in the past and employed till now successfully in a wide field of different classical ("non-exotic") and, with increasing intensity, non-classical applications as well (see, e.g. [1]). These theoretical tools are the classical nucleation theory and its modern reformulations including quantum nucleation (for an overview see e.g. [2, 3, 4]), the theory of spinodal decomposition [5] and the theory of coagulation processes from its foundations by Smoluchowski [6] up to recent developments like diffusion-limited aggregation phenomena and fractal cluster formation (e.g. [7]). When one discusses "exotic" clustering phenomena, thus the question arises whether above

CP644, *Exotic Clustering: 4ᵗʰ Catania Relativistic Ion Studies*, edited by S. Costa, A. Insolia, and C. Tuvè
© 2002 American Institute of Physics 0-7354-0099-7/02/$19.00

mentioned theoretical tools may give some key to the understanding of clustering in non-classical applications as well.

In analyzing the respective literature on exotic clustering, one gets the impression that in these applications spinodal decomposition approaches are preferred [8]. By this reason, here first the attention is directed to the limitations of this approach as compared with nucleation-growth models of clustering. As it will be shown, in most cases of interest, when a system is transferred into thermodynamic states, where a phase transition may occur, the evolution to the new phase is governed by nucleation and growth and not by spinodal decomposition.

In a next step, basic ideas of nucleation theory are sketched briefly. It is shown then that, as one of the possible applications, dynamic nucleation-growth models may give a key for the understanding of basic features of multifragmentation in heavy-ion collisions. In the theoretical approach outlined here, multi-fragmentation is treated as a consequence of a first-order (and not a second-order) phase transition in an expanding fluid-like system.

The results are compared then with statistical interpretations of multifragmentation experiments and attempts to interpret the respective results in terms of Fisher's statistical droplet model as a critical phenomenon. The limits of applicability of Fisher's droplet model are analyzed proving the advantages of dynamical nucleation-growth models as compared with Fisher's model in the interpretation of the experimental results.

Finally, some further possible developments of the dynamical nucleation-growth model to exotic clustering, in general, and to nuclear multifragmentation, in particular, are anticipated briefly.

NUCLEATION VERSUS SPINODAL DECOMPOSITION

When analyzing liquid-vapor phase transitions, the thermodynamics of the system is usually described by isotherms qualitatively similar to the isotherms of a van der Waals fluid shown in Fig. 1a in reduced variables ($\Pi = p/p_c$, $\omega = v/v_c$ and $\theta = T/T_c$). Here p, v and T are pressure, volume and absolute temperature, the subscript c specifies the respective values of these parameters at the critical point. The region in in the ($\Pi - \theta$) - space, where a transition to a new phase may occur, is enclosed by the binodal (full) curves, while the spinodal (dashed curve) separates (in the employed mean-field approach) thermodynamically metastable from thermodynamically unstable (inside the spinodal curve) states (see Fig. 1b). In thermodynamically metastable states (located between binodal and spinodal curves), the formation of the new phase proceeds via nucleation and growth while in the

thermodynamically unstable region, enclosed by the spinodal, the transformation
is governed by spinodal decomposition.

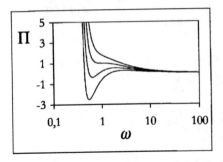 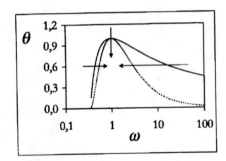

FIGURE 1. *(a)* Isotherms of a van der Waals fluid. Π is the reduced pressure, ω the
reduced volume (left side); *(b)* Binodal (full) and spinodal (dotted) curves of a van der
Waals fluid in the temperature vs. volume (θ, ω) - plane (right side). The arrows indicate
different possibles ways of entering the unstable region (see text).

The arrows in Fig. 1b refer to different possibilities how the system may be
transferred into metastable or unstable states. As it is evident, in most cases, in
order to reach unstable states, thermodynamically metastable states have to be
passed first (e.g. along the ways indicated by horizontal arrows in Fig. 1b). However, with increasing penetration into the metastable region, the intensity of clustering processes increases dramatically, in general. This conclusion can be derived,
for example, from one of the basic expressions of the classical theory of nucleation,
the dependence of the steady-state nucleation rate, J (number of clusters formed
per unit time in a unit volume of the ambient phase) on supersaturation $\Delta\mu$ (or the
degree of penetration of the system into the metastable region). It can be written
as

$$J = J_0 \exp\left\{-\frac{\Delta G_{cr}(A)}{k_B T}\right\} \tag{1}$$

with

$$\Delta G(A) \cong -A\Delta\mu + \alpha a^{2/3} + k_B T \tau \ln(A) . \tag{2}$$

Here J_0 is a kinetic prefactor to J, which only slightly depends on the thermodynamic parameters of the system as compared with the exponential term, it is
determined in its form by the type of kinetics of cluster growth. $\Delta G(A)$ is the
so-called work of formation of a cluster consisting of A particles, the subscript *(cr)*
specifies here its value for the critical cluster size (corresponding to the maximum
of $\Delta G(A)$ in dependence on A; see Fig. 2a). Such maximum for finite values of A
occurs only for $\Delta\mu = \mu_\beta - \mu_\alpha > 0$, i.e., when the chemical potential per particle
μ_α in the newly evolving phase is smaller as compared with the value μ_β in the
ambient phase. k_B is the Boltzmann constant and α a parameter proportional to

the surface tension, σ, between both considered phases ($\alpha = \alpha_0 \sigma$). The last term in the expression for the work of cluster formation was proposed by Fisher [9]. Here a new parameter τ appears, which should have - according to thermodynamic considerations - a value in the range $2 \leq \tau \leq 2.5$. In most cases, this term is of minor importance except in the vicinity of the critical point [2, 3, 9, 10].

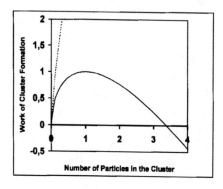 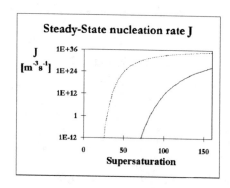

FIGURE 2. *(a)* Dependence of the work of cluster formation, $(\Delta G / \Delta G_{cr})$, on the number, (A/A_{cr}), of basic constituents of the newly evolving phase (left side). The dotted curve corresponds to $\Delta \mu < 0$, while the full curve refers to $\Delta \mu > 0$; *(b)* Typical dependence of the steady-state nucleation rate, J, as a function of supersaturation (right side). The full curve is drawn with some given value of the surface tension $\sigma = \sigma_1$, while the dotted curve refers to a value $\sigma_2 = 0.5 \sigma_1$ (for the details see [11]).

Beyond the vicinity of the critical point, the dependence of the steady-state nucleation rate J on supersaturation ($\Delta \mu$ or equivalent expressions) is of a form as shown in Fig. 2b (for the details see e.g. [11]). The two different curves refer to two different values of the parameter α (or the surface tension σ) illustrating simultaneously the sensitive dependence of J on σ. As it is evident, only after a certain value of the supersaturation is reached, measurable nucleation rates are observed. With a further increase of $\Delta \mu$, the nucleation rate rapidly increases by many orders of magnitude. As a consequence, only for very low values of the kinetic prefactor in the expression for the steady-state nucleation rate, the system may be transferred into thermodynamically unstable states passing the metastable region. In most cases of phase formation, homogeneous unstable states cannot be reached thus due to the intensive cluster formation in the course of the transfer of the system into the final states.

For completeness we have to note that there exist also (a comparatively small) set of pathes in the space of thermodynamic state parameters allowing the transfer immediately from stable to unstable states (indicated by a vertical arrow in Fig. 1b) via the critical point. The respective number of realizations of such kind of transfer into the unstable states is much less as compared with above mentioned alternative ways via metastable states. This way, in most cases of interest, nucleation-growth processes will determine the transition to the two-phase states.

Note that, even in the case, when the critical point is passed, the intensive fluctuations in the vicinity of the critical point can be expected to affect the state of the system considerably and, consequently, its further evolution into the two-phase system. Such history-effects should have to be incorporated into the description then as well. As it seems, such kind of problems has not been tackled so far.

DYNAMIC NUCLEATION-GROWTH MODELS

Model Assumptions and Basic Equations

In standard nucleation theory (see [2, 3, 4]), the time-evolution of the cluster size distribution function, $N(A, t)$, is modelled by a set of kinetic equations allowing to determine the number of clusters $N(A, t)$ consisting of A particles in dependence on time, t. Hereby it is supposed in a commonly good approximation that the cluster growth proceeds by incorporation or emission of single particles only. The resulting set of equations reads (for $A \geq 2$)

$$
\frac{\partial N(A, t)}{\partial t} = w^{(+)}(A - 1, t) \left\{ N(A - 1, t) - N(A, t) \exp \left[\frac{\Delta G(A) - \Delta G(A - 1)}{k_B T} \right] \right\}
$$

$$
- w^{(+)}(A, t) \left\{ N(A, t) - N(A + 1, t) \exp \left[\frac{\Delta G(A + 1) - \Delta G(A)}{k_B T} \right] \right\} . \tag{3}
$$

The coefficients $w^{(+)}(A, t)$ have the meaning of the number of particles incorporated into a cluster of size A per unit time. As initial and boundary conditions, we will assume here

$$
N(A = 1, t = 0) = N_0 = \text{constant} , \qquad N(A \geq 2, t = 0) = 0 . \tag{4}
$$

The missing so far function $N(1, t)$ can be determined then via the mass balance equation $\sum_{A=1}^{\infty} A N(A, t) = N_0$.

It is evident that specific thermodynamic properties of the systems under consideration enter the description mainly via the work of cluster formation, $\Delta G(A)$, while kinetic properties are reflected exclusively via the coefficients of aggregation, $w^{(+)}$.

Although the description is developed in classical terms, in the determination of above mentioned specific properties of the systems under consideration quantum-mechanical approaches could be utilized as well. However, in the numerical solutions given below classical expressions will be used exclusively.

Results of Solution of the Kinetic Equations

In a preceding analysis [10], above formulated set of kinetic equations was solved for two different modes of growth, the so-called diffusion limited growth and the kinetic or ballistic growth. Moreover, Fisher's term $k_B T \tau \ln(A)$ in the work of cluster formation was neglected.

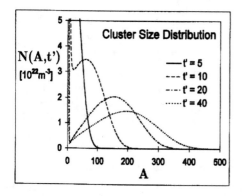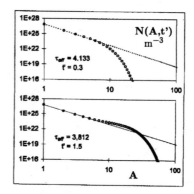

FIGURE 3. *(a)* Cluster size distribution $N(A, t')$ for different moments of time (left). *(b)* Cluster size distribution at different moments of time t' in logarithmic coordinates (right).

Here we employ more general expressions for the kinetic coefficients, reproducing the above mentioned modes of growth in limiting cases (see [2, 3, 10] for details). Moreover, we include the Fisher term into the work of cluster formation. Some of the results are shown in Figs. 3-5. As it turns out, the resulting dependencies, as discussed below, are qualitatively not changed by the introduction of the mentioned more sophisticated description as compared with the approach outlined in [10].

In Fig. 3a, the cluster size distribution function is shown for different moments of time (in a reduced time scale t' (see [2, 3])). Starting with a monodisperse distribution ($N(A = 1, t = 0) = N_0$, $N(A \geq 2, t = 0) = 0$), initially monotonously decreasing distributions are developed until at later stages of the process first a shoulder and then a second peak are formed. As evident from Fig. 3b, the part of the resulting distributions, corresponding to cluster sizes $1 \leq A \leq A_{max}$, can be well-approximated by dependencies of the form $N(A) \propto A^{-\tau_{eff}}$. Hereby both quantities A_{max} (giving the upper limit of the range, where the approximation $N(A) \propto A^{-\tau_{eff}}$ fits the distributions quite well) and τ_{eff} depend both on the initial supersaturation and the time the system has evolved in the course of the nucleation-growth process. The values of A_{max} and τ_{eff} are shown in Figs. 4a and 4b for different initial supersaturations and as functions of time of evolution. As it turns out, A_{max} may reach values of the order 10^2 while τ_{eff} commonly varies in the range $2 \leq \tau_{eff} \leq 6$. In all cases, both parameters change suddenly at some

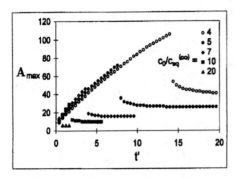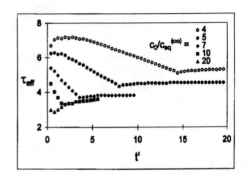

FIGURE 4. In the range $1 \leq A \leq A_{max}$, the distribution can be well-approximated by a dependence of the form $N(A) \propto A^{-\tau_{eff}}$. The values of A_{max} [(a), left side] and τ_{eff} [(b), right side] are determined kinetically and depend on the initial supersaturation (expressed here via the ratio c_0/c_{eq}^{∞}) and the time the system has evolved in the nucleation-growth process.

given moment of time. This effect is connected with the evolution of a shoulder in the distribution, restricting, after its formation, the validity of the considered exponential approximation to smaller cluster sizes.

Fragment size distributions of the form $N(A) \propto A^{-\tau}$ are widely observed in experiments on multifragmentation in heavy-ion collisions with values of τ varying in the range $2 \leq \tau \leq 6$. This way, the considered dynamic nucleation-growth model gives a qualitatively surprising adequate description of these results. This coincidence allows to formulate the hypothesis that nuclear multifragmentation may be interpreted as a result of a nucleation-growth process in a first-order phase transition.

So far we have investigated the evolution of the cluster size distribution function in nucleation-growth processes starting with metastable initial states but outside the immediate vicinity of the critical point. In the critical point, the quantites $\Delta\mu$ and σ become equal to zero and Fisher's term dominates the work of critical cluster formation (while, as already, mentioned, beyond the critical region it is of minor importance). The results of solution of the set of kinetic equations for this particular case are shown in Fig. 5. In this case, in the course of time the distribution $N(A) \propto A^{-\tau}$ is established. Here τ is not determined kinetically but by the respective value occuring in the expression for the work of critical cluster formation.

Note, however, that, due to critical slowing down, the time scales required to establish such dependencies for initial states near the critical point become very large. Moreover, in the vicinity of the critical point, coagulation processes become of major importance. Such processes, as mentioned, are not accounted for in the kinetic equations employed. Thus, even if the dynamic nucleation-growth model

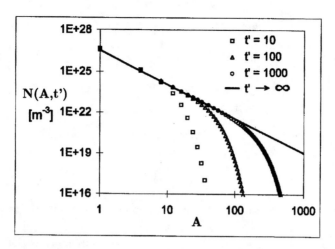

FIGURE 5. Cluster size distribution $N(A, t')$ for different moments of time as obtained by the solution of the kinetic equations starting from initial states in the immediate vicinity of the critical point.

leads to dependencies of the form $N(A) \propto A^{-\tau}$ for the critical point, this is not a support of a treatment of multifragmentation as a second-order phase transition. A more detailed discussion of this topic will be given in connection with an analysis of Fisher's statistical droplet model in the next section.

Limits of Validity of Fisher's Droplet Model

Fisher's droplet model represents a particular form of the so-called equilibrium distribution of cluster sizes $N^{(e)}(A)$ employed widely in classical nucleation theory. In this application, it refers to a very artificial model situation (so-called Szilard's model [3, 4]) not realized in nature but helpful to a certain degree for the formulation of the theoretical concepts.

Fisher derived his expression by methods of equilibrium statistical physics. He arrived at

$$N^{(e)} \cong N_0 \exp\left\{-\frac{\Delta G(A)}{k_B T}\right\} . \tag{5}$$

This result can be derived also immediately from the basic ideas of the theory of fluctuations in thermodynamic equilibrium states (see [13]). It gives, by its derivation, the number of clusters consisting of A particles in an equilibrium system. Already by this reason it becomes evident that the model cannot be employed to describe real cluster size distributions evolving in thermodynamically metastable or unstable initial states. But such kind of states one is interested in once one is attempting to describe phase transitions, in general, or multifragmentation as a consequence of a phase transition, in particular.

Substituting Eq. (2) into this relation, we obtain

$$\frac{N^{(e)}}{N_0} \cong A^{-\tau} \left\{ \exp\left(\frac{\Delta\mu}{k_B T}\right) \right\}^A \left\{ \exp\left(-\frac{\alpha_0 \sigma}{k_B T}\right) \right\}^{(A^{2/3})} \tag{6}$$

In the critical point ($\Delta\mu = 0$, $\sigma = 0$), we get then

$$\frac{N^{(e)}}{N_0} \cong A^{-\tau} . \tag{7}$$

This way, for initial states at the critical point Fisher's model is in agreement with the results of the solution of the kinetic equations employed commonly for the description of nucleation-growth processes (but remember the limitations of mentioned set of kinetic equations). This correlation is widely lost for any other metastable or unstable initial states.

In a finite closed system, as the final state of nucleation-growth processes, one large cluster is formed supplemented by a distribution of monomers, dimers. etc. [14]. The supersaturation becomes nearly equal to zero and Fisher's model can be employed then to describe the part of the distribution for small cluster sizes (but not the whole distribution). However, to repeat, in general, real cluster size distributions evolving in nucleation-growth processes, are different from the predictions of Fisher's model.

DISCUSSION

In addition to the arguments outlined in the preceding section, the widely met application of Fisher's model to an interpretation of nuclear multifragmentation, considering it as the result of clustering in an expanding gas of nucleons, is questionable also from another point of view. Multifragmentation is a dynamic process, where the conditions for clustering are changing with time. By this reason, a dynamic description of clustering is required which allows a self-consistent description of both clustering and the evolution of the state of the system. The interplay of these both processes determines finally the fragmentation pattern.

A theory of clustering in expanding fluids was developed in preceding papers. It allows to explain the existence of a broad spectrum of τ-values in the distributions $N(A) \propto A^{-\tau}$ as a result of the interplay between the rates of aggregation and expansion. This way, essential qualitative features of the experiments on multifragmentation may be explained straightforwardly without the necessity of employing additional assumptions like the introduction of such parameters like freeze-out volumes etc. [3, 4, 15]. Note that, in the approach developed, constancy of entropy is not assumed. As soon as aggregation processes take place, this condition is not fulfilled any more.

Dynamic nucleation-growth models allow a description of multifragmentation also in terms of an alternative approach considering it as the consequence of bubble formation in an expanding fluid. Respective work is in progress.

REFERENCES

1. Insolia, A., Costa, S. and Tuve, C., *Phase Transitions and Strong Interactions*, Proceedings 3rd Catania Conference on Relativistic Ion Studies, Acicastello, Italy, 22-26 May, 2000; Elsevier Publishers, Amsterdam, 2001.

2. Schmelzer, J. W. P., Röpke, G., and Mahnke, R., *Aggregation Phenomena in Complex Systems*, WILEY-VCH Publisher, Weinheim, 1999.

3. Schmelzer, J. W. P., Röpke, G., and Priezzhev, V. B. (Eds.), *Nucleation Theory and Applications*, 2 volumes, Joint Institute for Nuclear Research Publishing House, Dubna, Russia, 1999, 2002.

4. Gutzow, I., and Schmelzer, J., *The Vitreous State: Thermodynamics, Structure, Rheology, and Crystallisation*, Springer, Berlin, 1995.

5. Cahn, J. W., and Hilliard, J. E., J. Chem. Phys. **28**, 258 (1958); **31**, 688 (1959).

6. Smoluchowski, M., Phys. Z. **17**, 557 (1916); Z. Phys. Chem. **92**, 129 (1917).

7. Feder, J., *Fractals*, Plenum Press, New York, 1988.

8. Nörenberg, W., Papp, G., and Rozmej, P., Eur. Phys. J. **A 9**, 327 (2000) and references therein.

9. Fisher, M. E., Physics **3**, 255 (1967).

10. Schmelzer, J., Röpke, G., and Ludwig, F.-P., Phys. Rev. **C 55**, 1917 (1997).

11. Schmelzer, J. W. P., and Schmelzer, J. Jr., J. Colloid Interface Sci. **215**, 345 (1999).

12. Slezov, V. V., and Schmelzer, J., J. Phys. Chem. Sol. **55**, 243 (1994).

13. Landau, L. D., and Lifshitz, E. M., *Statistical Physics*, Pergamon Press, New York, 1969.

14. Schmelzer, J., and Schweitzer, F., Z. Phys. Chem. **271**, 565 (1990).

15. Schmelzer, J., Labudde, D., and Röpke, G., Physica **A 254**, 389 (1998).

Exotic shapes in ^{40}Ca and ^{36}Ar studied with Antisymmetrized Molecular Dynamics

Y. Kanada-En'yo*, M. Kimura† and H. Horiuchi**

*Institute of Particle and Nuclear Studies
†The Institute of Physical and Chemical Research, Wako 351-0198, Japan
**Department of Physics, Kyoto University, Kyoto 606-01, Japan

Abstract. The structures of the ground and excited states of sd-shell nuclei: ^{40}Ca and ^{36}Ar were studied with a method of antisymmetrized molecular dynamics. Recently observed rotational bands were described with deformed intrinsic states which are dominated by excited configurations such as $4p-4h$, $8p-8h$ and $4p-8h$ states. The results showed the coexistence of various kinds of exotic shapes with oblate, prolate and large prolate deformatios in the low-energy region of ^{40}Ca. Possible cluster aspects in heavy sd-shell nuclei were implied in these exotic shapes.

INTRODUCTION

Owing to the experiments of gamma-ray measurements, many excited bands in the nuclei near ^{40}Ca have been recently observed [1, 2]. The rotational bands with large moments of inertia are hot subjects relating with super deformations in sd-shell nuclei. In ^{40}Ca, there exist many low-lying bands, which imply the coexistence of various shapes. It is interesting problem why the shape coexistence occurs even in the doubly magic nucleus. The shape coexistence problem and the mechanism of deformations in these sd-shell nuclei is one of the attractive subjects.

Our interests are possible exotic shapes and cluster aspects in the deformed excited states. It has been already known that clustering is one of the essential features in light nuclei . Exotic shapes are formed due to cluster structures in some light nuclei, For example, a famous cluster structure in ^{12}C is three α clusters which form a triangle shape. Parity doublets of the rotational bands in ^{20}Ne are described by a parity asymmetry shape with a ^{16}O+α cluster structure. It is a long problem whether or not cluster aspects appear in heavier nuclei. Many of the familier cluster structures are those in light stable nuclei. In the recent experimental and theoretical research [3, 4, 5, 6, 7], cluster structures were suggested in light unstable nuclei such as ^{10}Be and ^{12}Be, where new type clusters such as ^{6}He and ^{8}He are proposed. Our aim in this paper is systematic study of the sd-shell nuclei near ^{40}Ca to find possible exotic states with cluster aspects.

For the systematic study of ground and excited states of sd-shell nuclei, theoretical difficulties exist in the coexistence of cluster and mean-field aspects. Needless to say, cluster aspects are important in light nuclei, while the mean-field is an essential feature in heavy nuclei. Since the sd-shell is an inter-mediate mass-number region, where both aspects should be significant, we need a miscroscopic method beyond the traditional clsuter models. Namely, we adopt a method of antisymmetrized molecular dy-

CP644, *Exotic Clustering: 4th Catania Relativistic Ion Studies*, edited by S. Costa, A. Insolia, and C. Tuvè

namics(AMD) [8, 9, 10, 11, 5]. The method AMD was applied for light nuclei and has been proved to be a powerful approach to describe the ground and excited states of unstable nuclei as well as stable nuclei. In the pioneering study with the AMD method, the structure changes on the ylast line of ^{20}Ne are explained in terms of alteration of the cluster structure [9]. In the recent studies of light unstable nuclei with AMD, a new cluster aspect in unstable nuclei was suggested that cluster cores develop in the deformed neutron mean-field in neutron-rich nuclei.

In the present work, we adopt a new effective force with a finite-rage three-body term, with which the binding energies and the radii of ^{4}He, ^{16}O and ^{40}Ca are systematically reproduced, because the usual effective forces used in the AMD study, such as Minnesota forces, Volkov forces and Modified Volkov forces, have serious problems in reproducing these basic properties in the wide mass-number region.

In this paper, the structures of the ground and excited states of ^{40}Ca and ^{36}Ar are studied by the AMD method. We investigate the shape coexistence problem of these nuclei. The mechanism of deformations are discussed, focusing on cluster aspects. We also introduce the AMD study of neuton-rich sd-shell nuclei near ^{32}Mg.

FORMULATION

In this section, the formulation of AMD for the nuclear structure study of ground and excited states is briefly explained. For more detailed descriptions of the AMD framework, the reader is referred to Refs [9, 10, 11].

The wave function of a system is written by AMD wave functions,

$$\Phi = c\Phi_{AMD} + c'\Phi'_{AMD} + \cdots. \qquad (1)$$

An AMD wave function of a nucleus with a mass number A is a Slater determinant of Gaussian wave packets;

$$\Phi_{AMD}(\mathbf{Z}) = \frac{1}{\sqrt{A!}}\mathscr{A}\{\varphi_1, \varphi_2, \cdots, \varphi_A\}, \qquad (2)$$

$$\varphi_i = \phi_{\mathbf{X}_i}\chi_{\xi_i}\tau_i : \begin{cases} \phi_{\mathbf{X}_i}(\mathbf{r}_j) \propto \exp\left[-\nu\left(\mathbf{r}_j - \frac{\mathbf{X}_i}{\sqrt{\nu}}\right)^2\right], \\ \chi_{\xi_i} = \begin{pmatrix} \frac{1}{2} + \xi_i \\ \frac{1}{2} - \xi_i \end{pmatrix}, \end{cases} \qquad (3)$$

where the ith single-particle wave function φ_i is a product of the spatial wave function $\phi_{\mathbf{X}_i}$, the intrinsic spin function χ_{ξ_i} and the iso-spin function τ_i. The spatial part $\phi_{\mathbf{X}_i}$ is presented by variational complex parameters X_{1i}, X_{2i}, X_{3i}. χ_{ξ_i} is the intrinsic spin function defined by ξ_i, and τ_i is the iso-spin function which is fixed to be up(proton) or down(neutron) in the present calculations. The values $\mathbf{Z} \equiv \{X_{ni}, \xi_i\}$ ($n = 1, 2, 3$ and $i = 1, \cdots, A$) are the variational parameters which express an AMD wave function.

In order to obtain the wave functions of ground and excited states, we perform a generator coordinate method in the framework of AMD. First we vary the energy of the parity eigen state projected from an AMD wave function under a constraint that the total oscillator quanta must equal to a given number $\mathcal{N}$ as $\langle aa^\dagger \rangle = \mathcal{N}$. The energy variation is numerically calculated by a frictional cooling method [9]. An energy curve is obtained as a function of the coordinate $\mathcal{N}$ in the constraint. In the second step, the spin-parity eigen states projected from the obtained AMD wave functions are superposed by diagonalizing Hamiltonian and Norm matrices, $\langle P_{MK'}^{J\pm} \Phi_{AMD}(\mathbf{Z}_i) | H | P_{MK''}^{J\pm} \Phi_{AMD}(\mathbf{Z}_j) \rangle$ and $\langle P_{MK'}^{J\pm} \Phi_{AMD}(\mathbf{Z}_i) | P_{MK''}^{J\pm} \Phi_{AMD}(\mathbf{Z}_j) \rangle$, We call the present calculations as VBP(variation before projection) because the variation is performed before the spin projection.

INTERACTIONS

The ordinary effective forces with no three-body term such as the Volkov force or with a zero-range three-body term like the MV1 force are not appropriate to describe the binding energies and radii of nuclei covering wide mass number region from α to ^{40}Ca. Hence we use an interaction containing a finite-range three-body term in the present calculations. The central part of the interaction is explained by a conbination of the two-body and three-body terms. The interaction parameters used in the present paper are as follows,

$$V_{central} = \Sigma_{i<j} V^{(2)} + \Sigma_{i<j<k} V^{(3)}, \tag{4}$$

$$V^{(2)} = (1 - m + bP_\sigma - hP_\tau - mP_\sigma P_\tau)\left\{ V_a \exp[-(\tfrac{r_{12}}{r_a})^2] V_b \exp[-(\tfrac{r_{12}}{r_b})^2] \right\} \tag{5}$$

$$+ V_c \exp[-(\tfrac{r_{12}}{r_c})^2], \tag{6}$$

$$V^{(3)} = V_d \exp[-d(r_{12}^2 + r_{23}^2 + r_{31}^2)^2], \tag{7}$$

$$V_a = -198.34 \text{ MeV}, V_b = 300.86 \text{ MeV}, V_c = 22.5 \text{ MeV}, \tag{8}$$

$$r_a = 1.2 \text{ fm}, r_b = 0.7 \text{ fm}, r_c = 0.9 \text{ fm}, V_d = 600 \text{ MeV}, d = 0.8 \text{ fm}^{-2} \tag{9}$$

$$m = 0.193, b = -0.185, h = 0.37, \tag{10}$$

where the width and strength parameters are chosen so as to reproduce reasonably the sizes of α and ^{40}Ca, and the binding energies of α, ^{16}O and ^{40}Ca. In choosing parameters, $\alpha + \alpha$ phase shifts and the saturation property of symmetric nuclear matter have been also taken into consideration. The adopted interaction is a sum of the central force, the G3RS spin-orbit force [12] with the strength $u_{ls} = 2500$ MeV and the Coulomb force.

RESULTS

The excited deformed bands in the ^{40}Ca and ^{36}Ar have been recently observed in the gamma-ray measurements [1, 2], which reveal the existence of many low-lying bands in these nuclei. In the experimental level scheme of ^{40}Ca, the rotational band with a large moment of inertia (No.1 in Fig.1) is a hot topic relating with a possible super deformation.

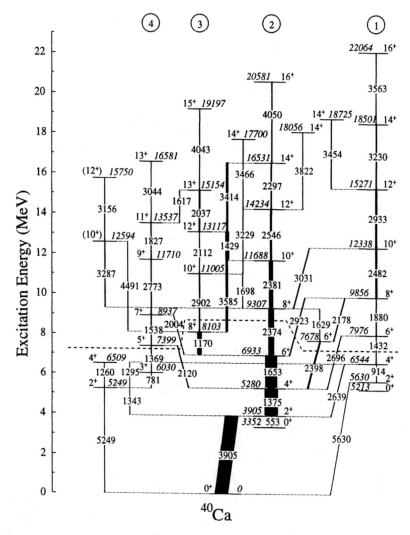

FIGURE 1. The experimental data of the energy levels of ^{40}Ca. The figure is taken from [1].

We study the ground and excited states of ^{40}Ca and ^{36}Ar, with the method of a generator coordinate method in the framework of AMD with respect to the total oscillator quanta. In the energy curves obtained after spin-projection as a function of the oscillator quanta, we can not necessarily find local minima except for the absolute minimum. However, after diagonalization of the Hamiltonian and Norm matrices, the energy curves for the excited states have other local minima due to the conditions orthogonal to the lower states. As a result, we obtain many low-liying rotational bands (Fig.2), each of which is dominated by the spin-parity projected states from an intrinsic AMD wave function.

By analyzing deformations of these dominant intrinsic states, it is found that various

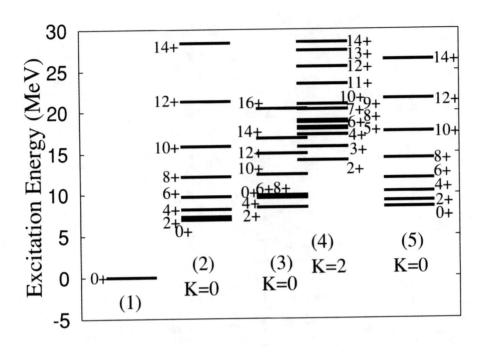

FIGURE 2. The theoretical energy levels of ^{40}Ca obtained after the diagonalization with respect to the spin-parity projected states.

shapes coexist in ^{40}Ca. The interesting point is that each of the intrinsic states in ^{40}Ca is well described by either of $0p - 0h$, $4p - 4h$ or $8p - 8h$ configurations. First we obtain the normal spherical ground state with the doubly closed sd-shell configuration. With the increase of the oscillator quanta, the deformed excited states appear in the low-energy region. In the region concerning $4p - 4h$ states, there exist opposite shapes as two local minimum states. One is an oblate shape with the deformation parameter $\beta = 0.2$, while the other is a prolate deformation with $\beta = 0.2$. The oblate state provides the excited $K^\pi = 0^+$ band (2), and the bands (3) and (4) with $K^\pi = 0^+, 2^+$ are dominated by the states projected from the prolate intrinsic state. The reason for the higher 0^+ state than the 2^+ state in the $K^\pi = 0^+$ band (3) is the mixing with the ground state. With a larger oscillator quanta, we find another largely prolate deformation which is dominated by $8p - 8h$ configurations. The energy levels of the band (5) obtained from this large deformation indicate a large moment of inertia of this rotational band. It is very surprising that the excited band with such highly excited configurations as $8p - 8h$ exists in low-energy region. We identify this band as the experimentally observed super-deformed band which starts from the 0^+ state at 5.2 MeV.

As shown in Table 1, the theoretical values of the $E2$ transition strength in the band (5) are extremely large comparing with those in other excited band. According to a quantitative comparison of the strength in the band (5) with the experimental data, the

TABLE 1. The theoretical values (W.u.) of $E2$ transition strength in the excited $K^\pi = 0^+$ bands.

transition	band(2)	band(3)	band(5)
$2^+ \to 0^+$	11	2	30
$4^+ \to 2^+$	16	21	53
$6^+ \to 4^+$	16	13	61
$8^+ \to 6^+$	17	14	78

theoretical $E2$ strength for $4^+ \to 2^+$ is smaller than the corresponding experimental data $B(E2; 6.54\text{MeV} \to 5.63\text{MeV}) = 100(W.u.)$. Therefore, it is conjectured that the intrinsic deformation $\beta = 0.4$ of the band (5) in the present results must be still small compared with the value $\beta = 0.6$ for a typical super deformation. The underestimation of the deformations in the present theory will be improved by the superposition along generator coordinates and also the extension of the single-particle wave functions.

As mentioned above, the point in the present results of ^{40}Ca is the coexistence of various shapes: the spherical state, the oblate shape, the normal prolate deformation and the large prolate deformation. By analysing the intrinsic structures of these deformed states, exotic shapes and cluster aspects are revealed. The density distributions of the dominant intrinsic states of the excited bands are presented in Fig.3. As shown in Fig.3, the oblate state (2) has a hexagon structure like a three-leaf clover which consists of three ^{12}C cores surrounding an α at the center. In the super deformation, the system has a parity-asymmetry shape like a pear because of a ^{28}Si+^{12}C-like clustering. We focus the characteristics of the ^{28}Si+^{12}C-like cluster structures in the super deformation. The first point is that the clusters are not weak coupling but are strong coupling ones. It is interesting that the ^{12}C cluster consists of four nucleons in sd-shell and eight nucleons in pf-shell. Namely, it is a inter-shell cluster which lies over different shells. Because of the strong coupling feature, it is expected that the deformed mean-field effect can be important as well as cluster aspects in the state. In second, if such a cluster structure indeed develops, it forms a parity asymmetric deformation, which may provide the parity doublets made of positive and negative-parity bands. Thirdly, it is reasonable that both clusters are the sub-shell closed nuclei. In other words, the stability of these cluster cores is based on the shell effects of the sub-shell closure. For further analysis of the cluster aspects, we need advanced calculations where the residual effects such as the coupling of other cluster channels, the tail of inter-cluster motion, and the core deformations are more carefully taken into consideration.

It should be pointed that the present results imply the possible exotic shapes due to the cluster effects such as ^{12}C cores, though there are few experimental evidences of the existence of clusters in ^{40}Ca. The investigation of the level structures of the negative-parity states is one of the probes, because the parity-doublets of positive and negative-parity bands support a cluster structure as the origin of a parity-asymmetry intrinsic state. The present calculations predict a $K^\pi = 0^-$ band with the band-head 1^- state at 11.7 MeV , which is about 3 MeV higher than theoretical band-head energy of the positive-parity super deformation. The deformation of the dominant intrinsic state is rather large as $\beta = 0.3$ and the intrinsic structure is similar to that of the super deformation in the

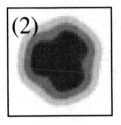 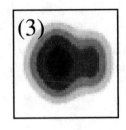 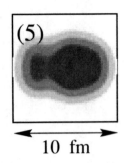

10 fm

FIGURE 3. The density distributions of the intrinsic states before spin-parity projections. The notations (2),(3),(5) are same as those in Fig.2. In each figure, the density of the dominant AMD wave function of each band is presented. The matter density is integrated along the axis with the largest moment of inertia.

positive-parity states. Therefore we consider this negative-parity band as a candidate of the parity doublet. Strictly speaking, in negative parity states, the rotational levels fragments because the state mixing between bands is rather strong. As a result, the intra-band $E2$ transitions in the $K^\pi = 0^-$ band are not so strong as those in the positive-parity band of the super deformation.

In the theoretical results of ^{36}Ar, we find excited bands with a large prolate deformation $\beta = 0.3$ which are dominated by $4p - 8h$ configurations. This band starts from a 4 MeV excitation energy, and well corresponds to the newly observed rotational band with a large moment of inertia in the experimental data [2].

At the end of this section, I briefly report the AMD study on neutron-rich sd-shell nuclei near ^{32}Mg calculated by one of the authors(M. K). The structures of the ground and excited states of ^{32}Mg are hot subjects concerning the island of inversion of a neutron magic number $N = 20$. The ground and excited states of Mg isotopes are calculated by an extended version of AMD with the Gogny force. The calculations well reproduce the energy levels of ^{32}Me and the features of the island of inversion. In the neutron-rich sd-shell nuclei, the deformed neutron mean-field plays an important role. The unique predictions in this work are the low-lying negative parity states in ^{32}Mg. Since the nuclei ^{32}Mg and ^{40}Ca have the same neutron number, it is interesting future problem to compare the intruder states of ^{32}Mg with the excited states of ^{40}Ca.

SUMMARY

The structures of the ground and excited states of sd-shell nuclei: ^{40}Ca and ^{36}Ar were studied with the method of antisymmetrized molecular dynamics. The recently observed excited bands were well described with the deformed intrinsic states in the present calculations. It was found that various kinds of shape coexist in the low-energy region. In ^{40}Ca, it was suggested in the present results that the oblate shape, the normal prolate deformation and the large prolate deformation coexist in the low-lying excited bands besides the spherical ground state. The oblate, normal prolate, and the large prolate states are explained by the excited configurations with $4p - 4h$, $4p - 4h$, and $8p - 8h$, respectively. It is a mystery that the rotational band with the highly excited configurations($8p - 8h$) starts from the low-energy region. The exotic shapes due to the cluster effects were implied in the intrinsic structures of these deformed states. The present results suggested possible cluster aspects in heavy sd-shell nuclei.

ACKNOWLEDGMENTS

The authors would like to thank Professors A. Tohsaki, Y. Akaishi and K. Ikeda for helpful discussions and comments. They are also thankful to Dr. N. Itagaki and E. Ideguchi for many discussions. This work was partially performed in the "Research Project for Study of Unstable Nuclei from Nuclear Cluster Aspects" sponsored by Institute of Physical and Chemical Research (RIKEN). The computational calculations of this work are supported by the Supercomputer Project No.58, No.70 of High Energy Accelerator Research Organization(KEK), and Research Center for Nuclear Physics(RCNP) in Osaka University.

REFERENCES

1. E. Ideguchi et al., Phys. Rev. Lett. **87**, 222501(2001)
2. C.E.Svensson et al., Phys. Rev. Lett. **85**, 2693(2000)
3. M. Freer, et al., Phys. Rev. Lett. **82**, 1383 (1999); M. Freer, et al., Phys. Rev. C **63**, 034301 (2001).
4. A. Saito, et al., *Proc. Int. Sympo. on Clustering Aspects of Quantum Many-Body Systems*, eds A. Ohnishi, N. Itagaki, Y. Kanada-En'yo and K. Kato, (World Scientific Publishing Co.) (to be published).
5. Y. Kanada-En'yo, H. Horiuchi and A. Doté, Phys. Rev. C **60**, 064304(1999).
6. N. Itagaki and S. Okabe, Phys. Rev. C **61**, 044306 (2000);
7. Y. Kanada-En'yo, Phys. Rev. C **66**, 011303 (2002)
8. A. Ono, H. Horiuchi, T. Maruyama, and A. Ohnishi, Prog. Theor. Phys. **87**, 1185 (1992).
9. Y. Kanada-En'yo and H. Horiuchi, Prog. Theor. Phys. **93**, 115 (1995).
10. Y. Kanada-En'yo, A. Ono, and H. Horiuchi, Phys. Rev. C **52**, 628 (1995); Y. Kanada-En'yo and H. Horiuchi, Phys. Rev. C **52**, 647 (1995).
11. Y. Kanada-En'yo, Phys. Rev. Lett. **81**, 5291 (1998).
12. N. Yamaguchi, T. Kasahara, S. Nagata, and Y. Akaishi, Prog. Theor. Phys. **62**, 1018 (1979); R. Tamagaki, Prog. Theor. Phys. **39**, 91 (1968).

Nonlinear Approach in Nuclear Dynamics

K.A. Gridnev,*, V.G. Kartavenko[†] and W. Greiner**

*Institute of Physics, St.–Petersburg State University, 198504, Russia
[†]Bogoliubov Laboratory of Theoretical Physics, JINR, Dubna, 141980, Russia
**Institut für Theoretische Physik, JWG Universität, Frankfurt/Main, D-60054, Germany

Abstract. Attention is focused on the various approaches that use the concept of nonlinear dispersive waves (solitons) in nonrelativistic nuclear physics.
The problem of dynamical instability and clustering (stable fragments formation) in a breakup of excited nuclear systems are considered from the points of view of the soliton concept. It is shown that the volume (spinodal) instability can be associated with nonlinear terms, and the surface (Rayleigh-Taylor type) instability, with the dispersion terms in the evolution equations. The both instabilities may compensate each other and lead to stable solutions (solitons).
A static scission configuration in cold ternary fission is considered in the framework of mean field approach. We suggest to use the inverse mean field method to solve single-particle Schrödinger equation, instead of constrained selfconsistent Hartree–Fock equations. It is shown, that it is possible to simulate one-dimensional three-center system in the approximation of reflectless single-particle potentials.
The soliton-like solutions of the Korteweg-de Vries equation are using to describe collective excitations of nuclei observed in inelastic alpha-particle and proton scattering. The analogy between fragmentation into parts of nuclei and buckyballs has led us to the idea of light nuclei as quasi-crystals. We establish that the quasi-crystalline structure can be formed when the distance between the alpha-particles is comparable with the length of the De Broglia wave of the alpha-particle. Applying this model to the scattering of alpha-particles we obtain that the form factor of the clusterized nucleus can be factorized into the formfactor of the cluster and the density of clusters in the nucleus. It gives possibility to study the distribution of clusters in nuclei and to resolve what kind of distribution we are dealing with: a surface or volume one.

INTRODUCTION

Nonlinear dynamic phenomena in different complex systems are currently a topic of considerable interest in modern physics. It is mainly caused by great progress in the development of methods to solve exactly nonlinear partial differential equations. A hole class of these equations admits solutions in a form of so-called nonlinear dispersive solitary waves - solitons [1].

The present report is devoted to the various approaches that use the concept of nonlinear dispersive waves (solitons) in nonrelativistic nuclear physics. In Section 2 we discuss a possible method to analyse a static scission configuration in cold ternary fission in the framework of mean field approach. The inverse mean field method is applied to solve

---

[1] Work supported in part by Deutsche Forschungsgemeinschaft, Russian Foundation for Basic Research and the Heisenberg-Landau program.

single-particle Schrödinger equation, instead of constrained selfconsistent Hartree–Fock equations. It is shown, that it is possible to simulate one-dimensional three-center system in the approximation of reflectless single-particle potentials.

In Section 3, the problem of dynamical instability and clustering (stable fragments formation) in a breakup of excited nuclear systems are considered from the points of view of the soliton concept. It is shown that the volume (spinodal) instability can be associated with nonlinear terms, and the surface (Rayleigh-Taylor type) instability, with the dispersion terms in the evolution equations. The both instabilities may compensate each other and lead to stable solutions (solitons).

Section 4 is devoted to the short analysis of the collective collective excitations observed in inelastic alpha-particle and proton scattering. The analogy between fragmentation into parts of nuclei and buckyballs has led us to the idea of light nuclei as quasi-crystals. We establish that the quasi-crystalline structure can be formed when the distance between the alpha-particles is comparable with the length of the De Broglia wave of the alpha-particle. Applying this model to the scattering of alpha-particles we obtain that the form factor of the clusterized nucleus can be factorized into the formfactor of the cluster and the density of clusters in the nucleus. It gives the possibility to study the distribution of clusters in nuclei and to resolve what kind of distribution we are dealing with: a surface or volume one.

SCISSION CONFIGURATION

Motivation

Recent progress in modern experimental techniques ($\gamma - \gamma - \gamma$ and $x - \gamma - \gamma$ triple coincidence, which allow the fine resolution of the mass, charge and angular momentum content of the fragments) gives new experimental data (See e.g. Proceedings [2, 3]) confirming that the following collective modes: heavy cluster radioactivity, bimodal fission, cold fission, and inverse processes, such as (subbarier) fusion, could belong to the general phenomena of cold nuclear fragmentation [4]. Cluster like models [5, 6] were used successfully to reproduce general features of the cold ternary fragmentation. However the scission configuration has been built in fact by hands.

Cold fragmentation indicates the formation of a binary, ternary and possibly multi-center quasistationary nuclear systems, which lives $\sim 10^{-21}$ sec without reaching statistical equilibrium. Since the lifetime of this nonequilibrium system is an order of magnitude greater than the time of the nucleon traveling at the Fermi velocity to pass through the total system, a mean field can be formed in composite system. Therefore it is actual to develop microscopical or semi-microscopical approach to this scission-point concept of nuclear fragmentation.

There are well developed methods to calculate, in the framework of many-body self–consistent approach, static properties of a well isolated nucleus in its ground state [7, 8]. There also exists a well developed two-center shell model [9]. However, a three-center shell model has not been developed yet, except for very early steps [10]. Three-center shapes are practically not investigated, in comparison with the two-center ones. There exists the generalizations of mean–field models to the case of two-centers [11, 12], but a ternary configuration is out of consideration, because of uncertainties to select a peculiar set of constraints.

There exist a number of calculations for nucleus-nucleus collisions in the frame of time-dependent mean–field methods, but an evolution of the cold fragmentation has not been investigated yet. Therefore, although the principal way to describe nuclear fragmentation in the framework of many-body self-consistent approach exists, it is interesting to develop other mean-field approaches to analyse these phenomena from different points of view.

Framework

Methods of nonlinear dynamics gave yet the possibility to derive for nuclear physics unexpected collective modes, which can not be obtained by traditional methods of perturbation theory near some equilibrium state (see e.g. review [13]). The most important reason is that the fragmentation and clusterization is a very general phenomenon. There are cluster objects in subnuclear and macro physics. Very different theoretical methods were developed in these fields. However, there are only few basic physical ideas, and most of the methods deal with nonlinear partial differential equations. One of the most important part of soliton theory is the inverse scattering method [14, 15, 16] and its applications to the integration of nonlinear partial differential equations [17]. The inverse methods to integrate nonlinear evolution equations are often more effective than a direct numerical integration. Let us demonstrate this statement for the following simple case. The type of systems under consideration are slabs of nuclear matter [18], which are finite in the z coordinate and infinite and homogeneous in two transverse directions. The wave function for the slab geomethry is

$$\psi_{\mathbf{k}_\perp n}(\mathbf{x}) = \frac{1}{\sqrt{\Omega}} \psi_n(z) \exp(i\mathbf{k}_\perp \mathbf{r}), \qquad \varepsilon_{\mathbf{k}_\perp n} = \frac{\hbar^2 k_\perp^2}{2m} + e_n, \tag{1}$$

where $\mathbf{r} \equiv (x, y), \mathbf{k}_\perp \equiv (k_x, k_y)$, and Ω is the transverse normalization area.

$$-\frac{\hbar^2}{2m} \frac{d^2}{dz^2} \psi_n(z) + U(z) \psi_n(z) = e_n \psi_n(z), \tag{2}$$

A direct method to solve the single-particle problem (2) is to assign a functional of interaction $\mathscr{E}$ (usually an effective density dependent Skyrme force), to derive the ansatz for the one-body potential, as the first variation of a functional of interaction in density $U(z) = U[\rho(z)] = \delta\mathscr{E}/\delta\rho$. Then to solve the Hartree-Fock problem under the set of constraints, which define the specifics of the nuclear system. In the simplest case of a ground state, one should conserve the total particle number of nucleons (A), which is related to the "thickness" of a slab, via $A \Longrightarrow \mathscr{A} = (6A\rho_N^2/\pi)^{1/3}$, which gives the same radius for a three-dimensional system and its one-dimensional analogue. As a result, one obtains the energies of the single particle states e_n, their wave functions $\psi_n(z)$, the density profile $\rho(\mathbf{x}) \Longrightarrow \rho(z)$

$$\rho(z) = \sum_{n=1}^{N_0} a_n \psi_n^2(z), \qquad \mathscr{A} = \sum_{n=1}^{N_0} a_n, \qquad a_n = \frac{2m}{\pi\hbar^2}(e_F - e_n), \tag{3}$$

and the corresponding single-particle potential. a_n are the occupation numbers, N_0 is the number of occupied bound orbitals. The Fermy-energy e_F controls the conservation of the total number of nucleons. The energy (per nucleon) of a system is given by

$$\frac{E}{A} \Longrightarrow \frac{\hbar^2}{2m\mathscr{A}}\left(\sum_{n=1}^{N_0} a_n \int_{-\infty}^{\infty}\left(\frac{d\psi_n}{dz}\right)^2 dz + \frac{\pi}{2}\sum_{n=1}^{N_0} a_n^2\right) + \frac{1}{\mathscr{A}}\int_{-\infty}^{\infty}\mathscr{E}[\rho(z)]dz. \tag{4}$$

Finally, the set of formulas (1–4) completely defines the direct self-consistent problem. Following the inverse scattering method, one reduces the main differential Schrödinger equation (2) to the integral Gel'fand-Levitan-Marchenko equation [14, 15]

$$K(x,y) + B(x+y) + \int_x^{\infty} B(y+z)K(x,z)dz = 0. \tag{5}$$

for a function $K(x,y)$. The kernel B is determined by the reflection coefficients $R(k)(e_k = \hbar^2 k^2/2m)$, and by the N bound state eigenvalues

$$B(z) = \sum_{n=1}^{N} C_n^2(\kappa_n) + \frac{1}{\pi}\int_{-\infty}^{\infty} R(k)\exp(ikz)dk, \qquad e_n = -\hbar^2\kappa_n^2/2m.$$

N is the total number of the bound orbitals. The coefficients C_n are uniquely specified by the boundary conditions and the symmetry of the problem under consideration. The general solution, $U(z) = -(\hbar^2/m)(\partial K(z,z)/\partial z)$, should naturally contain both, contributions due to the continuum of the spectrum and to its discrete part. There seems to be no way to obtain the general solution $U(z)$ in a closed form. Eqs. (2,5) have to be solved only numerically. In Ref. [19] we used the approximation of reflectless $(R(k) = 0)$, symmetrical $(U(-z) = U(z))$ potentials. This gave the possibility to derive the following set of relations

$$U(z) = -\frac{\hbar^2}{m}\frac{\partial^2}{\partial z^2}\ln(\det\|M\|) = -\frac{2\hbar^2}{m}\sum_{n=1}^{N}\kappa_n\psi_n^2(z),$$

$$\psi_n(z) = \sum_{n=1}^{N}(M^{-1})_{nl}\lambda_l(z), \qquad \lambda_n(z) = C_n(\kappa_n)\exp(-\kappa_n z),$$

$$M_{nl}(z) = \delta_{nl} + \frac{\lambda_n(z)\lambda_l(z)}{\kappa_n + \kappa_l}, \qquad C_n(\kappa_n) = \left(2\kappa_n\left|\prod_{l\neq n}^{N}\frac{\kappa_n + \kappa_l}{\kappa_n - \kappa_l}\right|\right)^{1/2}. \tag{6}$$

Consequently, in the approximation of reflectless potentials $(R(k) = 0)$, the wave functions, the single-particle potential and the density profiles are completely defined by the bound state eigenvalues via formulas (1,6).

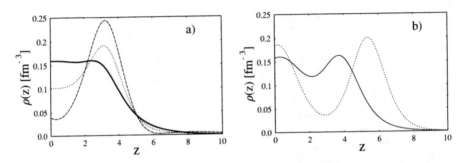

FIGURE 1. The density profiles of the light ($A \approx 20$, $\mathscr{A} \approx 1.0$) three-levels ($N = 3$, $N_0 = 3$) model system calculated in the frame of inverse mean field method. a) the ground state (solid line); a fragmented two-center configuration (dotted and dashed lines); b) a ternary fragmentation of the system into three fragments.

Results and discussion

We have carried out the series of calculations for the different layers, imitating nuclear systems in their ground state. For a direct part of the calculations by the Hartree-Fock method, the interaction functional was chosen in the form of effective Skyrme forces. The calculated spectrum of bound states was fed into the scheme of the inverse scattering method, and the relations were used to recover the wave functions of the states, the single-particle potentials, and the corresponding densities. Recently [20] we generalize this method to the case of fragmented configuration, trying to imitate two- and three-center nuclear systems. We use here only the inverse mean-field scheme (6). In Fig.1 we present the results of the calculations of three-level ($N = 3$, $N_0 = 3$) light ($A \approx 20$, $\mathscr{A} \approx 1.0$) model system simulating the ground state (Fig. 1(a) solid line), and fragmentation of the system into two fragments (Fig. 1(a), the dotted and dashed lines). In the same figure (Fig. 1(b)) we present the fragmentation of the system into three fragments (solid and dotted lines). One can see that it is possible to simulate a one-dimensional three-center system via inverse scattering method. The following conclusions can be drawn [20].

- The density profiles, calculated in the framework of inverse method, are practically identical to the results of calculation by SHF method. These results are valid for the ground state and for the system in the external potential field.
- The global properties of single-particle potentials (the depth and an effective radius) have been reproduced quite well, but the inverse method yields the quite strongly pronounced oscillations of the potential distributions within the inner region, and slightly different asymptotic tails of potential. In the framework of inverse scattering method, all bound states are taken into account in the calculation of the potential (6), but for the density distribution only the occupied states are taken into account (see Eqs. (3)). Therefore, the slope of the tails of the potential and of the density distributions will we different.

- We used, the approximation of reflectionless potentials, which gave us the possibility to obtain a simple closed set of relations (6), to calculate wave functions, density distributions and single particle potentials. The omitted reflection terms $(R(k) = 0)$ are not important for the evaluation of the density distributions, due to the fact that only the deepest occupied states are used to evaluate density distribution (see Eq. (3)). The introduction of these reflection terms will lead to a smoothing of the oscillations in the inner part of the potential and to a correction of its asymptotic behaviour.

- The presented method gives a tool to simulate the various sets of the static excited states of the system. This method could be useful to prepare in a simple way an initial state for the dynamical calculations in the frame of mean-field methods.

FRAGMENTATION

Motivation

An existence of solitary waves is determined by two essential factors, namely, nonlinearity and dispersion. Both the factors, which are responsible for the stability of a wave, are connected in their turn with two different types of instability. A localized pulse will tend to spread out due to dispersion terms of the equations of motion. The nonlinearity which is responsible for the formation of solitary waves, on the other hand, automatically leads to their destruction, if it is alone. Both instabities may compensate each other and lead to stable solutions (solitons).

Let us look from these points of view to multifragmentation phenomena [21]. Twenty years ago multifragmentation has been associated with the onset of the spinodal instability[22]. This instability is associated with the transit of a homogeneous fluid across a domain of negative pressure, which leads to its breaking up into droplets of denser liquid embedded in a lower density vapor. Since the spinodal instability can occur in an infinite system, it can be called the bulk or volume instability. On the other hand, it physically means that pressure depends on density, that is just a nonlinearity in terms of density.

Ten years ago [23], it has been pointed out that a new kind of instability (sheet instability) may play an important role in multifragmentation. This new instability can be assigned to the class of surface instabilities of the Rayleigh-Taylor kind[24]. System escapes from the high surface energy of the intermediate complex by breaking up into a number of spherical fragments with less overall surface. At the same time, it physically means the existence of the gradient terms of the equations of motion, i.e the dispersion. The spinodal instability and the Rayleigh-Taylor instability may compensate each other and lead to stable quasi-soliton type objects. In the next Sec. we will present a simple model to illustrate this physical picture[21].

201

Inverse mean-field method

The dynamical description will be done in the framework of the inverse mean-field method. One can found the details of this approach in [25, 26, 27]. The evolution of a system is given by the famous hydrodynamical Korteweg-de Vries equation (KdV) for the mean-field potential $U(z,t)$

$$\sum_{n=1}^{N} \frac{\partial U}{\partial (S_n t)} = 6U \frac{\partial U}{\partial z} - \frac{\hbar^2}{2m} \frac{\partial^3 U}{\partial z^3}, \tag{7}$$

where S_n are constants which are determined by the initial conditions.

General solution of KdV Eq. (7) can be derived in principle via direct methods numerically. This way is to assign a functional of interaction $\mathscr{E}$ (as usual an effective density dependent Skyrme force), a total number of particles (or a thickness of a slab $\mathscr{A}$) and to solve Hartree-Fock equations to derive a spectrum of the single particle states e_n and wave functions $\psi_n(z,0)$, the density profile $\rho(z,0)$ and the one-body potential $U(z,0)$ for the initial compressed nucleus. Then, one calculate an evolution of the one-body potential with the help of Eq. (7).

On the other side, there exists an inverse method to solve KdV Eq. (7). Following this method one derives, in the case of reflectless initially (t=0) symmetrical potentials, absolutely the same set relations, as Eqs. (6) with the trivial substitution:

$$z \rightarrow z + 2\hbar^2 \kappa_n^2 S_n t / m, \tag{8}$$

following from KdV Eq. (7).

Discussion

The following conclusions can be drawn.

- The wave functions, potential and the density profile are completely defined by the bound state eigenvalues. The first step is to solve the Schrödinger equation for the initial potential $U(z,0)$, which is suitable to simulate compressed nuclear system or to simulate this state with the help of spectrum.
- Then one calculates the evolution of $\rho(z,t)$ and $U(z,t)$ with the help of Eqs. (6,8). For large z and t the time - dependent one - body potential and the corresponding density distributions are represented by a set of stable solitary waves.
- The energy spectrum of an initially compressed system completely determines widths, velocities and the phase shifts of the solitons.
- The number of waves is equal to the number of occupied bound orbitals. Thickness ('number' of particles) of an n - wave is equal to a_n.
- Reflecting terms ($R(k) \neq 0$) of $U(z,t)$ cause ripples (oscillating waves of a small amplitude) in addition to the solitons in the final state.
- The initially compressed system expands so that for large times one can observe separate density solitons and ripples ('emissions'). This picture is in accordance

with the TDHF simulation of the time evolution of a compressed O^{16} nucleus [28]. The disassembly shows collective flow and clusterization.

- It is important to note that the clusterization was not observed in the absence of the self-consistent mean-field potential, i.e this confirmes our supposition that the nonlinearity is very important for the clustering.

The presented model is too primitive in order to describe a real breakup process. However this model can be used to illustrate an inverse mean-field method scheme, a nonlinear principle of superposition and the idea that nonlinearity and dispersion terms of the evolution equation can lead to clusterization in the final channel.

SUMMARY AND OTHER PROBLEMS

Attention is focused on the various approaches that use the concept of nonlinear dispersive waves (solitons) in nonrelativistic nuclear physics.
The problem of dynamical instability and clustering (stable fragments formation) in a breakup of excited nuclear systems are considered from the points of view of the soliton concept. It is shown that the volume (spinodal) instability can be associated with nonlinear terms, and the surface (Rayleigh-Taylor type) instability, with the dispersion terms in the evolution equations. The both instabilities may compensate each other and lead to stable solutions (solitons).
We suggest the method to analyse a static scission configuration in cold ternary fission in the framework of mean field approach. The inverse mean field method is applied to solve single-particle Schrödinger equation, instead of constrained selfconsistent Hartree–Fock equations. It is shown, that it is possible to simulate one-dimensional three-center system via inverse scattering method in the approximation of reflectless single-particle potentials. These models may be useful as a guide to understand the general properties of fragmented systems and to formulate the suitable set of constraints for the realistic three-dimensional mean field calculations of the three-center nuclear system.
The presented above results are the only few points in the wide field of nonlinear problems of the modern nuclear physics (see e.g. [29] for the recent references). Short comments concerning on the few another possible applications.
Recent experiments on the scattering of alpha-particles from light nuclei demonstrated the appearance of bands belonging to the same angular momenta [30, 31]. They occur within the rotational bands, which consists of even and odd states. It turns out that at the high energy excitations of alpha-clusterized nuclei, when the shell structure is loosing, one has a deal with Bose-Einstein condensate of alpha-particles, which is governed by Gross-Pitaevsky equation [32]

$$-\frac{\hbar^2 \nabla^2}{2m_\alpha} \psi_\alpha(r) + V_{ex}(r)\psi_\alpha(r) + N_\alpha \cdot g|\psi_\alpha(r)|^2 \psi_\alpha(r) = \mu \psi_\alpha(r), \qquad (9)$$

where m_α is the mass of α-particle, V_{ex} - an effective (oscillator) potential. N_α is the number of α-particles, g - the coupling constant and μ is the chemical potential. Neglecting the kinetic energy term in (9), one arrives to the well known Thomas-Fermi solution [33]

$$\rho_\alpha(r) \equiv N_\alpha \cdot |\psi_\alpha(r)|^2 \approx (\mu - V_{ex}(r))/g. \tag{10}$$

This solution shows that at high energy excitations the fragmentation of cluster states takes place. The number of these fragmented states is proportional to the number of clusters N_α.

In the case of repulsive interaction ($g \geq 0$) and if $(\mu - V_{ex}) \gg g$ one can expand the local momentum

$$
\begin{aligned}
p_\alpha(r) &\equiv \sqrt{2m_\alpha (\mu - V_{ex} - g\rho_\alpha)} \\
&\approx \sqrt{2m_\alpha (\mu - V_{ex})} \cdot \left(1 - \frac{g\rho_\alpha}{2(\mu - V_{ex})} + \ldots \right).
\end{aligned} \tag{11}
$$

This standard semi-classical WKB procedure leads to the conclusion that discretization of momentum depends upon the number of α-particles

$$\int_{r_1}^{r_2} p(r)dr = (2N+1)\pi/2 + GN_\alpha, \tag{12}$$

where N is the radial quantum number. Relying on this result we can present the band structure of cluster states by the following formula for the spectrum

$$E = A + BL(L+1) + CN + DN^2 + GN_\alpha, \tag{13}$$

where the standard formula describing rotational molecular spectra was modified by adding of the new term which is proportional to the number of alpha-particles.

The moment of inertia deduced from the band structure of the rotational states indicates that there are a few alpha particles orbiting a core [30]. The analogy between fragmentation into parts of nuclei and buckyballs has led us to the idea of light nuclei as quasi-crystals [34]. We establish that the quasi-crystalline structure can be formed when the distance between the alpha-particles is comparable with the length of the De Broglia wave of the alpha-particle [35]. Applying this model to the scattering of alpha-particles [36] we obtain that the form factor of the clusterized nucleus can be factorized into the formfactor of the cluster and the density of clusters in the nucleus. It gives us the possibility to study the distribution of clusters in nuclei and to resolve what kind of of distribution we are dealing with: a surface or volume one. Similar circumstances occur by the scattering of electrons from metal clusters and fullerenes [37].

REFERENCES

1. G.B. Whitham, *Linear and Nonlinear Waves*, Wiley, NY, 1974.
2. *Perspectives in Nuclear Physics*, Procs. Int. Conf., Eds. J.H. Hamilton, H.K. Carter, and R.B. Piercey, World Scientific, Singapour, 1999.
3. *Fusion Dynamics at the Extremes*, Int. Workshop Dubna, Russia, May 2000, Eds. Yu.Ts. Oganessian, and V.I. Zagrebaev, World Scientific, Singapoure, 2001.
4. P. Armbruster, *Rep. Prog. Phys.* **62** (1999) 465.
5. A. Săndulescu, A. Florescu, and W. Greiner, *J. Phys. G: Nucl. Part. Phys.* **22** (1989) 1815.

6. F. Gönnenwein, and B. Borsig, *Nucl.Phys.* **A530** (1991) 27.
7. D. Vautherin, M. Vénéroni, and D. M. Brink, *Phys. Lett.* **B33** (1970) 381.
8. T. Burvenich, D.G. Madland, J.A. Maruhn, and P.G. Reinhard, *Phys. Rev.* **C65** (2002) C65:044308.
9. J. Maruhn, and W. Greiner, *Z. Physik* **251** (1972) 431.
10. J. Hahn, H.–J. Lustig, and W. Greiner, *Z. Naturf.* **32a** (1977) 215.
11. M. Bender,K. Rutz, P.G. Reinhard, J.A. Maruhn, and W. Greiner, *Phys. Rev.* **C58** (1998) 2126.
12. J.F. Berger, M. Girod, and D. Gogny, *Nucl. Phys.* **A502** (1989) 85c.
13. V.G. Kartavenko, *Phys. Part. Nucl.* **24** (1993) 619.
14. I.M. Gel'fand, and B.M. Levitan, *Izv. Akad. Nauk SSSR, Ser. Mat.* **15** (1951) 309.
15. V.A. Marchenko, *Dokl. Akad. Nauk SSSR* **104** (1955) 695.
16. L.D. Faddev, *Sov. Phys. Dokl.* **3** (1959) 747.
17. *Theory of Solitons: Inverse Scattering Method*, Ed. by S. P. Novikov, Nauka, Moscow, 1970.
18. P. Bonche, S. Koonin, and J.W. Negele, *Phys. Rev.* **C13** (1976) 1226.
19. V.G. Kartavenko, and P. Mädler, *Izv. Akad. Nauk SSSR, Ser. Fiz.* **51** (1987) 1973.
20. V.G. Kartavenko, A. Săndulescu and W. Greiner, *Int. J. Mod. Phys.* **E8** (1999) 381.
21. V. Kartavenko, K. Gridnev, W. Greiner W., *Int.J.Mod.Phys.* **3E** (1994) 1219.
22. P.J. Siemens, *Nature* **305** (1983) 410.
23. L.G. Moretto, and G.J. Wozniak, *Ann. Rev. Nucl. Part. Sci.* **43** (1993) 379.
24. Lord Rayleigh, *Scientific Papers*, Article 58, Dover 1964), NY, p.361.
25. E.F. Hefter, *Journ. de Phys.* **45**, C6:67 (1984).
26. E.F. Hefter, and K.A. Gridnev, *Prog. Theor. Phys.* **77** (1984) 549.
27. E.F. Hefter, and V.G. Kartavenko, *JINR Rapid. Comm.* **29** (1987) 3.
28. A. Dhar, S. Das Gupta, *Phys. Rev.* **C30** (1984) 1545.
29. V.G. Kartavenko, K.A. Gridnev, and W. Greiner, *Sov. J. Nucl. Phys.* **65** (2002) 669.
30. M. Brenner et al., *Heavy Ion Physics* **7** (1998) 355.
31. U. Abbondano, N. Cindro, and P.M. Milazzo, *Nuov. Chim.* **A110** (1997) 955.
32. Gridnev K.A., *Z. Phys.* **A349** (1994) 269.
33. V.G. Kartavenko, K.A. Gridnev, and W. Greiner, *Int. J. Mod. Phys.* **E7** (1998) 287.
34. K.A. Gridnev, M.W. Brenner, S.E. Belov, K.V. Ershov, V.G. Kartavenko, and W. Greiner, XVth Int. Workshop High Energy Physics and Quantum Field Theory, September 7–13, 2000, Tver, Russia, Eds. M. Dubinin, V. Savrin, Moscow, 2001, p. 406.
35. K.A. Gridnev, M.W. Brenner, S.E. Belov, E. Indola, V.G. Kartavenko, and W. Greiner, *I.EC 2000*, Procs. 1st Eurasia Conference on Nuclear Science and Its Application, October 23–27, 2000, Izmir, Turkey, pp.897–902.
36. K.A. Gridnev, M.W. Brenner, K.V. Ershov, V.G. Kartavenko and W. Greiner, Proc. of the Int. Symposium on *Clustering Aspects of Quantum Many-Body Systems*, Kyoto, Japan, November 12–14, 2001, World Scientific, Singapour, in press.
37. K.A. Gridnev, S.N. Fadeev, V.G. Kartavenko, and W. Greiner, *Physics of Unstable Nuclei*, Yukawa Int. Seminar Kyoto, Japan, November 5–10, 2001, *Prog. Theor. Phys.*, in press.

Dynamics and Exotic Cluster Production in the system Sn+Ni at 35 MeV/A

Massimo Papa*, Aldo Bonasera† and Toshiki Maruyama**

*Istituto Nazionale di Fisica Nucleare Sezione di Catania, Via S. Sofia 64, Catania 95123, Italy
†Istituto Nazionale di Fisica Nucleare Laboratorio Nazionale del Sud, Via S. Sofia 44, Catania 95123, Italy
**Japan Atomic Energy Research Institute, Tokai, Ibaraki 319-1195, Japan

Abstract. The $^{112}Sn + ^{58}Ni$ and $^{124}Sn + ^{64}Ni$ systems at 35 MeV/nucleon have studied by means of the recently proposed Constrained Molecular Dynamics model (CoMD). Different aspects concerning the central and mid-peripheral collisions are put in evidence. In particular asymmetry distributions of the produced cluster are discussed.

INTRODUCTION

In this contribution different aspects of the $^{112}Sn + ^{58}Ni$ and $^{124}Sn + ^{64}Ni$ systems at 35 MeV/nucleon have been studied with the recently developed Constrained Molecular Dynamics approach [1]. These collisions have been performed in Catania at the LNS in the framework of the REVERSE experiment which uses the forward part of the CHIMERA detector [2]. In Section 2 we discuss on the possibility to find out signals of the Isospin dynamics for central collision by looking to the symmetry variable distribution $Y = \frac{N-Z}{A}$ for cluster produced in the mass A range 30-69. In general this signal can be masked by the secondary evaporation process, but the detection of exotic fragments with particular values of Y could give interesting informations. The Section 3 is devoted to the study of the mid-peripheral collisions, which show the possibility to observe the intriguing phenomenon of the target multi-break-up. The summary is reported in Section 3.

SYMMETRY AND COULOMB EFFECTS IN CENTRAL COLLISION

One of the recent interesting aspects in studying heavy ion collision in the Fermi energy domain is to explore the role played by the symmetry term in EOS. In this section we want to show some results obtained by means of the CoMD model. In this new approach the occupation number in phase-space $\overline{f}$ for identical particles is bounded to be less or at most equal to one. This constraint coupled with a modified cooling procedure allows to obtain nuclear stable configurations (up to 1500-2000 fm/c) in which the average kinetic energy of nucleons is in good agreement with the Thomas Fermi model. More details can be found in [1]. In particular in this approach the symmetry part of the interaction

CP644, *Exotic Clustering: 4th Catania Relativistic Ion Studies*, edited by S. Costa, A. Insolia, and C. Tuvè
© 2002 American Institute of Physics 0-7354-0099-7/02/$19.00

is taken into account through the Pauli principle (kinetic term) and trough the following term in the effective interaction:

$$V_{sym} = \frac{a_{sym}}{2\rho_0} \sum_{j \neq j} \rho_{i,j} ([2\delta_{\tau_i,\tau_j} - 1)$$ (1)

In this equation $\rho_{i,j}$ represents the particle-particle overlap integral, τ_i indicates the isospin third component of the generic nucleon and a_{sym} is equal to 32 MeV.

In the following we will try to understand if a detailed study of the isotopic distributions can give signals reflecting the isospin dependent part of the effective interaction. In particular we will focus in this section to the central collision $b \leq 5$ fm). After the initialisation, in which the "ground state" configuration of the Sn and Ni isotopes have been found, the time evolution of the systems has been followed up to 2000 fm/c.

At this time, taking as reference a binding energy of 8 MeV/nucleon, the hot sources have an average excitation energy of about 1-1.5 MeV/nucleon . To evaluate to which extent the secondary evaporation process can affect the final results, a second stage of statistical sequential decay has been applied to the excited sources by means of the GEMINI code. Concerning the reaction dynamics it results that, for the range of impact parameter considered, the reaction mechanisms are mainly incomplete fusion and processes in which several IMFs are produced.

In Fig. 1 we show the charge distribution and the IMF multiplicity distribution for the $^{124}Sn + ^{64}Ni$ system after the dynamical stage. The charge distribution shows that globally there is no indication for a phase transition even if a more deep analysis put in evidence that a small percentage of events are located in the bending region of the Campi plot.

In Fig. 2 we also show as function of the atomic number Z the average ratio $\langle N/Z \rangle$ observed in a preliminary analysis of data [3] compared to the results of the CoMD calculations. The agreement is good. It shows the general tendency for the neutron-rich system to produce more neutron-rich light fragments.

Now we focus our attention on the charge/mass distribution expressed through the variable Y (see Sec.1) for medium-heavy ion fragments. The main reasons are at least two:

-i) the production of very light fragments can be heavily affected by the decay of the hot primary fragments. Moreover their binding energies depend on the details of the effective interaction and on quantal effects. Both and especially the second are still not included at the present development stage of the semi classical dynamical-many body approaches.

-ii) on the contrary the same approaches are better suited to describe the production of heavier clusters because the effective interaction used reflects mainly the bulk properties of the nuclear matter. We also observe that heavier clusters are formed on the average in short time (less than 200 fm/c). Finally the related Y distribution densely spans a wide range of values and therefore it can be usefully experimentally evaluated also within an uncertainty of the order of 30% with the recently developed charge-mass multi-detectors.

In Fig. 3a) we show the Y-distribution for masses in the range 30-69 as obtained from the sequential decay code GEMINI applied to the decay of the sources corresponding to the studied system. The excitation energy is about 7.8 MeV/nucleon. An angular

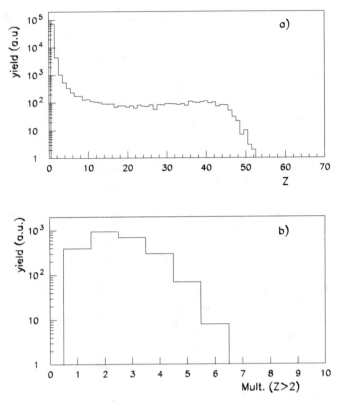

FIGURE 1. a) - Charge distribution for the $^{124}Sn + ^{64}Ni$ system. b) - related IMF multiplicity distribution.

momentum triangular distribution has been taken in to account according to the impact parameter window considered ($J = 0 \simeq 5\hbar$). In Fig. 3b) we compare for the two system the Y-distribution predicted by the CoMD calculations (empty and full circles) using the same cut on the fragment mass. In these figures the yields are in arbitrary units but the comparison is performed by normalizing to the same number of events. We observe that the value 30 for the lower cut on the mass has to be considered only indicative of a region of mass for which, according to our present experience on CoMD calculations (binding energies and nuclear radii, the item ii) previously exposed can be reasonably satisfied. Large scale calculations for ground state configurations of lighter nuclei are in progress to lower this threshold. The error bars in Fig. 3b are related to the statistics of the calculations.

As already observed the dynamical evolution has been followed for 2000 fm/c. The related Y-distribution are shown by points in Fig. 3b). At this stage each kind of dynamical effects have been already manifested. We can observe that the distributions for the studied system look differently being shifted on the average by an amount of about 0.04. On the Y axes the arrows point to the asymmetry of the targets (T), projectiles (P) and

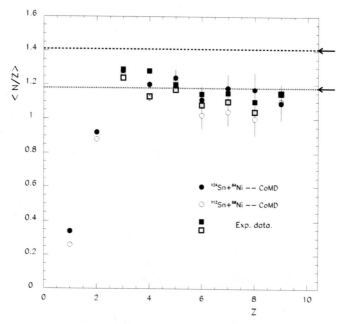

FIGURE 2. $\langle N/Z \rangle$ as function of the atomic number Z. Comparison between experimental data and CoMD calculations. The arrows and the horizontal lines indicate the N/Z value for the total systems.

total system (CS). The longer ones refer to the rich neutron system.

On can observe that there is no simple scaling operation, by using the relative Y values of T,P or CS, able to overlap the two distribution. This can be also deduced by looking at the differences of the average value of Y respect to the value corresponding to the total system. The system which displays the largest shift is $^{112}Sn + ^{58}Ni$ The sign of this difference indicates that, on the average, this systems has the strongest tendency in the dynamical stage to give up more protons and light proton-rich particles compared to neutron or neutron-rich particles. For $^{124}Sn + ^{64}Ni$ system on the contrary the average value of Y displays a smaller shift trough lower value with respect the one related to the total system. This means that the above system has on the average a weaker tendency to give up neutrons or neutron-rich fragments.

These effects and in particular the difference between the distribution shown in Fig. 3a) and the ones displayed in Fig. 3b) by means of circles are of dynamical nature. They are mainly connected with the interplay between the Coulomb and the symmetry energy per nucleon in the compression stage.

The next step in the calculations includes the estimation of the remaining in-flight evaporation of the residues. The residues can have excitation energy of the order of 1.5 MeV/nucleon after 2000 fm/c. The spectra shown trough histograms represent the final results obtained after this last stage. In particular the average values and variances approach to the value predicted by the statistical model calculations. Concerning this result we note that an overestimation of the excitation energy of about 0.7 MeV/nucleon,

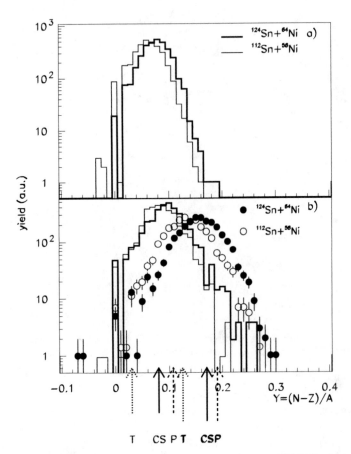

FIGURE 3. Y-distribution for the system under study: a) - results from GEMINI calculations applied to the composite nuclei b) - results predicted by CoMD (points). The final results obtained after a second stage reflecting the last part of the secondary emission, simulated by means of GEMINI code, is also displayed trough histograms.

arising from the assumption of a binding energy of 8 MeV/nucleon for all the fragments, is quite reasonable in our calculations. We note that without this overestimation the final Y-distribution should be much more similar to those obtained from the dynamical calculations. Therefore the results shown with histogram in this work may be considered as the worse case concerning the visibility of the dynamical effects.

Notwithstanding this, the final distribution (histograms in Fig. 3b)) are very different from the one predicted by the statistical sequential decay if one looks to the tails. In particular we observe that the relative yield for Y=0.2 compared to the maximum value is about 0.5% for the CoMD calculations (including also the secondary evaporation) while is 0.05% in the case of sequential decay. Moreover for the most neutron-rich system the region for $Y \geq 0.2$ shows a total yield which is about a factor ten higher than the case of the neutron-poor system. Experimentally this large difference represents a

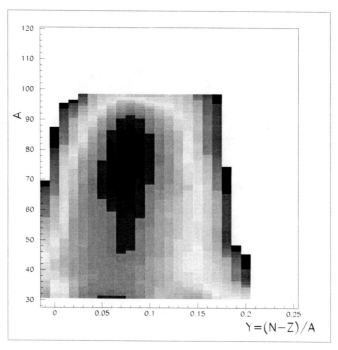

FIGURE 4. A-Y correlation for the system $^{112}Sn + ^{58}Ni$ and $A \geq 30$.

detectable signal also with a moderate statistics.

By analysing the mass-Y correlation, we find also that the involved masses, for the Y interval under study, belong, for a fraction of about 80% to a mass range between 30 and 60 units. Therefore these clusters are rather exotic having a very large excess of neutron. This is clearly visible through the scatter-plot shown in Fig. 4 and Fig. 5.

From the Fig. 4 and Fig. 5 is also visible that for Y values corresponding to the maximum in the final Y-distributions shown in Fig. 3b) the average produced mass is around 80 which strongly suggest a dynamics dominated by the incomplete fusion. In Fig. 6 we finally show the Dalitz plot for the 3 biggest fragments. By comparing the upper and bottom parts it is evident that the conditions $A_2 > 29$ (A_2 second biggest fragment) and $Y > 0.19$ select multifragmentation events in which at least three fragments are produced with approximately the same size. This fragments become relatively cold within a time interval of 2000 fm/c but they are produced in a shorter time of about 150 fm/c. For this reason they can be considered like the "messengers" of the dynamics of such short time scale. In this direction this kind of preliminary analysis (with more statistics an analysis using higher order moments of the distribution could be possible) can be more deeply investigated performing also a more fine selection of the reaction mechanism.

We finally remark that the quantitative results shown in this work strong depend on the kind of the effective interaction (see ref.[1] and eq. (1) in the text). However, for the systems under study, we can assume as a more general result that in a dynamical

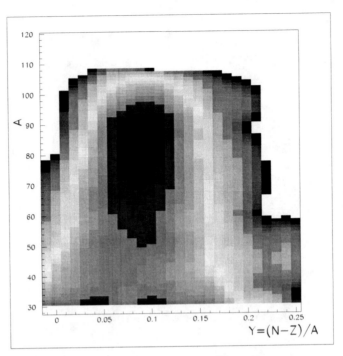

FIGURE 5. A-Y correlation for the system $^{124}Sn + ^{64}Ni$ and $A \geq 30$.

approach the tails of the Y distributions for the so called "liquid part", being determined by the fluctuations in the dynamics itself, contain selected information on the first moment of the disassembly of the hot sources and therefore are strongly sensitive to the compression phase of the nucleus-nucleus collision.

MID-PERIPHERAL REACTIONS AND TARGET MULTI-FRAGMENTATION

The system $^{124}Sn + ^{64}Ni$ at 35 MeV/nucleon has been studied also in the range of impact parameter $b = 6 \simeq 9$ fm. On the average the reaction mechanisms is dominated in this case by a quasi binary processes in which a target like (TLF) and projectile like (PLF) fragments are produced. In our study we concentrate on a reaction mechanisms in which a PLF fragment is still present but not the TLF one. The latter in fact can disassembly in to several IMFs. One of the reasons of interest is due to the fact that such a class of events, if detected in a quasi-complete way (total charge and momentum detected for a fraction greater than 80%) can be reconstructed with a small uncertainty. This is due to the focusing effect on the particles emitted from the quasi-projectile arising from the inverse kinematical condition. These kind of subjects were discussed in a series of meeting [4]. In particular for REVERSE kind experiment a reconstruction efficiency of

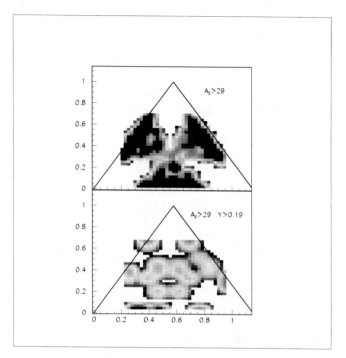

FIGURE 6. Dalitz plot for mass fragment produced through the collision $^{124}Sn + ^{64}Ni$ under the conditions expressed in the figure.

the order of $5 - 10\%$ was estimated. In the same meetings the TLF multiple break-up was related to the approximately equal sharing between PLF and TLF leading to a TLF with an excitation energy above 5 MeV/nucleon. In the following we show preliminary results of our calculations performed by means of the CoMD approach. In Fig. 7 we show a typical picture of the events we are discussing.

From Fig. 8 it is possible to note that the events related to the TLF fragmentation fill the kinetic energy spectrum, shown in the upper part of the figure, corresponding to the condition $A_1 \geq 70$ and $A_2 \leq 20$ on the masses of the two biggest fragments computed at 500 fm/c. The same figure evidence that the mechanism we are discussing selects a window in the impact parameter distribution. We observe that in a simply binary framework the kinetic energy of the PLF is related to the total kinetic energy dissipated in the collision. Therefore looking at the probability of the process as function of the PLF kinetic energy and taking into account the correction for mid-rapidity emission, it is possible to perform an excitation function of the process which will contains important informationŠ on the nature of the process. Moreover this process, which show clearly a dynamical nature, if studied more deeply can reveal interesting connection between the sharing of excitation energy, the dynamics of the nucleon exchange and the macroscopic condition to obtain the multifragmentation process.

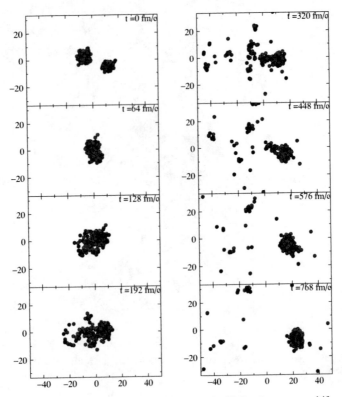

FIGURE 7. Typical time evolution of an event in which the TLF undergoes a multifragmentation.

SUMMARY

Different aspects of the $^{124}Sn + ^{64}Ni$ and $^{112}Sn + ^{58}Ni$ collision at 35 MeV/nucleon have been illustrated by means of the CoMD model. In central collision it has been shown that, notwithstanding the strong masking effect of the secondary evaporation process, the study of the high order moments of the Y-distribution for heavy cluster with mass smaller than 70 can reveal, with high visibility, effects connected to the difference in the Coulomb and symmetry energies per nucleon experienced by the studied systems in the compression phase.

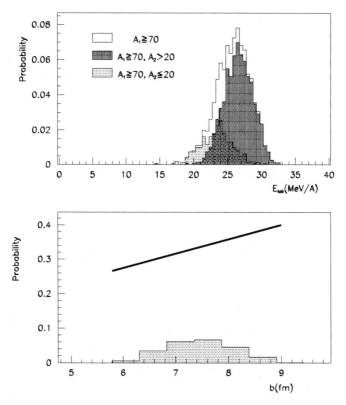

FIGURE 8. In the upper panel we show the spectra of the laboratory kinetic energy of PLF (A_1) for different conditions on the mass of the two biggest fragments A_1 and A_2. In the bottom panel it is shown the probability distribution for TLF fragmentation as function of the impact parameter b. The line represents the total impact parameter distribution considered in the simulations.

For mid-peripheral reaction about 15% of the reaction cross section is related to events in which PLF fragments are produced in the forward direction while the target like fragments disappear. This kind of process is clearly of dynamical origin and we found interesting to study both the mechanism able to determine the condition of TLF multi break-up and the way in which the fragmenatation happens.

REFERENCES

1. M. Papa, T.Maruyama and A.Bonasera, Phys. Rev. C 64, 024612 (2001).
2. A. Pagano et al Nucl. Phys. A681, 331 (2001).
3. S.Pirrone et al Proc. on Nuclear Physiscs at Border Line, May 21-24, 2001 Messina (Italy), Ed. by Word Scientific.
4. M.Papa, 1°, 2° and 3° meeting of the Reverse Collaboration. Isituto Nazionale di Fisica Nucleare -Laboratorio Nazionale del Sud Catania (Italy) 1998-2000.

Strange "Large" Clusters: Neutron Stars

3d Supernovae Collapse Calculations

Tobias Bollenbach* and Wolfgang Bauer*

*National Superconducting Cyclotron Laboratory and Department of Physics and Astronomy,
Michigan State University
East Lansing, MI 48824 - 1116, USA

Abstract. We give a brief overview of the current understanding of the explosion mechanism of core collapse supernovae. Our main focus is the impact of rotation on the explosion. Recent observations of the polarization of the light emitted by supernova explosions indicate that there are large deviations from spherical symmetry in the very heart of the explosion the origin of which is unknown.

We use the new approach of a three dimensional test particle based simulation to simulate the infall phase of a supernova event. The underlying microphysics is simplified to make this computationally possible. A systematic study of the influence of rotation mainly during the infall phase of the collapse of a typical iron core is performed. Indications for significant deviations from spherical symmetry are found in our very rapidly rotating models.

CORE COLLAPSE SUPERNOVAE

As an introduction, we briefly review the explosion mechanism of a typical core collapse supernova which occurs at the end of a massive star's main sequence evolution.

Collapse

The production of iron in nuclear reactions that occurs during the late stages of the stellar evolution essentially goes on until the iron core that is formed in the center of the star reaches a mass that results in gravitational forces which can no longer be supported by the pressure of the present degenerate electron gas. A typical mass of the iron core at this time is $1.35 M_\odot$.

Electron Capture and Photodisintegration

As soon as collapse begins, two instabilities are of importance. Ongoing electron capture, $p^+ + e^- \longrightarrow n + \nu_e$, reduces the electron fraction in the core, thus obviously reducing the pressure created by the electron gas. In the beginning of collapse the neutrinos created in the electron capture reactions are radiated away almost freely which ultimately results in a temperature decrease in the core – a phenomenon known as *neutrino cooling*. This helps the ongoing collapse further as a reduction of temperature implies a pressure decrease.

CP644, *Exotic Clustering: 4th Catania Relativistic Ion Studies,* edited by S. Costa, A. Insolia, and C. Tuvè
© 2002 American Institute of Physics 0-7354-0099-7/02/$19.00

The second instability is due to a process called *photodisintegration* that is possible at the extremely high temperatures now present in the core: heavy nuclei are fragmented to their constituents by extremely high energetic photons which requires huge amounts of energy. Therefore the temperature increase due to the compression of the core during its collapse is intensely weakened resulting again in a pressure decrement: gravity can no longer be compensated by pressure.

Inner Core, Outer Core, and Neutrino Trapping

Once core collapse has begun, the iron core matter falls almost at free-fall velocity towards the center of the star. Hence, the time scale for collapse is merely ≈ 100ms.

Two regions in the iron core must be separated. In the *inner core*, the infall velocity of the matter is proportional to the distance from the center which obviously causes all the matter in the inner core to finally arrive at the center of the star simultaneously. The inner core ends at the distance where the infall velocity of the matter exceeds the local sound velocity. Beyond that (time-dependent) distance the *outer core* falls towards the center at supersonic velocities. It has decoupled from the inner core and arrives at the center of the star later.

The collapse does not stop before nuclear density is exceeded. It is believed that roughly three times the density of isospin symmetric nuclear matter is reached at maximum compression of the core [24, 26].

During the core collapse neutrinos are produced copiously; about 10^{58} are emitted during a supernova explosion. An important phenomenon called *neutrino trapping* occurs at densities higher than $\approx 10^{11}$g/cm^3: neutrinos can no longer escape the core freely but are trapped in there since at these densities elastic scattering of the neutrinos by the nuclei becomes relevant. The mean free path for the neutrinos is so small now that they can only diffuse slowly (compared to the collapse time scale) through this high density region – they are virtually trapped.

Core Bounce

As soon as a density higher than nuclear is reached, the equation of state (EOS) of the nuclear matter becomes repulsive as a consequence of the Pauli principle for neutrons. At a density of roughly three times nuclear this repulsion becomes so strong that the matter stiffens, the collapse is halted, and the inner core rebounds. In doing so, shockwaves are sent out to the infalling outer core. This event is known as *core bounce*.

Theoretical predictions of the maximum density reached at bounce and its vigorousness strongly depend on the nuclear EOS which is still largely unknown for these extremely high densities and temperatures. The probably most realistic EOS currently available for these conditions is the one developed by Lattimer and Swesty (LS EOS) [1, 2].

Delayed Shock Mechanism

During collapse an enormous amount of gravitational energy is released. The main question is by which mechanism a fraction of $\gtrsim 1\%$ (that would be sufficient to explain the observed supernova explosion energies of $\approx 10^{51}$ erg) of this energy can be coupled to the mantle and the envelope of the star in order to eject them.

An appealing mechanism for this was favored till the mid 1980s: after core bounce the created shockwave moves outward through the infalling matter of the outer core, reaches the outer layers of the star, and ejects them – the star explodes.

Unfortunately, things turned out to be not quite this simple because the shock loses enormous amounts of energy while it beats its way through the infalling matter of the outer core. The densities and temperatures in the shock region are so high that electron capture resulting in neutrino losses and photodisintegration occur massively. Large amounts of heavy nuclei are broken up completely to free nucleons which costs gigantic amounts of energy that is taken from the shock. Consequently, if the outer core is sufficiently large, the shock eventually completely stalls and never even reaches the outer layers.

Once the shock has stalled, there is still an enormous amount of energy in the core, mainly in the form of thermal excitations and neutrino and electron chemical potentials. The neutrinos start to diffuse outward bathing the whole matter above the *neutrino sphere*[1] in a very intense neutrino flux. This helps to revive the stalled shock: the shocked matter above the neutrino sphere now contains a lot of free nucleons which can absorb a small fraction of the neutrinos (e.g. by the inverse of the electron capture reaction). This ultimately leads to the heating and expansion of the matter (*neutrino heating*) so that the shock can resume its way outward.

This *delayed shock mechanism* (suggested by Wilson in 1985 [3]) is the commonly accepted theory for core collapse supernova explosions. Still, it depends crucially on such input as the cross sections for neutrino capture on nucleons, the neutrino production rates, the details of the neutrino transport, and other factors many of which are not known with great certainty. Depending on these input parameters, simulations of the delayed mechanism (performed by several different groups) yielded successful explosions [4, 5] while in other cases the explosions failed [6]. A very good summary of the delayed mechanism and the skepticism about its relevance was recently provided by Janka [7].

Convection

Several two dimensional simulations have been carried out in the last ten years in which especially the impact of *convection* has been studied [4, 5, 8, 9]. Convection will certainly occur in the region between the shock and the neutrino sphere where the shock leaves behind an entropy profile that is instable to convection. It is widely believed today that convection in this region helps the success of the explosion by transporting energy

---

[1] The surface of the neutrino sphere is given by the distance at which the optical depth for neutrinos is 1. The neutrinos can be considered to radiate away freely approximately from there.

from above the neutrino sphere (where the matter can be heated by neutrino absorption in a relatively easy way) to the outer layers of the star. Two dimensional simulations done by Herant et al. [4, 8] for example resulted in successful explosions, while their one dimensional ones (that did not include convection but used the same microphysics otherwise) failed.

Still, this is not necessarily the final solution to the supernova problem. In particular, it is conceivable that phenomena that can only occur in three dimensions play a role.

Rotation

It is well known that stars carry angular momentum. The impact of this rotation on the core collapse and the explosion mechanism has been studied relatively little. Some studies were made, however, in which indications have been found that rotation does not affect the explosion mechanism dramatically [5, 9]. It appears that explosions are slightly delayed and a little bit weaker if rotation is included.

It must be mentioned that often rotation is mimicked in questionable ways necessary because most simulations are performed in two dimensions. Well known phenomena from earth such as vortices in flowing liquids or tornadoes in the atmosphere could hardly occur in less than three dimensions. Rotation can become extremely rapid in the late stages of core collapse, because the inner core contracts progressively, enforcing higher angular velocities due to global angular momentum conservation.

Another argument for studying the impact of rotation are recent observations of the polarization of the light emitted by supernovae made by Wang and Wheeler [10, 11, 12]. Using a method in which they observe the polarization and the wavelength of the light simultaneously ("*spectropolarimetry*"), they found that the light emitted by core collapse supernovae is significantly polarized. They draw the conclusion that there must be large asymmetries in the explosion mechanism to cause this polarization. Rotation is a very good candidate to explain such deviations from spherical symmetry.

OUR THREE DIMENSIONAL SIMULATION

The basic idea of our own simulation is to simplify the underlying microphysics by using input from one and two dimensional simulations in order to be able to follow the core's dynamics in three dimensions during collapse and bounce. The main focus is put on the impact of rotation during collapse. Our hope was to thus find deviations from spherical symmetry that are so significant that they may deliver alternatives to the currently favored complicated convection-driven explosion mechanism just described.

Test Particle Method

The method we use is similar to the so called *test particle method* (or *pseudoparticle method*) that has been used extensively in nuclear physics [13, 14, 15].

While in nuclear physics usually up to several thousand test particles represent one nucleon, the scales are opposite in our model for the collapse of an iron core with a mass around $1.3M_\odot$: assuming that we use 10^6 test particles, one test particle represents a mass which is just a little less than that of the entire earth!

In our model [16] all N_{tp} particles have a mass $M_{IC}/N_{tp} =: m_{tp}$, where M_{IC} denotes the mass of the whole iron core. For each individual particle, position $\vec{r}_j$ and momentum $\vec{p}_j$ are tracked (as classical three vectors). The equations of motion for the test particles are the relativistic versions of the Newtonian ones known from classical mechanics:

$$\dot{\vec{r}}_j = \frac{d\vec{r}_j}{dt} = \frac{\vec{p}_j}{\sqrt{m_{tp}^2 + (\frac{\vec{p}_j}{c})^2}} \tag{1}$$

$$\dot{\vec{p}}_j = \frac{d\vec{p}_j}{dt} = \vec{F}_{G,j}(\vec{r}_1,\ldots,\vec{r}_{N_{tp}}) + \vec{F}_{EOS,j}(\vec{r}_j) \tag{2}$$

$$j = 1,\ldots,N_{tp},$$

where $\vec{F}_{G,j}$ denotes the force on particle j due to gravity and $\vec{F}_{EOS,j}$ the force due to the equation of state. Gravity is modeled using the Newtonian monopole approximation:

$$\vec{F}_{G,j} = -G\,\frac{m_{tp}^2\,\#\left\{i \in \{1,\ldots,N_{tp}\} : |\vec{r}_i| < |\vec{r}_j|\right\}}{|\vec{r}_j|^3}\vec{r}_j. \tag{3}$$

This approximation is obviously only appropriate as long as the deviations from spherical symmetry are sufficiently small. The calculation of $\vec{F}_{EOS,j}$ will be explained below.

Grid

In order to be able to locally define thermodynamic quantities, most notably the local mass density, we introduced a spherical coordinate grid.

The boundaries between the cells of this grid are defined by the surfaces of constant r, constant ϕ, and constant θ in spherical coordinates (r,ϕ,θ) using the standard notation $(r = |\vec{x}|,\ \phi = $ azimuth angle, $\theta = $ polar angle). For the ϕ coordinate the boundaries are located at

$$0,\ 1 \times \frac{2\pi}{N_\phi},\ 2 \times \frac{2\pi}{N_\phi},\ldots,\ (N_\phi - 1) \times \frac{2\pi}{N_\phi},$$

where N_ϕ denotes the total number of boundaries (for the ϕ coordinate). For the θ coordinate the boundaries are chosen so that the difference between $\cos\theta$ of two arbitrary neighboring boundaries is constant, i.e.

$$|\cos\theta_1 - \cos\theta_2| = const.$$

if θ_1, θ_2 are neighboring θ-boundaries. We made this so to make the volume of a grid cell independent of θ. Finally, for the r coordinate we chose equidistant boundaries so these are located at

$$1 \times \frac{R}{N_r}, \ 2 \times \frac{R}{N_r}, \ldots, \ N_r \times \frac{R}{N_r} = R,$$

where R denotes the radial location of the edge of the spherical grid and N_r the number of boundaries for the r coordinate.

The grid cells are conveniently labeled by three integers $n_r \in \{1, \ldots, N_r\}$, $n_\phi \in \{1, \ldots, N_\phi\}$, $n_{\cos\theta} \in \{1, \ldots, N_{\cos\theta}\}$ that will be referred to as *grid coordinates* from now on. Here increasing n_X corresponds to increasing X for $X \in \{r, \phi, \cos\theta\}$.

Using this, the volume of a grid cell as a function of its grid coordinates is

$$Vol(n_r, n_\phi, n_{\cos\theta}) = Vol(n_r) = \frac{4\pi R^3}{3 N_\phi N_{\cos\theta}} \left\{ \left(\frac{n_r}{N_r}\right)^3 - \left(\frac{n_r - 1}{N_r}\right)^3 \right\}. \tag{4}$$

During a core collapse simulation an enormous change in the length scale occurs: the inner iron core contracts roughly by a factor of 10^2 to 10^3. Even if N_r is chosen very large[2] the inner core at core bounce would still be in a few innermost cells if the grid were fixed in space destroying an appropriate resolution of core bounce and other phenomena. That is why we chose to scale down the grid simultaneously with the core using a simple method described in [16].

Calculation of Densities

A mass density $\rho(n_r, n_\phi, n_{\cos\theta})$ can be calculated for each cell of the grid using the following evident way:

$$\rho(n_r, n_\phi, n_{\cos\theta}) = \frac{N_{tp}(n_r, n_\phi, n_{\cos\theta}) \, m_{tp}}{Vol(n_r)}, \tag{5}$$

where $N_{tp}(n_r, n_\phi, n_{\cos\theta})$ denotes the number of test particles in the cell.

In order to minimize errors due to fluctuations and to smooth the density distribution, we used a slightly more sophisticated method [16] in which the test particles' mass is smeared over the cell it is in and the seven (well-defined) neighboring cells which are located nearest to the test particle.

Calculation of Derivatives

To follow the dynamics of core collapse (more precisely: to calculate $\vec{F}_{EOS,j}$ from equation (2)), the calculation of (spatial) derivatives of thermodynamic quantities is necessary.

---

[2] Note that the total number of cells is clearly limited by the condition $N_{cells} \ll N_{tp}$.

Let Ω be a thermodynamic quantity defined on the grid (meaning that for each cell $(n_r, n_\phi, n_{\cos\theta})$ a real number $\Omega(n_r, n_\phi, n_{\cos\theta}) \in \mathbb{R}$ is defined). To approximate the gradient $\nabla\Omega(r, \phi, \theta)$, we use a modification of a standard technique for calculating numerical derivatives [17] and the well known expression for the gradient in spherical coordinates:

$$\nabla\Omega(r,\phi,\theta) = \frac{\partial\Omega}{\partial r}\Big|_{(r,\phi,\theta)}\vec{e}_r + \frac{1}{r\sin\theta}\frac{\partial\Omega}{\partial\phi}\Big|_{(r,\phi,\theta)}\vec{e}_\phi + (-\sin\theta)\frac{1}{r}\frac{\partial\Omega}{\partial(\cos\theta)}\Big|_{(r,\phi,\theta)}\vec{e}_\theta, \quad (6)$$

where $\vec{e}_r, \vec{e}_\phi, \vec{e}_\theta$ denote the orthonormal basis vectors for spherical coordinates. What remains to be done is the numerical definition of

$$\frac{\partial\Omega}{\partial r}\Big|_{(r,\phi,\theta)}, \quad \frac{\partial\Omega}{\partial\phi}\Big|_{(r,\phi,\theta)}, \quad \text{and} \quad \frac{\partial\Omega}{\partial(\cos\theta)}\Big|_{(r,\phi,\theta)}.$$

We will exemplarily describe our technique for $\frac{\partial\Omega}{\partial r}$. Let (r, ϕ, θ) be located in the cell with grid coordinates $(n_r, n_\phi, n_{\cos\theta})$. Then two obvious approximations for $\frac{\partial\Omega}{\partial r}$ are

$$\left(\frac{\partial\Omega}{\partial r}\Big|_{(r,\phi,\theta)}\right)_{right} = \frac{\Omega(n_r+1, n_\phi, n_{\cos\theta}) - \Omega(n_r, n_\phi, n_{\cos\theta})}{R/N_r} \quad (7)$$

$$\left(\frac{\partial\Omega}{\partial r}\Big|_{(r,\phi,\theta)}\right)_{left} = \frac{\Omega(n_r, n_\phi, n_{\cos\theta}) - \Omega(n_r-1, n_\phi, n_{\cos\theta})}{R/N_r}. \quad (8)$$

We decided to interpolate the derivative linearly between these two values:

$$\frac{\partial\Omega}{\partial r}\Big|_{(r,\phi,\theta)} = \frac{r-r_S}{r_B-r_S}\times\left(\frac{\partial\Omega}{\partial r}\Big|_{(r,\phi,\theta)}\right)_{right} + \frac{r_B-r}{r_B-r_S}\times\left(\frac{\partial\Omega}{\partial r}\Big|_{(r,\phi,\theta)}\right)_{left}, \quad (9)$$

where r_S and r_B denote the r-coordinates of the cell boundaries of cell $(n_r, n_\phi, n_{\cos\theta})$ with $r_S \le r \le r_B$. $\frac{\partial\Omega}{\partial\phi}$ and $\frac{\partial\Omega}{\partial\cos\theta}$ are in principle interpolated the same way.

Equation of State, Calculation of Thermodynamic Quantities

The equation of state (EOS) we used in this work needs the density ρ, temperature T, electron fraction Y_e, and the composition of the matter as input (the latter is given by the average charge and mass numbers of the present nuclei). A realistic calculation of these quantities is not nearly as simple in our model as that of the density.

We circumvented these problems by mapping data from the one dimensional simulations performed by Cooperstein and Wambach [18, 19] on our model. Their tabulated data enables us to define the temperature $T(\rho)$ and electron fraction $Y_e(\rho)$ as a function of density and should yield a good approximation for the state of the matter until core bounce.

We interpolated the data from Cooperstein's work using appropriate fit functions. The relation $T(\rho)$ that was finally used in our simulations is shown in figure 1 together with Cooperstein's values. The electron fraction $Y_e(\rho)$ was obtained using the same procedure.

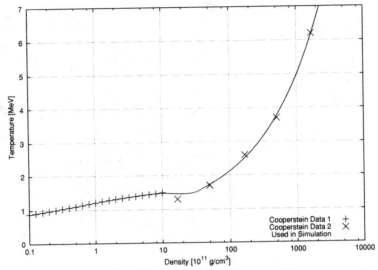

FIGURE 1. $T(\rho)$ for the infall phase

Helmholtz EOS and Lattimer & Swesty EOS, Calculation of $\vec{F}_{EOS,j}$

As a realistic EOS for core collapse conditions we used a combination of the nuclear EOS by Lattimer & Swesty [2, 1] and the Helmholtz EOS by Timmes (which is an EOS for the electron/positron gas) [20, 21]. The former is used for $\rho \geq 10^{11} \text{g/cm}^3$, the latter for $\rho < 10^{11} \text{g/cm}^3$ where the nuclear contribution to the pressure is negligible.

The input for the LS EOS is ρ, T, and Y_e. The Helmholtz EOS gets the composition of the matter instead of Y_e.

We mainly use the EOS to obtain $\vec{F}_{EOS,j}$ as introduced in equation (2). This is done as follows: the internal energy per baryon u_{int} is returned by the EOS for every grid cell. This is easily converted into an internal energy per test particle by multiplying with $m_{tp}/m_B =: \nu$, the number of baryons per test particle. We then calculate a gradient of this quantity at the location of test particle j using the technique described above. Then

$$\vec{F}_{EOS,j} = -\nu \, \nabla u_{int} \big|_{(r,\phi,\theta)_j} \qquad (10)$$

where $(r,\phi,\theta)_j$ are the spherical coordinates of test particle j's position.

This technique is justified by energy conservation: during the infall phase it is a good approximation to neglect energy losses due to neutrinos (and photons) radiating away from the core because the magnitude of these losses is relatively small. Thus the sum of the core's kinetic energy E_{kin}, gravitational energy E_G, and internal energy E_{int}

$$E_{tot} = E_{kin} + E_G + E_{int} \qquad (11)$$

should be approximately constant during collapse (after bounce this is no longer valid as neutrino losses cannot be neglected anymore). It can be shown [16] that our method of calculating $\vec{F}_{EOS,j}$ implies the (approximate) conservation of E_{tot}.

In our simulations, energy (and angular momentum) is kept track of to assure that the method just described works properly.

Symmetry Assumptions and Numerical Problems

To make the simulation code as efficient as possible, we decided to assume both equatorial and cylindrical symmetry on the grid (but not for the test particles) after not finding significant deviations from these in simulation runs in which these symmetries were not enforced.

Some other subtleties required our attention. For example, very low (and zero) densities are certain to appear in the grid at the outer edge of the iron core which is completely unrealistic because of the presence of the star's outer layers. Therefore a minimum density ρ_{min} is enforced throughout the grid by setting ρ to ρ_{min} wherever $\rho < \rho_{min}$.

Time Development

The time development of the collapsing core is obtained by numerically integrating the system of $2 \times 3 \times N_{tp}$ coupled first-order ordinary differential equations (1) and (2). This is done using a *fourth-order Runge-Kutta* algorithm, a standard method precisely described e.g. in [17]. The whole simulation program is written in C++.

Advantages and Weaknesses of this Method

A big advantage of the method described in the preceeding sections is its simplicity: it enables us to simulate the collapse of a rotating iron core in three space dimensions until core bounce on a quite ordinary home computer. Our code also conserves angular momentum. If the step size is chosen sufficiently small, energy conservation is given in good approximation.

The model described here is work in progress. At present, we have not yet implemented test particle collisions that lead to heating, stopping, and neutrino production. Thus our simulation is currently limited to the infall phase.

SIMULATION RUNS

Initial Conditions

It is desirable to use a realistic density profile of a presupernova core like that calculated with the stellar evolution code of Woosley and Weaver [22, 23]. However, due to problems caused by the enormous scale change that occurs during core collapse and because the effects of rotation should mainly play a role during the late stages of collapse (when angular momentum conservation leads to the highest angular velocities) we decided to use the following approach: the simulation is started when collapse is already going on and the (inner) core has already contracted by a factor of 5. Therefore we used a contracted variant of the mass distribution of the Woosley and Weaver progenitor as published in [22]. The fact that the matter is already falling inward is mimicked by imprinting an initial velocity profile $v(r) = k_v r$ on the test particles, where a realistic value $k_v = 89.1 \text{ s}^{-1}$ (estimated using the results of previous simulations [24]) is chosen. Since this is a good model only for the inner core, M_{IC} is chosen to be $0.7 M_\odot$ – a typical value for the inner core mass.

Since centrifugal forces are relatively weak during the early collapse stages the initial mass distribution is still assumed to be spherically symmetric.

Stellar evolution calculations for a rotating progenitor done by Heger [25] indicate that it is a very good approximation to assume that the inner core (initially) rotates like a rigid body. Therefore we used a constant initial angular velocity ω_0.

$N_{tp} = 10^6$ test particles, the grid parameters $N_r = 110$, $N_{\cos\theta} = 100$, and a background density $\rho_{min} = 1.3 \times 10^{11} \text{g/cm}^3$ were applied in all runs of the series. Table 1 shows the remaining data ($|\frac{E_{rot}}{E_G}|_{init}$ is the ratio of the initial rotational to the initial gravitational energy).

TABLE 1. Simulation runs using the combination of the Helmholtz and the LS EOS. The abbreviations are explained in the text.

Name of Run	ω_0 [$\frac{rad}{s}$]	$|\vec{L}_0|$ [10^{41}Js]	$|\frac{E_{rot}}{E_G}|_{init}$	t_{bounce} [ms]	ρ_{max} [ρ_0]
HLSEOS01	10	0.466	0.027%	3.44	2.99
HLSEOS07	70	3.26	1.3%	3.61	2.35
HLSEOS10	100	4.67	2.7%	3.76	1.63
HLSEOS13	130	6.06	4.5%	3.89	1.03
HLSEOS16	160	7.46	6.8%	4.03	0.56
HLSEOS19	190	8.86	9.6%	4.17	0.34
HLSEOS22	220	10.3	13%	4.31	0.21

Results

As the data obtained from the simulation runs of this series in table 1 shows, the time of core bounce t_{bounce} and the maximum density at bounce ρ_{max} depend on the initial angular momentum $|\vec{L}_0|$ in a strictly monotonous way. More rapid rotation obviously leads to a later core bounce and a reduced maximum compression of the core.

Figure 2 shows the development of the mass density distribution during the infall phase till shortly after bounce for models HLSEOS07 (slowly rotating), HLSEOS16 (moderately rotating), and HLSEOS22 (rapidly rotating).

The apparently different sizes of the density profiles are due to the grid scaling mentioned above (the same radius is used in all plots in this figure): the abrupt change between the light gray and the white regions is located at the radial edge of the grid.

An oblate shape (due to centrifugal forces) of the more rapidly rotating models is obvious in (b) and (c). It can also be seen that the mass distributions at core bounce get

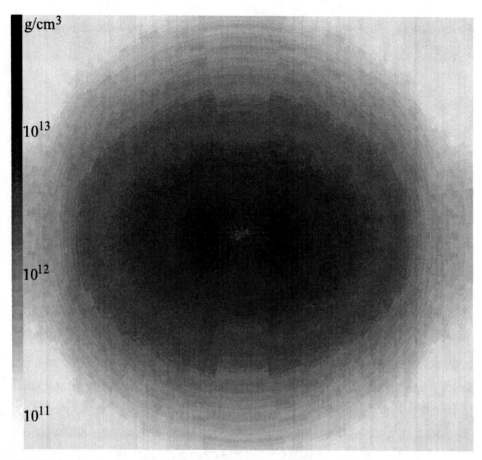

FIGURE 2. Mass density in a slice in the x-z-plane at five different times for models HLSEOS07 (left), HLSEOS16 (center), and HLSEOS22 (right). Events at the respective times: (a) onset of simulation, (b) after 2 ms, (c) presence of centrifugal forces becomes apparent (after 3 ms), (d) core bounce, (e) shortly after core bounce. The density scale is only approximately valid: the highest density (indicated by black) decreases from left to right. All plots have the same radius (123.42 km) indicated by the black line in the top left. See table 2 for more data.

more diffuse with increasing rotation (d). A slightly prolate shape shortly after bounce is apparent in model HLSEOS07. In the very rapidly rotating model HLSEOS22 a density depletion along the z-axis (vertical in the plot) can be seen shortly after bounce (figure 3 shows this in a magnified way[3]).

Another feature that can only be seen in the videos showing the full motion of core collapse for the rapidly rotating model HLSEOS22 (and also HLSEOS19 and

---

[3] Color versions of figures 2 and 3 can be found in [16] and illustrate the descibed feature much more clearly.

TABLE 2. Key for figure 2. t_a through t_e are the times (in ms) corresponding to the density profiles labeled (a) through (e) in the figure. $\rho_{sc.min}$ and $\rho_{sc.max}$ are the densities (in g/cm^3) corresponding to the bottom end (white) and top end (black) of the density key.

	HLSEOS07	HLSEOS16	HLSEOS22
t_a	0	0	0
t_b	2.00	2.00	2.00
t_c	3.00	3.00	3.00
t_d	3.61	4.03	4.31
t_e	3.80	4.51	5.33
$\rho_{sc.min}$	5.13×10^{10}	5.30×10^{10}	5.30×10^{10}
$\rho_{sc.max}$	6.15×10^{14}	1.48×10^{14}	5.93×10^{13}

HLSEOS16) is that significantly more mass rebounds from the core near the equatorial plane than along the rotation axis after bounce.

Vortices along the rotation axis that appeared during the infall phase in other models in which different EsOS were used [16] cannot be identified in this series.

POSSIBLE IMPLICATIONS OF THESE RESULTS

Our simulation succeeds in reproducing several results of previous ones which used different techniques of modeling rotating core collapse: we can confirm that faster rotation delays core bouce and an approximate value of $\rho_{max} \approx 3\rho_0$ (mentioned e.g. in [24, 26]) for the maximum density at bounce (for the slowly rotating models). More interesting is the validation that increasing rotation leads to lower densities at bounce (mentioned e.g. in [9, 27]). Like Zwerger [27] we find centrifugal bounces (occuring before nuclear matter density is reached) in our very rapidly rotating models.

An interesting new feature of the rapidly rotating models is the density depletion along the rotation axis shortly after core bounce and the fact that most of the mass is shot out near the equatorial plane after bounce. A highly speculative conclusion out of this is that the shockwave created after core bounce might be more vigorous in the equatorial plane. Thus the shock might not stall there and lead to an explosion that starts in this plane.

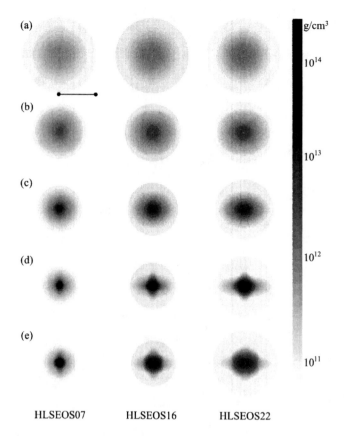

FIGURE 3. Density depletion along the z-axis after bounce in model HLSEOS22 at $t = 5.21$ms. The radius in this plot (measured from the center horizontally to the right) is 51.43km.

The density depletion along the rotation axis after bounce is extremely relevant for neutrino transport: it is evident that neutrinos trapped in the central core would be able to escape much easier by diffusing through the low density region along the rotation axis. Thus such a large neutrino flux might be created there that neutrino heating could lead to a jet-like explosion along this axis.

If neutrinos are predominantly emitted along the rotation axis, the effects of parity violation in the neutrino emission might lead to an asymmetry in the number of neutrinos emitted from the "north pole" relative to those emitted from the "south pole". This would lead to a recoil of a neutron star resulting from the explosion. Such a macroscopic *kick* has indeed been observed [28], and interpreted in terms of parity violation in neutrino emission [29, 30]. However, these calculations were performed for spherically symmetric systems, and the vortex effects noted by us will amplify this effect.

However, a lot more work needs to be done to underpin these preliminary results. Of particular importance is the implementation of collisions between individual test particles in order to follow the development after bounce. The inclusion of neutrino transport in our model is also most desirable if serious conclusions about the effects of neutrinos shall be drawn. After bounce also a more sophisticated way of mapping data obtained from lower dimensional simulations on our model is needed.

Both phenomena we just suggested are possible explanations for the polarization observations by Wang and Wheeler [10, 11, 12] because it is known that the light scattered from or through asymmetric surfaces is likely to be polarized.

ACKNOWLEDGMENTS

This work was supported by US National Science Foundation under grants PHY-0070818 and INT-9981342. W.B. acknowledges support from the A.v.Humboldt Foundation, and T.B. is grateful for support from the Studienstiftung des Deutschen Volkes.

REFERENCES

1. Lattimer, J., and Swesty, F., *Nucl. Phys. A*, **535**, 331–367 (1991).
2. Lattimer, J., and Swesty, F., Equation of state version 2.7 (ls eos v2.7), http://www.ess.sunysb.edu/dswesty/lseos.html.
3. Wilson, J., *Numerical Astrophysics*, Jones and Bartlett, Boston, 1985, p. 422.
4. Herant, M., Benz, W., Hix, W., Fryer, C., and Colgate, S., *Astrophys. J.*, **435**, 339–361 (1994).
5. Fryer, C., and Heger, A., *Astrophys. J.*, **541**, 1033–1050 (2000).
6. Mezzacappa, A., Liebendörfer, M., Messer, O., Hix, W., Thielemann, F.-K., and Bruenn, S., *Phys. Rev. Lett.*, **86**, 1935 (2001).
7. Janka, H.-T., *Astron. Astrophys.*, **368**, 527–560 (2001).
8. Herant, M., Benz, W., and Colgate, S., *Astrophys. J.*, **395**, 642–653 (1992).
9. Yamada, S., and Sato, K., *Astrophys. J.*, **434**, 268–276 (1994).
10. Wang, L., and Wheeler, J., *Sky and Telescope* (2002).
11. Wang, L., Howell, D., Höflich, P., and Wheeler, J., *Astrophys. J.*, **550**, 1030–1035 (2001).
12. Wang, L., Wheeler, J., Li, Z., and Clocchiatti, A., *Astophys. J.*, **467**, 435–445 (1996).
13. Wong, C.-Y., *Phys. Rev. C*, **25**, 1460–1475 (1982).
14. Bauer, W., Bertsch, G.F., Cassing, W., and Mosel, U., *Phys. Rev. C*, **34**, 2127 (1986).
15. Gong, W.G., Bauer, W., Gelbke, C.K., and Pratt, S., *Phys. Rev. C*, **43**, 781 (1991).
16. Bollenbach, T., *Numerical Study of Rotating Core Collapse Supernovae*, Master's thesis, Michigan State University, East Lansing, MI (2002).
17. Press, W., Teukolsky, S., Vetterling, W., and Flannery, B., *NUMERICAL RECIPES in C*, Press Syndicate of the University of Cambridge, Cambridge, UK, 1988, second edn.
18. Cooperstein, J., and Wambach, J., *Nucl. Phys. A*, **420**, 591–620 (1984).
19. Cooperstein, J., *Nucl. Phys. A*, **438**, 722–739 (1985).
20. Timmes, F., The helmholtz eos, http://flash.uchicago.edu/ fxt/eos.shtml.
21. Timmes, F., and Swesty, F., *Astrophys. J. Suppl. S.*, **126**, 501–516 (2000).
22. Woosley, S., and Weaver, T., *Ann. Rev. Astron. Astrophys.*, **24**, 205–253 (1986).
23. Weaver, T., Zimmermann, G., and Woosley, S., *Astrophys. J.*, **225**, 1021–1029 (1978).
24. Bethe, H., *Rev. Mod. Phys.*, **62**, 801–866 (1990).
25. Heger, A., Langer, N., and Woosley, S., *Astrophys. J.*, **528**, 368–396 (2000).
26. Baron, E., Cooperstein, J., and Kahana, S., *Nucl. Phys. A*, **440**, 744–754 (1985).
27. Zwerger, T., and Müller, E., *Astron. Astrophys.*, **320**, 209–227 (1997).
28. Lyne, A.G., and Lorimer, D.R., *Nature*, **369**, 127 (1994).
29. Horowitz, C. J., Gang Li, *Phys. Rev. Lett.*, **80**, 3694-3697 (1998); Erratum-ibid. **81**, 1985 (1998).
30. Horowitz, C. J., Piekarewicz, J., *Nucl. Phys. A*, **640**, 281-290 (1998).

Hyperon ordering in neutron star matter

L. Mornas*, J.P. Suárez Curieses* and J. Diaz Alonso*†

*Departamento de Física, Universidad de Oviedo, E-33007 Oviedo (Asturias) Spain
†LUTH, FRE2462 CNRS, Observatoire de Paris-Meudon, F92195 Meudon , France

Abstract. We explore the possible formation of ordered phases in neutron star matter. In the framework of a quantum hadrodynamics model where neutrons, protons and Lambda hyperons interact via the exchange of mesons, we compare the energy of the usually assumed uniform, liquid phase, to that of a configuration in which di-lambda pairs immersed in an uniform nucleon fluid are localized on the nodes of a regular lattice. The confining potential is calculated self-consistently as resulting from the combined action of the nucleon fluid and the other hyperons, under the condition of beta equilibrium. We are able to obtain stable ordered phases for some reasonable sets of values of the model parameters. This could have important consequences on the structure and cooling of neutron stars.

INTRODUCTION

The equation of state in the interior of neutron stars is usually considered to be that of an interacting Fermi liquid in a uniform phase. Under certain conditions however, a crystallized phase may be energetically more favorable. This in turn can have interesting consequences on the structure and evolution of the neutron stars. We may quote as examples thereof *(i)* triaxial configurations with emission of gravitational waves [1], *(ii)* modification of the oscillation modes of the star [2] or *(iii)* of the neutrino transport properties [3].

In a previous paper [4] we presented the results pertaining to a simplified model where we considered three baryonic species: the neutron, proton and Lambda hyperon, interacting with each other through the exchange of σ and ω mesons. We found that it was possible to find some sets of the model parameters such that the energy of a configuration where pairs of Λ hyperons with antiparallel spins located at the nodes of a cubic lattice, could be energetically more favorable than the corresponding liquid phase.

The model presented in [4] was a strictly minimal one. Here we would like to perform again these calculations with a more realistic description of the nuclear interaction. In order to reproduce the correct values of the incompressibility modulus and the asymmetry energy, we choose an approach based on the Density Dependent Hadron Field theory (DDRH) developped by the Giessen group [5, 6, 7]. In this model, the nucleon-meson vertices have a functional dependence on the density operator. In this way, some basic features of the short-range correlations are taken into account.

CP644, *Exotic Clustering: 4th Catania Relativistic Ion Studies*, edited by S. Costa, A. Insolia, and C. Tuvè
© 2002 American Institute of Physics 0-7354-0099-7/02/$19.00

MODEL OF BARYONIC MATTER

Lagrangian

We describe the nuclear interaction in the framework of a phenomenological quantum relativistic model. The interaction piece of the Lagrangian density in the Density Dependent Hadron Field Theory (DDRH) reads:

$$\mathscr{L}_{\text{int}} = \sum_{B=N,H} \Gamma_{\sigma B} \overline{\Psi}_B \, \sigma \, \Psi_B - \Gamma_{\omega B} \overline{\Psi}_B \, \gamma^\mu \, \omega_\mu \, \Psi_B + \Gamma_{\delta B} \overline{\Psi}_B \, \delta \, \Psi_B - \Gamma_{\rho B} \overline{\Psi}_B \, \gamma^\mu \rho_\mu \, \Psi_B \quad (1)$$

where the couplings $\Gamma_{\alpha B}$ are density-dependent through their functional dependence on the baryonic current J^μ.

$$\Gamma_{\alpha B} = \Gamma_{\alpha B}[\hat{J}^\mu \hat{J}_\mu] \quad \text{with} \quad \hat{J}^\mu = \sum_B \overline{\hat{\Psi}}_B \gamma^\mu \hat{\psi}_B \quad (2)$$

We will consider only one species of hyperons, the Λ. In a realistic model, it can be expected that the Σ hyperon will also play a role. However, the interaction of this hyperon with nuclear matter is less well known, so that it has not yet been implemented in the framework of a DDRH-type model.

Liquid phase

In the liquid phase we impose the usual conditions for charge neutrality and chemical equilibrium under β decay among the baryons (neutron, proton, Λ hyperon) and leptons (electron, muon) taken into account in our model.

$$n_e + n_\mu = n_p, \qquad \mu_n = \mu_\Lambda, \qquad \mu_p - \mu_n = \hat{\mu} = \mu_e = \mu_\mu$$

These equations determine the relative fraction of each species of particles. The equation of state is then obtained from the energy momentum tensor. For details we refer the reader to [5, 6, 7]. In particular we will be interested in comparing the energy density of the liquid phase $\rho_L = T^{00}$ to its counterpart in the crystallized phase.

Self consistent confining potential

We now assume that the hyperons are ordered on a lattice and are localized in gaussian clouds. We further assume that there are two hyperons with antiparallel spins per lattice site . This is partly in order to avoid complexities related to spin-spin interaction, but we can also justify this choice by arguing that the Λ-Λ interaction, as inferred from data on double hypernuclei, is attractive and may even favor a bound state ($\sim$ H-dibaryon). In the spirit of the Sommerfeld aproximation, we neglect the redistribution of the surrounding nucleons.

The potential energy of an hyperon around a lattice site is calculated as the sum of the interaction potential energies generated by hyperons at other lattice sites $\vec{r}_i = a(l\vec{i} + m\vec{j} + n\vec{k})$.

$$U_{\text{sup}}(\vec{r}) = \sum_i U(\vec{r} - \vec{r}_i) \tag{3}$$

In this expression, the potential energy of each hyperon is obtained by taking the convolution

$$U(\vec{r}) = \frac{2\pi}{r} \int_0^\infty R\, n(R)\, dR \int_{|r-R|}^{|r+R|} V^{OBE}(x)\, x\, dx \tag{4}$$

of a gaussian distribution

$$n(r) = 2\Psi_\Lambda^*(r)\Psi_\Lambda(r) = 2\left(\frac{M_\Lambda v_0}{\pi}\right)^{3/2} e^{-M_\Lambda v_0 r^2} \tag{5}$$

with the elementary one boson exchange potential $V^{OBE}(x)$ corrected for the finite size of hyperons by form factors (with cutoffs Λ_σ, Λ_ω) and for relativistic effects in the momentum expansion. Note that in the present configuration, only the central and spin-spin components contribute.

The potential calculated in this way is then approximated by a parabola U_{par} parameterized by its depth U_0 and the frequency of oscillation of the hyperons in the potential well $v_0 = \sqrt{\nabla^2 U_{\text{par}}(\vec{r})|_{\vec{r}=0}}/M_\Lambda$. The width $\Gamma = 1/\sqrt{M_\Lambda v_0}$ of the gaussian distribution (5) is determined self-consistently by the frequency of this harmonic oscillator.

Equation of state of the ordered phase

We again impose the conditions for β equilibrium

$$\mu_n = \mu_p + \mu_e, \quad \mu_n = \mu_\Lambda, \quad \mu_e = \mu_\mu \tag{6}$$

with the chemical potentials given by

$$
\begin{aligned}
\mu_p &= \sqrt{p_{fp}^2 + M_p^2} + \Gamma_{\omega N}(<\omega^0> + <\omega_\Lambda>_s) + \Gamma_{\rho N}<\rho^0> + \Sigma_{\text{rearr}}^N \\
\mu_n &= \sqrt{p_{fn}^2 + M_n^2} + \Gamma_{\omega N}(<\omega^0> + <\omega_\Lambda>_s) - \Gamma_{\rho N}<\rho^0> + \Sigma_{\text{rearr}}^N \\
\mu_\Lambda &= M_\Lambda + U_0 + \frac{5}{2}v_0 + \Gamma_{\omega\Lambda}<\omega^0> + \Sigma_{\text{rearr}}^\Lambda
\end{aligned}
\tag{7}
$$

The nucleons evolve in constant mean fields consisting of two contributions: To the field produced by the homogeneous nucleon fluid (e.g. ω^0), we add the spatial average of the fields generated by the periodic hyperon distribution (e.g. $<\omega_\Lambda>_s$). The chemical potential of the hyperons is now determined by the parameters of the confining potential. The rearrangement terms Σ_{rearr} contain derivatives of the couplings $\Gamma_{\alpha B}$ with respect to the density.

We further impose the conditions of charge neutrality $n_\mu + n_e = n_p$, and express the baryonic density $n_B = n_n + n_p + n_\Lambda$ as the sum of the nucleon densities and the hyperon density, the latter being related to the lattice parameter a by $n_\Lambda = 2/a^3$. Finally, we write the defining equations for the effective baryon masses and the self consistent equations obeyed by the fields σ, ω^0, ρ^0 and δ.

$$M_p = m_N - \Gamma_{\sigma N}(<\sigma> + <\sigma_\Lambda>_s) - \Gamma_\delta <\delta>$$
$$M_n = m_N - \Gamma_{\sigma N}(<\sigma> + <\sigma_\Lambda>_s) + \Gamma_\delta <\delta>$$
$$M_\Lambda = m_\Lambda - \Gamma_{\sigma\Lambda} <\sigma>$$

$$<\sigma> = \Gamma_{\sigma N}(n_p^{(s)} + n_n^{(s)}) = \frac{\Gamma_{\sigma N}}{m_\sigma^2} \sum_{i=p,n} \frac{M_i}{2\pi^2} \left[p_{fi}\varepsilon_{fi} - M_i^2 \ln \left(\frac{p_{fi} + \varepsilon_{fi}}{M_i} \right) \right]$$

$$<\delta> = \Gamma_{\delta N}(n_p^{(s)} - n_n^{(s)}), \qquad <\sigma_\Lambda>_s = \frac{\Gamma_{\sigma\Lambda}}{m_\sigma^2} n_\Lambda$$

$$<\omega^0> = \frac{\Gamma_{\omega N}}{m_\omega^2}(n_n + n_p) = \frac{\Gamma_{\omega N}}{m_\omega^2} \left(\frac{p_{fp}^3}{3\pi^2} + \frac{p_{fn}^3}{3\pi^2} \right)$$

$$<\rho^0> = \frac{\Gamma_{\rho N}}{m_\rho^2}(n_p - n_n), \qquad <\omega_\Lambda>_s = \frac{\Gamma_{\omega\Lambda}}{m_\omega^2} n_\Lambda \qquad (8)$$

The equations for chemical equilibrium (6),(7),(8) again determine the chemical composition of the matter at a given baryonic density. They have to be solved self consistently with the equations for the confining potential (3), to which they are interrelated through the effective hyperon mass, M_Λ the lattice parameter a, the depth U_0 and the oscillation frequency ν_0.

NUMERICAL RESULTS

We will use the rational parametrization of the coupling constants suggested in Ref. [8] with the parameters of Ref. [5]

$$\Gamma_{\alpha N}(n_B) = a_\alpha \left[\frac{1 + b_\alpha \left(\frac{n_B}{n_{sat}} + d_\alpha \right)}{1 + c_\alpha \left(\frac{n_B}{n_{sat}} + e_\alpha \right)} \right], \qquad \Gamma_{\alpha\Lambda}(n_B) = x_{\alpha\Lambda}\Gamma_{\alpha N}(n_B) \qquad (9)$$

This function is chosen so as reproduce parameter free Dirac-Brueckner calculations in the nucleon sector. The parameters $x_{\alpha B}$ are adjusted in order to reproduce the binding energy of the Λ in hypernuclei. The coupling constants used in the present work correspond to the "model 1 with DD phenomenological" parametrization of Reference [7]. We would like to stress the fact that, in contrast to the calculation performed in [4], we have essentially no free parameters in the results presented in this work.

When numerical convergence is reached, we obtain the parameters U_0, ν_0, M_i, p_{Fi}, etc. which fully determine the thermodynamical state of the system. We now are in a position to calculate the energy density of the crystallized phase. It is obtained as

$$\rho_C = \rho_p + \rho_n + \rho_e + \rho_\mu + \rho_\Lambda + \rho_{fields} \qquad (10)$$

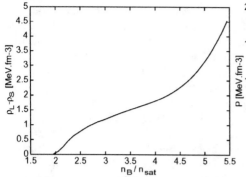

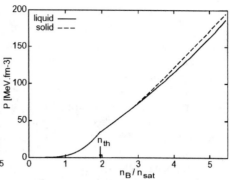

Figure 1. Energy difference between the liquid and crystallized phase, as a function of baryonic density

Figure 2. Pressure in the liquid and the crystallized phases, as a function of baryonic density.

$$\rho_\Lambda = \left(M_\Lambda + \frac{3}{2}v_0 + \Gamma_{\omega\Lambda} <\omega^0> + U_0\right) n_\Lambda$$

$$\rho_{\text{fields}} = \frac{\Gamma_{\sigma N}}{2}\left(<\sigma> + <\sigma_\Lambda>_s\right)\left(n_p^{(s)} + n_n^{(s)}\right) + \frac{\Gamma_{\sigma\Lambda}}{2} <\sigma><n_\Lambda^{(s)}>_s + \frac{\Gamma_{\delta N}}{2} <\delta>\left(n_p^{(s)} - n_n^{(s)}\right)$$

$$+ \frac{\Gamma_{\omega N}}{2}\left(<\omega^0> + <\omega_\Lambda>_s\right)\left(n_n + n_p\right) + \frac{\Gamma_{\omega\Lambda}}{2} <\omega^0><n_\Lambda>_s + \frac{\Gamma_{\rho N}}{2} <\rho^0>\left(n_p - n_n\right)$$

This energy is then compared to the energy density of the corresponding liquid phase ρ_L. The result is displayed in Figure 1. With the chosen model parameters we obtained an ordered phase which is energetically more favorable than the liquid one above the threshold for hyperon production $n_{\text{th}} = 1.95\,n_{\text{sat}}$. The shape of this curve is similar to the one which was obtained for parameter set C in [4], with however a somewhat smaller energy gain. The stability of the solid phase increases at high density.

Finally, we obtain the pressure from the thermodynamical relation

$$P = n_B^2 \frac{\partial(\rho/n_B)}{\partial n_B} \tag{11}$$

The result is displayed in Figure 2. In contrast with the results obtained for parameter set C in [4], the equation of state of the ordered phase is now slighlty harder than the liquid one.

CONCLUSION

We investigated the possible formation of an ordered phase of the baryonic matter present in the core of the neutron stars, in which the hyperons are located on the nodes of a cubic lattice. The model presented in this short contribution is a first upgrade of our original barebones model [4]. We now have a more realistic description of the nuclear interaction, in particular of the nucleon sector. Our main conclusion is that the findings of [4] are confirmed, namely the ordered phase can be energetically more favorable for some choices of the model parameters. As could be expected, our result is found to depend more strongly on the features of the – largely unknown – in-medium hyperon-hyperon potential than on the properties of the nucleon background such as incompressibility modulus and asymmetry. A more systematic study with several alternative parametrizations is underway and will be presented in a further publication.

The model which we have presented here can be improved in several ways. In order to be consistent with the DDRH picture, we described the hyperon-hyperon interaction by sigma and omega exchange with density dependent couplings in the form $\Gamma_{\alpha\Lambda}(n_N, n_\Lambda) = x_{\alpha\Lambda}\Gamma_{\alpha N}(n_N + n_\Lambda)$ with $\alpha \in \{\sigma, \omega\}$. On the other hand, some authors [9] also introduce additional $\sigma*$ and ϕ exchange in order to reproduce the attractive $\Lambda - \Lambda$ interaction inferred from the available data on double hypernuclei. Furthermore, parametrizations of the free hyperon-nucleon and hyperon-hyperon potentials consider the exchange of e.g. kaons or η which are also not present in our calculations.

Ideally one could extract an in-medium hyperon-hyperon potential from Brueckner-type calculations and test the result comparing with the available hypernuclei data. Whereas several non relativistic calculations exist in the litterature [10, 11], the corresponding calculations have not been performed in a relativistic framework. In particular a full fledged relativistic calculation including all hyperons (not only Λ but also Σ and Ξ) and a subsequent DDRH-type parametrization would be especially welcome. It would allow us to have a greater variety of lattice configurations involving in particular the Σ^- hyperon. Indeed the Σ^- is formed in the liquid phase at the same density as the Λ in most neutron star matter calculations.

An other direction of research would be to release the Sommerfeld approximation and take into account the redistribution of nucleons. This involves considering screening correlations of the RPA type to the potential.

Acknowledgment: This work was partially supported by project MCT-00-BFM-0357.

REFERENCES

1. Haensel, P., "," in *Relativistic Gravitation and Gravitational Radiation*, edited by J.-A. Marck and J.-P. Lasota, Proceedings of Les Houches School of Physics, Cambridge University Press, 1997.
2. Bonazzola, S., *Private communication* (2002).
3. Baiko, D.-A., and Haensel, P., *Acta Phys. Polon. B*, **30**, 1097 (1999).
4. Pérez-Garcia, M.-A., Corte-Rodríguez, N., Mornas, L., Suárez-Curieses, J.-P., and Diaz-Alonso, J., *Nucl. Phys. A*, **699**, 939 (2002).
5. Hofmann, F., Keil, C., and Lenske, H., *nucl-th/0007050* (2000).
6. Keil, C., Hofmann, F., and Lenske, H., *Phys. Rev. C*, **61**, 064309 (2000).
7. Hofmann, F., Keil, C., and Lenske, H., *Phys. Rev. C*, **64**, 025804 (2001).
8. Typel, S., and Wolter, H., *Nucl. Phys. A*, **656**, 331 (1999).
9. Schaffner, J., Dover, C., Gal, A., Millener, D., Greiner, C., and Stöcker, H., *Ann. Phys. (N.Y.)*, **235**, 35 (1994).
10. Baldo, M., Burgio, G.-F., and Schulze, H.-J., *Phys. Rev.*, **61**, 055801 (2000).
11. Vidaña, I., Polls, A., Ramos, A., Engvik, L., and Hjorth-Jensen, M., *Phys. Rev.*, **62**, 035801 (2000).

Quark Matter Formation in Neutron Stars and Implications for Gamma-Ray Bursts

Z. Bereziani*, I. Bombaci†, A. Drago**, F. Frontera‡ and A. Lavagno§

*Dipartimento di Fisica, Università di L'Aquila & INFN Sez. del Gran Sasso, I-67010 Coppito, Italy
†Dipartimento di Fisica "E. Fermi", Università di Pisa & INFN Sez. di Pisa, I-56127 Pisa, Italy
**Dipartimento di Fisica, Università di Ferrara & INFN Sez. di Ferrara, I-44100 Ferrara, Italy
‡Dipartimento di Fisica, Università di Ferrara, I-44100 Ferrara, Italy & IASF, CNR, I-40129 Bologna, Italy
§Dipartimento di Fisica, Università di Torino, Politecnico di Torino & INFN Sez. di Torino, I-44100 Ferrara, Italy

Abstract. We propose a model to explain how a Gamma Rays Burst can take place days or years after a supernova explosion. Our model is based on the conversion of a pure hadronic star (neutron star) into a star made at least in part of deconfined quark matter. The conversion process can be delayed if the surface tension at the interface between hadronic and deconfined-quark-matter phases is taken into account. The nucleation time (*i.e.* the time to form a critical-size drop of quark matter) can be extremely long if the mass of the star is small. Via mass accretion the nucleation time can be dramaticaly reduced and the star is finally converted into the stable configuration. A huge amount of energy, of the order of 10^{52}–10^{53} erg, is released during the stellar conversion and can produce a powerful Gamma Ray Burst. The delay between the supernova explosion generating the metastable neutron star and the new collapse can explain the delay proposed in GRB990705 [1] and in GRB011211 [2].

INTRODUCTION

The recent discovery of redshifted Fe K-lines in the X-ray afterglow of some Gamma Ray Bursts (GRBs) [1–5] gives a very strong indication in favour of an association between GRBs and Supernova (SN) explosions. Particularly, in the case of the gamma ray burst of July 5, 1999 (GRB990705) and in the case of GRB011211, it has been possible to estimate the time delay between the two events. For GRB990705 the supernova explosion is evaluated to have occurred about 10 years before the GRB [1], while for GRB011211 about four days before the burst [2].

The scenario which emerges from these findings is the following two-stage scenario: (i) the first event is the supernova explosion which forms a compact stellar remnant, *i.e.* a neutron star (NS); (ii) the second catastrophic event is associated with the NS and it is the energy source for the observed GRB.

These new observational data, and the scenario outlined above, poses severe problems for most of the current theoretical models for the central energy source (the so called "central engine") of GRBs. The main difficulty of all these models is to give an answer to the following questions: what is the origin of the second "explosion"? How to explain the long time delay between the two events?

CP644, *Exotic Clustering: 4th Catania Relativistic Ion Studies*, edited by S. Costa, A. Insolia, and C. Tuvè
© 2002 American Institute of Physics 0-7354-0099-7/02/$19.00

In the so called *supranova* model [6] for GRBs the second catastrophic event is the collapse to a black hole of a supramassive neutron star, *i.e.* a fast rotating NS with a baryonic mass M_B above the maximum baryonic mass for non-rotating configurations. In this model, the time delay between the SN explosion and the GRB is equal to the time needed by the fast rotating newly formed neutron star to get rid of angular momentum and to reach the limit for instability against quasi-radial modes where the collapse to a black hole occurs (see *e.g.* [7]). Questions concerning both the duration of the burst and its energy have been raised by several authors [8,9], suggesting that in the supranova model the burst would turn out to be too short and too weak.

In the present work we propose an alternative model to explain the GRB–SN association and in particular the long time delay inferred for GRB990705 and GRB011211. In our model the second explosion is related to the conversion from a metastable, purely hadronic star (neutron star) into a more compact star in which deconfined quark matter (QM) is present.

This possibility has already been discussed in the literature [10–12]. The new and crucial idea we introduce here, is the metastability of the purely hadronic star due to the existence of a non-vanishing surface tension at the interface separating hadronic matter from quark matter. The *mean-life time* of the metastable NS can then be connected to the delay between the supernova explosion and the GRB. As we shall see, in our model we can easily obtain a burst lasting a few tens of seconds, in agreement with the observations. The order of magnitude of the energy released is also the appropriate one to power a GRB.

COMPACT STARS

Quantum Chromodynamics (QCD), as the fundamental theory of strong interactions, predict the transition to a deconfined quark phase to occur at a density of a few times nuclear matter saturation density ($\rho_0 \sim 2.8 \times 10^{14} \text{g/cm}^3$). The search of this new phase of matter is one of the main goal in heavy ion physics [13]. Experiments at Brookhaven National Lab's Relativistic Heavy Ion Collider (RHIC) and at CERN's Large Hadron Collider (LHC), will hopefully clarify this issue in the near future. The core of a neutron star is one of the best candidate in the universe where a deconfined quark phase of matter could be found.

Various possibilities have been discussed in the literature, concerning the theoretical picture of the internal structure of the compact stars usually called "neutron stars". Depending on the microscopic properties of ultradense matter, and particularly on the possibility to have deconfined quark matter in the star, it is possible to have three different classes of compact stars: (a) purely Hadronic Stars (HS), in which below the usual neutron star crust one has a layer of neutron-rich nuclear matter in beta-equilibrium with electrons and muons, and possibly an inner core containing hyperons or a condensate of negative kaons in addition to the particles mentioned above. In these compact stars no fraction of quark matter is present; (b) Hybrid Stars (HyS), which possess a "quark core" either as a mixed phase of deconfined quarks and hadrons or as a pure quark matter phase; (c) Quark Stars (QS), which can be realized as pure

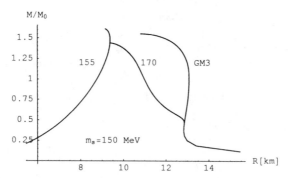

FIGURE 1. The mass-radius relation for different types of compact stars. The curve labelled GM3 is relative to a pure hadronic star described by the GM3 EOS with hyperons. The curve labelled 170 is relative to an hybrid star in which the quark matter sector is described by the MIT bag model with $B^{1/4} = 170$ MeV and $m_s = 150$ MeV. Finally the curve labelled 155 is a pure quark star. Stellar masses are plotted in unit of the solar mass.

u,d,s strange quark matter stars (the so called Strange Stars) satisfying the Bodmer–Witten hypothesis [14,15], or as stars which above a strange quark matter core possess a mantle of mixed quark–hadron phase [16]. Compact stars of each of the three classes could be endowed with strong magnetic fields and could manifest their presence in the universe as pulsars or as compact X-ray sources in binary systems. A sizeable amount of observational data collected by the new generations of X-ray satellites, is providing a growing body of evidence for the existence of quark stars [17–22].

QUANTUM NUCLEATION OF QUARK MATTER IN HADRONIC STARS

In our scenario, we consider a purely hadronic star whose central density (pressure) is increasing due to spin-down or due to mass accretion (from a companion star or from the interstellar medium). As the central density approaches the quark deconfinement critical density, a virtual drop of quark matter can be formed in the central region of the star. The quantum fluctuations of a spherical droplet of quark matter having a radius R are regulated by a potential energy of the form [23]

$$U(R) = \frac{4}{3}\pi R^3 n_q (\mu_q - \mu_h) + 4\pi\sigma R^2 \tag{1}$$

where n_q is the quark baryon density, μ_h and μ_q are the hadronic and quark chemical potentials at a fixed pressure P and σ is the surface tension for the surface separating the quark phase from the hadronic phase. For the sake of simplicity and to make our present discussion more transparent, in the previous expression we neglect the so called curvature energy [24] and the terms connected with the Coulomb energy [25,26]. The inclusion of these terms will not modify the general physical picture of our model and the conclusion of the present study [27]. The value of the surface tension σ is poorly known, and typical values used in the literature range within 10–50 MeV/fm^2 [25,26,28,29].

The process of formation of a bubble having a critical radius, can be computed using a semiclassical approximation. The procedure is rather straightforward. First one computes using the well known Wentzel–Kramers–Brillouin (WKB) approximation the ground state energy E_0 and the oscillation frequency ν_0 of the virtual QM drop in the potential well $U(R)$. Then it is possible to calculate in a relativistic framework the probability of tunneling as [26]

$$p_0 = \exp\left[-\frac{A(E_0)}{\hbar}\right]$$

(2)

where A is the action under the potential barrier

$$A(E) = \frac{2}{c}\int_{R_-}^{R_+}\left\{[2\mathcal{M}(R)c^2 + E - U(R)][U(R) - E]\right\}^{1/2}dR.$$

(3)

Here $R_\pm$ are the classical turning points and

$$\mathcal{M}(R) = 4\pi\rho_h\left(1 - \frac{n_q}{n_h}\right)^2 R^3$$

(4)

is the droplet effective mass, ρ_h being the hadronic mass density. n_h and n_q are the baryonic number densities at a same and given pressure in the hadronic and quark phase, respectively. The nucleation time is then equal to

$$\tau = (\nu_0 p_0 N_c)^{-1},$$

(5)

where N_c is the number of virtual centers of droplet formation in the star. N_c is of the order of 10^{48} [26].

In our analysis we have adopted rather common models for describing both the hadronic and the quark phase of dense matter. For the hadronic phase we used models which are based on a relativistic lagrangian of hadrons interacting via the exchange of pion, sigma, rho and omega mesons. The parameters adopted are the standard ones [30]. Hereafter we refer to this model as the GM equation of state (EOS). For the quark phase we have adopted a phenomenological EOS [31] which is based on the MIT bag model for hadrons. The parameters here are: the mass m_s of the strange quark, the so-called pressure of the vacuum B (bag constant) and the QCD structure constant.

RESULTS

We show in Fig. 1 the typical mass-radius relations for the three possible types of compact stars. As it appears, stars having a deconfined quark content (HyS or QS) are more compact than purely hadronic stars (HS). In our scenario a metastable HS having, e.g., a mass of $1.3\,M_\odot$ and a radius of ~ 13 km can collapse into an HyS having a radius of ~ 9.5 km or into a QS with radius ~ 8.5 km. The nature of the stable configuration reached after the stellar conversion (*i.e.* an HyS or a QS) will depend on the parameters of the quark phase EOS.

The nucleation time, *i.e.* the time needed to form a critical droplet of deconfined quark matter can be calculated for different values of the stellar central pressure P_c which enters in the expression of the energy barrier in eq. (1). The nucleation time can be plotted as a function of the gravitational mass M_{HS} of the HS corresponding to the given value of the central pressure. The result of our calculations for a specific EOS of dense matter are reported in Fig. 2, where each curve refer to a different value of the bag constant. As we can see, from the results in Fig. 2, an hadronic star can have a mean-life time many orders of magnitude larger than the age of the universe ($T_{univ} = 10$–20×10^9 yr $= 3.1$–6.3×10^{17} s). As the star accretes a small amount of mass, the consequential increase of the central pressure lead to a huge reduction of the nucleation time and, as a result, to a dramatic reduction of the HS mean-life time.

Once a critical-size droplet of quark matter is formed, it is possible to calculate its rate of growth [10,32]. The QM front absorbs hadrons liberating the constituent quarks. As the "chemical" equilibrium is re-established by the weak-interaction processes of the type $u + e^- \rightarrow s + \nu_e$, a large amount of neutrinos and antineutrinos are produced in the stellar core. Whether the conversion process occurs in a deflagration regime (either laminar or turbolent) or in a detonation regime (thus driving an explosive transient) has been debated in the literature [10,32], but detailed studies are still lacking. Available studies [10,32] indicate that the stellar conversion process occurs in a very short time, in the range of 10^{-2}–10^2 s, depending on the speed of the "burning front" of hadronic matter into quark matter.

Next we consider the total energy E^{conv} released in the conversion from a metastable hadronic star to an hybrid star or a quark star (the final state of such transition depends on the details of the quark matter EOS and in particular on the value of the bag parameter). The energy gain is calculated as a difference of the gravitational mass between the hadronic and hybrid-quark star at the same baryonic number [12]. In Table 1 we report $E^{conv}(\tau)$ and the corresponding "critical mass" $M_{cr}(\tau)$ of the metastable hadronic star, (taking $\tau = 1$ yr) when an hybrid star is formed in the conversion process. E^{conv} is a few 10^{52} erg. An energy of the order of 10^{53} erg is liberated when a quark star is formed [12]. Thus the total energy liberated in the stellar conversion process is in the range 10^{52}–10^{53} erg. This tremendous amount of energy, mainly originates from the energy liberated in the deconfinement phase transition [12].

To generate a strong GRB, an efficient mechanism to transfer the energy released in the stellar conversion into an electron-photon plasma is needed. In earlier work it was this difficulty that hampered the possibility to connect GRBs and the hadronic-quark matter phase transition in compact stars. Only more recently it was noticed [33] that near the surface of a compact stellar object, due to general relativity effects, the efficiency of the neutrino-antineutrino annihilation into e^+e^- pairs is strongly enhanced with respect to the Newtonian case. The efficiency of the neutrino pair annihilation into e^+e^- pairs could be as high as 10% [33]. The total energy deposited into the electron-photon plasma can therefore be of the order of 10^{51}–10^{52} erg. We must also mention that a more effective way to generate photons and/or e^+e^- pairs have been proposed in the literature, based on the decay of axion-like particles [34]. This mechanism, would

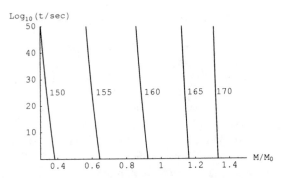

FIGURE 2. Quark matter nucleation time as a function of the gravitational mass of the hadronic star for $\sigma = 30$ MeV/fm^2. Each curve refer to a different value of $B^{1/4}$ (MeV). The strange quark mass is taken equal to 150 MeV. The hadronic sector is described by the GM3 EOS with hyperons.

TABLE 1. Critical mass M_{cr} of the metastable hadronic star (in unit of the mass of the sun $M_\odot = 1.989 \times 10^{33}$ g) and energy released E^{conv} in the conversion to hybrid star assuming the hadronic star mean life time τ equal to 1 year. Results are reported for various choices of the surface tension σ and of the bag constant B. The strange quark mass is taken equal to 150 MeV. For the hadronic matter EOS the GM3 model with hyperons has been used.

$B^{1/4}$ (MeV)	σ (MeV/fm^2)	$M_{cr}/M_\odot$	E^{conv} (10^{51}erg)
170	20	1.25	30.0
170	30	1.33	33.5
170	40	1.39	38.0
165	30	1.15	38.6
160	30	0.91	45.7

have an extremely large efficiency and would transfer into the GRB most of the energy released in the stellar conversion.

It is also important to stress that in our scenario the duration of the burst is related both to the speed of the "burning front" of hadronic matter to quark matter, discussed above, and to the time during which neutrinos are trapped in the dense stellar material. This *neutrino trapping time* is of the order of a few tens of seconds [35]. Therefore the duration of the burst turns out to be of the order of a few tens of seconds in a very natural way, in agreement with observations for "long" ($t_{GRB} > 10$ s) GRBs.

The strong magnetic field of the compact star will affect the motion of the electrons and positrons, and in turn could generate a moderate anisotropy in the GRB. Moreover, it has been recently shown [36] that the stellar magnetic field could influence the velocity of the "burning front" of hadronic matter into quark matter. This results in a strong geometrical asymmetry of the forming quark matter core along the direction of the stellar magnetic axis, thus providing a suitable mechanism to produce a collimated GRB [36]. Other anisotropies in the GRB could be generated by the rotation of the star.

DISCUSSION AND CONCLUSIONS

The nature of the compact star at birth (*i.e.* HS or HyS/QS) will depend on the value of its initial mass M_{in} with respect to the value of the critical mass M_{cr} for the stellar conversion. While the value of M_{cr} is set by the fundamental theory of strong interactions (EOS, QM-HM surface tension), the value of M_{in} is determined by stellar evolution and is related to the value of the mass of the progenitor star. Due to many uncertainties in the stellar evolution modelling, the so called *neutron star initial mass function* (*i.e.* the number of neutron stars as a function of their mass at birth) is not known with high accuracy. For those stars that explode as Type II supernovae, the models give [37] M_{in} in the range $1.2 - 1.8\ M_\odot$ (stellar remnants in the upper part of this range could directly form a black hole). If $M_{in} > M_{cr}$ then the quark deconfinement phase transition will occur during the stellar core bounce in the presupernova collapse, or within a few seconds after the core bounce. In this case the energy liberated during the quark deconfinement phase transition will help the supernova to explode [38]. An hybrid star or a quark star is formed directly in the supernova event. If $M_{in} < M_{cr}$ the compact remnant left by the supernova explosion is a pure HS. This star can live in a metastable state with a mean-life time which depends on the value of its central density. Eventually, as discussed in the present work, the star will be converted to a quark star or to an hybrid star and could originate a GRB.

In conclusion, we propose the following origin for (at least some of) the GRBs having a duration of tens of seconds. They can be associated with the conversion from a metastable pure hadronic star to a more compact hybrid star or quark star. The time delay between the supernova explosion originating the hadronic star and the GRB is regulated by those processes which increase the central pressure of the star (matter accretion and/or spin-down of the HS).

There are various specific signatures and interesting astrophysical consequences of the mechanism we are proposing. First of all, two different families of compact stars exist in nature: pure hadronic stars (metastable) which have radii in the range of $12 - 20$ km (as in the case [39] of the compact star 1E 1207.4-5209, assuming $M = 1.4\ M_\odot$), and hybrid or quark stars with radii in the range of $6 - 8$ km (see ref.s [17,19,20,22]. In other words, the existence of quark stars (or hybrid stars) does not conflict with the existence of pure hadronic stars. Second, all the GRBs generated by the present mechanism should have the same energy.

REFERENCES

1. L. Amati et al., Science 290 (2000) 953.

2. J.N. Reeves et al., Nature 414 (2002) 512.

3. J.S. Bloom et al., Nature 401 (1999) 453.

4. L. Piro et al., Science 290 (2000) 955.

5. L.A. Antonelli et al., Astrophys. J. 545 (2000) L39.

6. M. Vietri and L. Stella, Astrophys. J. 507 (1998) L45.

7. B. Datta, A.V. Thampan and I. Bombaci, Astron. and Astrophys. 334 (1998) 943.

8. M.J. Rees and P. Mészáros, Astrophys. J. 545 (2000) L73.

9. M. Böttcher and C.L. Fryer, Astrophys. J. 547 (2001) 338.

10. A. Olinto, Phys. Lett. B 192 (1987) 71.

11. K.S. Cheng and Z.G. Dai, Phys. Rev. Lett. 77 (1996) 1210.

12. I. Bombaci and B. Datta, Astrophys. J. 530 (2000) L69.

13. *Quark Matter '99*, edited by L. Riccati, M. Masera and E. Vercellin, Nucl. Phys. A 661 (1999).

14. A. R. Bodmer, Phys. Rev. D 4 (1971) 1601.

15. E. Witten, Phys. Rev. D 30 (1984) 272.

16. A. Drago and A. Lavagno, Phys. Lett. B 511 (2001) 229.

17. I. Bombaci, Phys. Rev. C 55 (1997) 1587.

18. K.S. Cheng, Z.G. Dai, D.M. Wei and T. Lu, Science 280 (1998) 407.

19. X.D. Li, I. Bombaci, M. Dey, J. Dey and E.P.J. van den Heuvel, Phys. Rev. Lett. 83 (1999) 3776.

20. X.D. Li, S. Ray, J. Dey, M. Dey and I. Bombaci, Astrophys. J. 527 (1999) L51.

21. R.X. Xu, Astrophys. J. 570 (2002) L65.

22. J.J. Drake et al., arXiv:astro-ph/0204159.

23. I.M. Lifshitz and Yu. Kagan, Zh. Eksp. Teor. Fiz. 62 (1972) 385 [Sov. Phys. JETP 35 (1972) 206].

24. J. Madsen, Phys. Rev. Lett. 70 (1993) 391, Phys. Rev. D 47 (1993) 5156.

25. H. Heiselberg, C.P. Pethick and E.F. Staubo, Phys. Rev. Lett. 70 (1993) 1355.

26. K. Iida and K. Sato, Phys. Rev. C 58 (1998) 2538.

27. Z. Berezhiani, I. Bombaci, A. Drago, F. Frontera and A. Lavagno, in preparation.

28. M.L. Olesen and J. Madsen, Phys. Rev. D (1994) 2698.

29. M.S. Berger and R.L. Jaffe, Phys. Rev. C 35 (1987) 213.

30. N.K. Glendenning and S.A. Moszkowski, Phys. Rev. Lett. 67 (1991) 2414.

31. E. Farhi and R.L. Jaffe, Phys. Rev. D 30 (1984) 2379.

32. J.E. Horvath and O.G. Benvenuto, Phys. Lett. B 213 (1988) 516.

33. J.D. Salmonson and J.R. Wilson, Astrophys. J. 517 (1999) 859.

34. Z. Berezhiani and A. Drago, Phys. Lett. B473 (2000) 281.

35. M. Prakash, I. Bombaci, M. Prakash, P.J. Ellis, J.M. Lattimer and R. Knorren, Phys. Rep. 280 (1997) 1.

36. G. Lugones, C.R. Ghezzi, E.M. de Gouveia Dal Pino and J.E. Horvath, Astrophys. J. (in press); arXiv:astro-ph/0207262.

37. F.X. Timmes, S.E. Woosley and T.A. Weaver, Astrophys. J. 457 (1996) 834.

38. O.G. Benvenuto and J.E. Horvath, Phys. Rev. Lett. 63, (1989) 716.

39. D. Sanwal, G.G. Pavlov, V.E. Zavlin and M.A. Teter, Astrophys. J. Lett. (in press); astro-ph/0206195.

Effects of quark deconfinement on the maximum mass of neutron stars

G. F. Burgio*, M. Baldo,* and H.-J. Schulze*

*INFN Sezione di Catania, 57 Corso Italia, I-95129 Catania, Italy

Abstract. We study the consequences of the hadron-quark phase transition for the values of the maximum mass of neutron stars (NS's). For the hadronic phase, we use an equation of state (EOS) derived within the Brueckner–Bethe–Goldstone formalism with realistic two-body and three-body forces. For quark matter we employ the MIT bag model with a density dependent bag parameter. We calculate the structure of NS interiors with the EOS comprising both phases, and we find that the NS's maximum masses fall in a relatively narrow interval, $1.4 M_\odot \lesssim M_{max} \lesssim 1.7 M_\odot$, the precise value being only weakly correlated with the value of the energy density at the assumed transition point in nearly symmetric nuclear matter.

INTRODUCTION

The fundamental input for building models of neutron stars is the equation of state of nuclear matter up to very large values of density [1]. In fact, as widely known, the matter in the NS's core possesses densities ranging from a few times ρ_0 (≈ 0.17 fm^{-3}, the normal nuclear matter density) up to one order of magnitude higher. In this range, a description of matter only in terms of nucleons and leptons may be inadequate, and one should take into account the appearance of several species of other particles, such as hyperons, or the transition to a quark-gluon plasma [2], although the exact value of the transition density to deconfined quark matter is still unknown.

Recently, we have proposed to constrain the maximum mass of neutron stars taking into account the phase transition from hadronic matter to quark matter inside the neutron star [3]. For this purpose, we describe the hadronic phase of matter by using a microscopic equation of state, obtained in the Brueckner–Bethe–Goldstone (BBG) theory [4], whereas the deconfined quark phase is treated within the popular MIT bag model [5]. The bag constant, B, which is a parameter of this model, is constrained to be compatible with the recent experimental results obtained at CERN on the formation of a quark-gluon plasma [6], recently confirmed by RHIC preliminary results [7]. However, whereas the possible quark-gluon plasma produced in heavy ion collisions is characterized by small baryon density and high temperature, the quark phase in neutron stars appears at high baryon density and low temperature. Within the original MIT bag model, if one adopts for the hadronic phase a non-interacting gas model of nucleons, antinucleons, and pions, then the deconfined phase is predicted to take place at an almost constant value of the quark-gluon energy density, irrespective of the thermodynamical conditions of the system [8]. For this reason, the transition line between the hadronic and quark phase is commonly drawn at a constant value of the energy density ε_Q. For that, CERN experi-

CP644, *Exotic Clustering: 4th Catania Relativistic Ion Studies*, edited by S. Costa, A. Insolia, and C. Tuvè
© 2002 American Institute of Physics 0-7354-0099-7/02/$19.00

ments reported $\varepsilon_Q \simeq 1.1 \, \text{GeV fm}^{-3}$.

We assume that the quark deconfinement takes place at this same value of the energy density even when correlations in the hadronic phase are present. Therefore, all our calculations can be limited to zero temperature. We will then study the predictions that one can draw from this hypothesis on NS's structure. Any observational data on neutron stars in disagreement with these predictions would give an indication on the accuracy of this assumption. In this way, we hope to directly relate the physics of the hadron-quark phase transition observed in high energy heavy ion collisions and the one in neutron star matter. For more details, the reader is referred to Refs[3, 9].

THEORETICAL FRAMEWORK

Hadronic phase

We start with the description of the hadronic phase. The Brueckner–Bethe–Goldstone (BBG) theory is based on a linked cluster expansion of the energy per nucleon of nuclear matter (see Ref. [4], chapter 1 and references therein). The basic ingredient in this many-body approach is the Brueckner reaction matrix G, which is the solution of the Bethe–Goldstone equation

$$G[n; \omega] = V + \sum_{k_a k_b} V \frac{|k_a k_b\rangle Q \langle k_a k_b|}{\omega - e(k_a) - e(k_b)} G[n; \omega], \qquad (1)$$

where V is the bare nucleon-nucleon (NN) interaction, n is the nucleon number density, and ω the starting energy. The single-particle energy $e(k)$ (assuming $\hbar=1$),

$$e(k) = e(k; n) = \frac{k^2}{2m} + U(k; n), \qquad (2)$$

and the Pauli operator Q determine the propagation of intermediate baryon pairs. The Brueckner–Hartree–Fock (BHF) approximation for the single-particle potential $U(k; n)$ using the *continuous choice* is

$$U(k; n) = \text{Re} \sum_{k' \leq k_F} \langle kk' | G[n; e(k) + e(k')] | kk' \rangle_a, \qquad (3)$$

where the subscript "a" indicates antisymmetrization of the matrix element. Due to the occurrence of $U(k)$ in Eq. (2), they constitute a coupled system that has to be solved in a self-consistent manner for several Fermi momenta of the particles involved. In the BHF approximation the energy per nucleon is

$$\frac{E}{A} = \frac{3}{5} \frac{k_F^2}{2m} + \frac{1}{2n} \sum_{k,k' \leq k_F} \langle kk' | G[n; e(k) + e(k')] | kk' \rangle_a. \qquad (4)$$

In this scheme, the only input quantity we need is the bare NN interaction V in the Bethe-Goldstone equation (1). In this sense the BBG approach can be considered as a microscopic one. The nuclear EOS can be calculated with good accuracy in the Brueckner two hole-line approximation with the continuous choice for the single-particle potential, since the results in this scheme are quite close to the calculations which include also the three hole-line contribution [10]. In the calculations reported here, we have used the Paris potential [11] as the two-nucleon interaction and the Urbana model as three-body force [12]. This allows the correct reproduction of the empirical nuclear matter saturation point ρ_0 [13]. Recently, we have included the hyperon degrees of freedom within the same approximation to calculate the nuclear EOS needed to describe the NS interior [14]. We have included the Σ^- and Λ hyperons. To this purpose, one needs also nucleon-hyperon (NY) and hyperon-hyperon (YY) interactions [14, 15]. However, because of a lack of experimental data, the hyperon-hyperon interaction has been neglected in the first approximation, whereas for the NY interaction the Nijmegen soft-core model [16] has been adopted. Once hyperons and leptons are introduced, the total EOS can be calculated for a given composition of the baryon components. Further details are given in Ref[14].

Quark phase

We begin with the thermodynamic potential of q quarks, where $q = u, d, s$ denote up, down, and strange quarks, expressed as a sum of the kinetic term and the one-gluon-exchange term [2, 17],

$$
\begin{aligned}
\Omega_q = & -\frac{3m_q^4}{8\pi^2}\left[\frac{\eta_q x_q}{3}(2x_q^2 - 3) + \ln(x_q + \eta_q)\right] \\
& + \frac{3m_q^4 \alpha_s}{4\pi^3}\left\{2\left[\eta_q x_q - \ln(x_q + \eta_q)\right]^2 - \frac{4}{3}x_q^4 + 2\ln(\eta_q)\right. \\
& \left. + 4\ln(\frac{\sigma_{\text{ren}}}{m_q \eta_q})\left[\eta_q x_q - \ln(x_q + \eta_q)\right]\right\},
\end{aligned}
\tag{5}
$$

where m_q and μ_q are the q current quark mass and chemical potential, respectively, and $x_q = \sqrt{\mu_q^2 - m_q^2}/m_q$, $\eta_q = \sqrt{1 + x_q^2} = \mu_q/m_q$. α_s denotes the QCD fine structure constant, whereas σ_{ren} is the renormalization point, $\sigma_{\text{ren}} = 313$ MeV. The number density ρ_q of q quarks is related to Ω_q via

$$
\rho_q = -\frac{\partial \Omega_q}{\partial \mu_q},
\tag{6}
$$

and the total energy density and pressure for the quark system are given by

$$
\varepsilon_Q = \sum_q (\Omega_q + \mu_q \rho_q) + B,
\tag{7}
$$

$$
P_Q = -\sum_q \Omega_q - B,
\tag{8}
$$

where B is the energy density difference between the perturbative vacuum and the true vacuum, i.e., the bag constant. In the original MIT bag model the bag constant

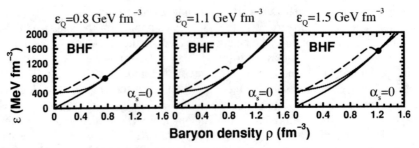

FIGURE 1. The energy density is displayed vs. the baryon density for almost symmetric matter (solid line). The dotted (dashed) lines represent calculations for u, d quark matter obtained within the MIT bag model, with B parametrized as a Gaussian (Woods-Saxon)-like function.

has the value $B \approx 55 \, \mathrm{MeV \, fm^{-3}}$, which is quite small when compared with the ones ($\approx 210 \, \mathrm{MeV \, fm^{-3}}$) estimated from lattice calculations [18]. In this sense B can be considered as a free parameter. The composition of β-stable quark matter is determined by imposing the condition of equilibrium under weak interactions, supplemented with the charge neutrality condition and the total baryon number conservation. It can easily be demonstrated that, in the case of massless u, d, and s quarks, the equilibrium solution reads

$$\rho_u = \rho_d = \rho_s, \quad \rho_e = 0, \tag{9}$$

and consequently the equation of state is

$$P_Q = \frac{1}{3}(\varepsilon_Q - 4B). \tag{10}$$

Here one should notice that the above expressions hold in the case of constant B. If the bag constant is density dependent, all thermodynamical relations must be reformulated [3, 19].

RESULTS AND DISCUSSION

We try to determine a range of possible values for B by exploiting the experimental data obtained at the CERN SPS, where several experiments using high-energy beams of Pb nuclei reported (indirect) evidence for the formation of a quark-gluon plasma [6]. According to the analysis of those experiments, the quark-hadron transition takes place at about seven times normal nuclear matter energy density ($\varepsilon_0 \approx 156 \, \mathrm{MeV \, fm^{-3}}$). As discussed above, in our analysis we assume that the transition to quark-gluon plasma is determined by the value of the energy density only (for a given asymmetry). With this assumption and taking the hadron to quark matter transition energy density from the CERN experiments we estimate the value of B and its possible density dependence in the following way.

First, we calculate the EOS for cold asymmetric nuclear matter characterized by a proton fraction $x_p = 0.4$ (the one for Pb nuclei accelerated at CERN-SPS energies) in the

BHF formalism with two-body and three-body forces as described earlier. The result is shown by the solid line in Fig. 1. Then we calculate the EOS for (u,d) quark matter using Eq. (7) (we assume massless u and d quarks). To be compatible with the experimental observation at the CERN-SPS, we have assumed that the hadron-quark transition takes place within a range of energy density values. For that we have considered three possible values of the transition energy density, i.e., $\varepsilon_Q = 0.8$, 1.1, and 1.5 GeV fm^{-3}. We find that at very low baryon density the quark matter energy density is higher than that of nuclear matter, while with increasing baryon density the two energy densities become equal at a certain point (indicated by the full dot), and after that the nuclear matter energy density remains always higher. We identify this crossing point with the transition density from nuclear matter to quark matter.

However, for no density independent value of B, the two EOS's cross each other, satisfying the experimental condition. Therefore, we have used a density dependent B. In the literature there are attempts to understand the density dependence of B [20]; however, currently the results are highly model dependent and no definite picture has come out yet. Therefore, we attempt to provide effective parametrizations for this density dependence. Our parametrizations are constructed in such a way that at asymptotic densities B has some finite value B_∞, which we fix in the following way. The energy density for u,d quark matter reads

$$\varepsilon_Q(\rho, x_p) = B(\rho) + \frac{3}{4}\left[\frac{\pi^2}{(1 - 2\alpha_s/\pi)}\right]^{1/3}\left[(1 + x_p)^{4/3} + (2 - x_p)^{4/3}\right]\rho^{4/3}. \tag{11}$$

B_∞ can be readily calculated at the transition energy density (known from experiments) which corresponds to a value of the baryonic number density $\bar{\rho}$ given by the hadronic equation of state, i.e.,

$$B_\infty = \varepsilon_Q(\bar{\rho}, x_p) - \frac{3}{4}\left[\frac{\pi^2}{(1 - 2\alpha_s/\pi)}\right]^{1/3}\left[(1 + x_p)^{4/3} + (2 - x_p)^{4/3}\right]\bar{\rho}^{4/3}. \tag{12}$$

Therefore we can determine a range of values for B_∞, $19 \leq B_\infty \leq 77$MeV fm^{-3}, depending on the value of α_s. We limit ourselves to consider only two possible values of α_s, i.e., $\alpha_s = 0$ and $\alpha_s = 0.1$. Although the values of B_∞ span a wide range, we have verified that our results do not change appreciably by varying this value, since at large densities the quark matter EOS is dominated by the kinetic term on the RHS of Eq. (11). With those values of B_∞ we then construct two parametrizations of B as function of the baryon density. We have used both a Gaussian and a Woods-Saxon-like parametrization. With the Gaussian parametrization B decreases monotonically with increasing the baryon density, whereas in the Woods-Saxon case B remains practically constant up to a certain density and then drops to B_∞ almost like a step function. This can be considered an extreme parametrization in the sense that it will delay the onset of the quark phase in neutron star matter as much as possible. For numerical details, the reader is referred to Ref[9].

With these parametrizations of the density dependence of B we now consider the hadron-quark phase transition in neutron stars. We calculate in the BHF framework the EOS of a conventional neutron star as composed of a chemically equilibrated and neutrally charged mixture of nucleons, leptons, and hyperons. The result is shown by the

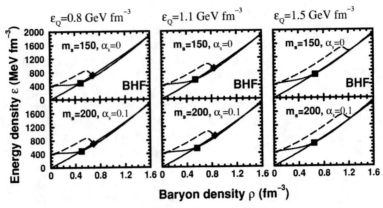

FIGURE 2. The energy density is displayed vs. the baryon density for beta-stable matter (solid line). The dotted (dashed) lines represent calculations for u, d, s quark matter obtained within the MIT bag model, with B parametrized as a Gaussian (Woods-Saxon)-like function. The full squares (diamonds) are the crossings between the hadron and the quark phase.

solid lines in Fig. 2. The other curves (with the same notation as in Fig. 1) represent the EOS's for beta-stable and charge neutral quark matter. In particular, the left-hand panels display calculations with a transition energy density $\varepsilon_Q = 0.8$ GeV fm^{-3}, whereas the central panels show the results for $\varepsilon_Q = 1.1$ GeV fm^{-3} and the right-hand panels for $\varepsilon_Q = 1.5$ GeV fm^{-3}. Two sets of values of the s-quark mass and the QCD coupling constant α_s are considered, namely $m_s = 150$ MeV, $\alpha_s = 0$ (upper panels) and $m_s = 200$ MeV, $\alpha_s = 0.1$ (lower panels). The full symbols represent the crossing points between the hadron and the quark phase, and lie inside the mixed phase region.

We determine the range of baryon density where both phases can coexist by following the construction from ref. [21]. In this procedure both hadron and quark phases are allowed to be charged, still preserving the total charge neutrality. The pressure is the same in the two phases to ensure mechanical stability, while the chemical potentials of the different species are related to each other to ensure chemical and beta stability. The resulting EOS for neutron star matter, according to the different bag parametrizations, is reported in Fig. 3, where the shaded area indicates the mixed phase region. A pure quark phase is present at densities above the shaded area and a pure hadronic phase is present below it. We notice that the mixed phase starts at values of the baryon density which increase with increasing ε_Q, and that those values turn out to be weakly dependent on the values of m_s and α_s. Moreover, the onset density of the mixed phase turns out to be slightly smaller than the density for hyperons formation in pure hadronic matter. Of course hyperons are still present in the hadronic component of the mixed phase. However, if $\varepsilon_Q = 1.5$ GeV fm^{-3}, no phase transition at all is present when the Woods-Saxon-like parametrization of B is adopted. In this case, neutron star matter always remains in the hadronic phase.

Finally, we solve the Tolman-Oppenheimer-Volkoff (TOV) equations [1] for the pressure P and the enclosed mass m with the EOS's of Fig. 3 as input. The TOV equations

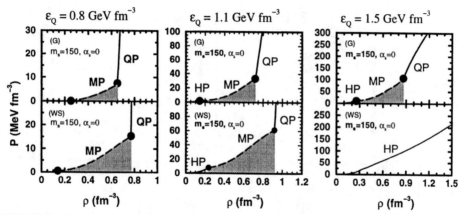

FIGURE 3. The total EOS including both hadronic and quark components is displayed for three transition energy densities $\varepsilon_Q = 0.8$, 1.1, and 1.5 GeV fm^{-3}. The upper (lower) panels show calculations obtained with the bag constant B parametrized as a Gaussian-like (Woods-Saxon-like) function. In all cases the shaded region, bordered by two dots, indicates the mixed phase MP, while HP and QP label the portions of the EOS where pure hadron or pure quark phases are present.

read

$$\frac{dP(r)}{dr} = -\frac{Gm(r)\varepsilon(r)}{r^2}\frac{\left[1+P(r)/\varepsilon(r)\right]\left[1+4\pi r^3 P(r)/m(r)\right]}{1-2Gm(r)/r}, \qquad (13)$$

$$\frac{dm(r)}{dr} = 4\pi r^2 \varepsilon(r), \qquad (14)$$

being G the gravitational constant. Starting with a central mass density $\varepsilon(r=0) \equiv \varepsilon_c$, we integrate out until the pressure on the surface equals the one corresponding to the density of iron. This gives the stellar radius R and the gravitational mass is then

$$M_G \equiv m(R) = 4\pi \int_0^R dr\, r^2 \varepsilon(r). \qquad (15)$$

For the description of the NS's crust, if present, we have joined the hadronic equations of state with the ones by Negele and Vautherin [22] in the medium-density regime, and the ones by Feynman-Metropolis-Teller [23] and Baym-Pethick-Sutherland [24] for the outer crust.

The calculated results, the NS's mass vs. radius (left panel) and central density (right panel), for all cases are shown in Fig. 4. The EOS with nucleons, leptons and hyperons gives a maximum mass of neutron stars of about $1.26 M_\odot$, below the "canonical value" of 1.44 solar masses. It is commonly believed that the inclusion of the quark component should soften the NS matter EOS. This does not hold true in the BHF case, where the EOS becomes, on the contrary, stiffer. Correspondingly, the inclusion of the quark component has the effect of increasing the maximum mass. As a consequence, the calculated maximum masses fall in any case in a relatively narrow range, $1.4 M_\odot \leq M_{\max} \leq 1.7 M_\odot$, slightly above the observational lower limit of $1.44 M_\odot$ [25].

As one can see from Fig. 4, the presence of a mixed phase produces a sort of plateau in the mass vs. central density relationship, which is a direct consequence of the smaller

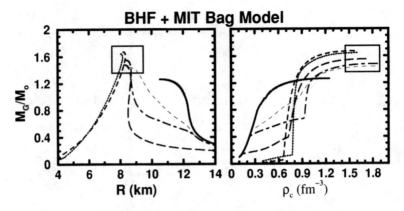

FIGURE 4. The mass-radius (left panel) and mass-central density (right panel) relations are displayed for $\varepsilon_Q = 0.8$, 1.1, and 1.5 GeV fm^{-3} and several parametrizations of the bag constant B. Calculations performed with the BHF EOS for the hadronic component are displayed by the solid lines.

slope displayed by all EOS in the mixed phase region, see Fig. 3. In this region, however, the pressure is still increasing monotonically, despite the apparent smooth behaviour, and no unstable configuration can actually appear. We found that the appearance of this slow variation of the pressure is due to the density dependence of the bag constant. Finally, it has to pointed out that the maximum mass value is dominated by the quark EOS at densities where the bag constant is much smaller than the quark kinetic energy.

CONCLUSIONS

In summary, under our hypothesis, we have found that a density dependent B is necessary to understand the CERN-SPS findings on the phase transition from hadronic matter to quark matter. Then, taking into account this experimental data, we calculated NS's maximum masses, using an EOS which combines a reliable EOS for hadronic matter and a bag model EOS for quark matter. The calculated NS's maximum masses lie in a narrow range in spite of using very different parametrizations of the density dependence of B. Other recent calculations of neutron star properties employing various RMF nuclear EOS' together with either effective mass bag model [26] or Nambu-Jona-Lasinio model [27] EOS' for quark matter, also give maximum masses of only about $1.7M_\odot$, even though not constrained to reproduce simultaneously the CERN-SPS data. The value of the maximum mass of neutron stars obtained according to our analysis appears robust with respect to the uncertainties of the nuclear EOS. Therefore, the experimental observation of a heavy neutron star, as claimed recently by some groups [28]($M \approx 2.2M_\odot$), if confirmed, would suggest mainly two possibilities. Either serious problems are present for the current theoretical modelling of the high-density phase of nuclear matter, or the working hypothesis that the transition to the deconfined phase occurs approximately at the same energy density, irrespective of the thermodynamic conditions, is substantially wrong. In both cases, one can expect a well defined hint on the high density nuclear matter EOS.

REFERENCES

1. S. L. Shapiro and S. A. Teukolsky, *Black Holes, White Dwarfs, and Neutron Stars* (John Wiley & Sons, New York, 1983).
2. E. Witten, Phys. Rev. **D30**, 272 (1984); G. Baym, E. W. Kolb, L. McLerran, T. P. Walker, and R. L. Jaffe, Phys. Lett. **B160**, 181 (1985); N. K. Glendenning, Mod. Phys. Lett. **A5**, 2197 (1990).
3. G. F. Burgio, M. Baldo, P. K. Sahu, A. B. Santra, and H.-J. Schulze, Phys. Lett. **B526**, 19 (2002).
4. M. Baldo, *Nuclear Methods and the Nuclear Equation of State* (World Scientific, Singapore, 1999).
5. A. Chodos, R. L. Jaffe, K. Johnson, C. B. Thorn, and V. F. Weisskopf, Phys. Rev. **D9**, 3471 (1974).
6. U. Heinz and M. Jacobs, nucl-th/0002042; U. Heinz, hep-ph/0009170.
7. See for instance "Theoretical Conference Summary", Quark Matter 2001, J. P. Blaizot, nucl-th/0107025.
8. J. Cleymans, R.V. Gavai, and E. Suhonen, Physics Rep. **130**, 217 (1986).
9. G. F. Burgio, M. Baldo, P. K. Sahu, and H.-J. Schulze, Phys. Rev. C, August 1 (2002), in press.
10. H. Q. Song, M. Baldo, G. Giansiracusa, and U. Lombardo, Phys. Rev. Lett. **81**, 1584 (1998), Phys. Lett. **B473**, 1 (2000); M. Baldo and G. F. Burgio, *Microscopic Theory of the Nuclear Equation of State and Neutron Star Structure*, in "Physics of Neutron Star Interiors", Eds. D. Blaschke, N. Glendenning, and A. Sedrakian, Lectures Notes in Physics, Springer, vol. 578 (2001), pp. 1-30.
11. M. Lacombe, B. Loiseau, J. M. Richard, R. Vinh Mau, J. Côté, P. Pirès, and R. de Tourreil, Phys. Rev. **C21**, 861 (1980).
12. J. Carlson, V. R. Pandharipande, and R. B. Wiringa, Nucl. Phys. **A401**, 59 (1983); R. Schiavilla, V. R. Pandharipande, and R. B. Wiringa, Nucl. Phys. **A449**, 219 (1986).
13. M. Baldo, I. Bombaci, and G. F. Burgio, Astron. Astrophys. **328**, 274 (1997).
14. M. Baldo, G. F. Burgio, and H.-J. Schulze, Phys. Rev. **C58**, 3688 (1998); Phys. Rev. **C61**, 055801 (2000).
15. I. Vidaña, A. Polls, A. Ramos, L. Engvik, and M. Hjorth-Jensen, Phys. Rev. **C62**, 035801 (2000).
16. P. Maessen, Th. Rijken, and J. de Swart, Phys. Rev. **C40**, 2226 (1989).
17. E. Fahri and R. L. Jaffe, Phys. Rev. **D30**, 2379 (1984).
18. H. Satz, Phys. Rep. **89**, 349 (1982).
19. G. X. Peng, H. C. Chiang, B. S. Zou, P. Z. Ning, and S. J. Luo, Phys. Rev. **C62**, 025801 (2000).
20. C. Adami and G. E. Brown, Phys. Rep. **234**, 1 (1993); Xue-min Jin and B. K. Jennings, Phys. Rev. **C55**, 1567 (1997).
21. N. K. Glendenning, Phys. Rev. **D46**, 1274 (1992).
22. J. W. Negele and D. Vautherin, Nucl. Phys. **A207**, 298 (1973).
23. R. Feynman, F. Metropolis, and E. Teller, Phys. Rev. **75**, 1561 (1949).
24. G. Baym, C. Pethick, and D. Sutherland, Astrophys. J. **170**, 299 (1971).
25. R. A. Hulse and J. H. Taylor, Astrophys. J. **195**, L51 (1975).
26. K. Schertler, C. Greiner, P. K. Sahu, and M. H. Thoma, Nucl. Phys. **A637**, 451 (1998); K. Schertler, C. Greiner, J. Schaffner-Bielich, and M. H. Thoma, Nucl. Phys. **A677**, 463 (2000).
27. K. Schertler, S. Leupold, and J. Schaffner-Bielich, Phys. Rev. **C60**, 025801 (1999).
28. P. Kaaret, E. Ford, and K. Chen, Astrophys. J. Lett. **480**, L27 (1997); W. Zhang, A. P. Smale, T. E. Strohmayer, and J. H. Swank, Astrophys. J. Lett. **500**, L171 (1998).

Strangeness and Strangelets

Experimental Searches for Strange Quark Matter

Reiner Klingenberg

Institut für Physik, Universität Dortmund, D-44221 Dortmund, Germany

Abstract. Strange quark matter (SQM) is a hypothetical multi-quark ensemble containing almost equal amount of up-, down- and strange quarks. Different experimental techniques looking for the possible existence of SQM as well as its production in the laboratory have been carried out and are reviewed in this contribution. It is focused on examples of terrestrial searches and recent experiments of relativistic particle and heavy-ion collisions, which try to produce SQM in the laboratory. An outlook on future experimental possibilities is given.

IDEAS ABOUT STRANGE QUARK MATTER

Strange quark matter (SQM) is a hypothetical state which is predicted by theoretical considerations but so far has not been confirmed in nature nor in experiments. A comprehensive introduction to theoretical issues of SQM can be found e. g. in several review articles [1, 2].

In particle physics hadrons are well established states, which show up either as mesons, which are pairs of quarks and antiquarks, or as hadrons which are composed of three quarks. Groups with a larger number of quarks are not known, but clusters of baryons like nuclei are well established in nature. There is no basic physical principle known which excludes the existence of larger hadrons. These hypothetical states are usually called *quark matter*, or SQM if they contain similar amounts of up, down and strange quarks. Smaller amounts of SQM are usually called *strangelets*. However the anticipated mass range for this kind of matter may lie between the masses of light nuclei and that of neutron stars $A \approx 10^{57}$, the latter one are called *strange stars*. As strange matter might be part of the cosmic radiation one is also speaking of strange quark *nuggets*.

First ideas about the existence of quark matter were published under the title "Collapsed Nuclei" [3]. According to those ideas collapsed nuclei have a higher density than ordinary nuclei and may have also a lower energy level. The spontaneous decay of nuclei into these collapsed states is inhibited by a saturation barrier which prolongs their lifetimes to more than 10^{14} times the age of the universe. Furthermore, it is presumed that such collapsed states have been created in the initial extremely hot and dense stages of the universe and part of them may still exist in certain regions. Considerations about the creation of quark matter during the expansion of the early universe have later been picked up [4]. According to that scenario a first order phase transition from deconfined quarks and gluons to confined hadrons allows the creation of bubbles of low temperature in coexistence with the hot phase. In the course of the expansion the bubbles are

CP644, *Exotic Clustering: 4th Catania Relativistic Ion Studies,* edited by S. Costa, A. Insolia, and C. Tuvè
© 2002 American Institute of Physics 0-7354-0099-7/02/$19.00

enlarged and clustered and only small regions of high temperature are remaining. They consist of a large fraction of the baryonic matter in states of quark matter.

Possibilities for the production and detection of metastable strangelets in heavy-ion collisions have been expressed [5]. Strangelets could be created by the fragmentation and recombination of quarks in a quark-gluon plasma which might be formed in heavy-ion collisions. Identification of strangelets should be possible by their low charge to mass ratio in a spectrometer, because the strange quark carries a negative charge of $-1/3$ and roughly equal numbers of u, d and s quarks predict near neutrality. Thus, strangelets might be even neutral or negatively charged. In other theoretical models the stability of SQM is discussed. One distinguishes between absolute stable, metastable and non-stable possibilities. While in the last case it would not be appropriate to use the term matter at all, the question is which conditions might lead to a stability or which decay modes are favoured. Stability and decay channels have been widely investigated by several authors [6, 7]. Calculations indicate that SQM tends to be more tightly bound for increasing number of incorporated quarks, e. g. [8, 9]. On the other hand there are predictions that also small strangelets might gain stability due to shell effects [10]. A discussion on different possible descriptions of strangelets can be found in the contriubtion "Stability of Strange Quark Matter" by Wanda Alberico in this volume.

TERRESTRIAL SEARCHES

If SQM is absolute stable it might be part of our natural environment. This idea has been tested by terrestrial searches which look for SQM similar to heavy isotopes with an unusual high mass and low charge. Such SQM could be of primary origin or could have been brought in by cosmic radiation.

Various tests to look for terrestrial SQM have been carried out. An experiment using the Rutherford backscattering method which is sensitive to heavy nuggets reports on a null result for masses up to 10^7 amu with limits down to 10^{-13} [11]. A search for low-charged heavy nuclei containing massive stable particles using a technique to measure the energy loss of accelerated particles in a gas ionization detector is used to estimate the mass of the particle [12]. No evidence for anomalous isotopes of terrestrial H, Li, Be, B, C, O and F in the mass range from 100 to 10000 amu is found. Limits vary between 10^{-9} and 10^{-24} depending on mass and isotope. Compared to this the most sensitive search for heavy hydrogen is based on a spectrometer experiment investigating heavy water enriched by electrolysis [13]. This experiment gives a concentration limit of 10^{-28} per nucleon for masses up to 1200 amu. Moreover, a search for the doubly charged strangelets as heavy isotopes of helium, which is not covered by the previous experiments, has been carried out [14]. This experiment excludes heavy particles on the level of 2×10^{-11} per ordinary helium-4 for masses in the 42 to 60 amu range and 2×10^{-12} for masses between 60 and 82 amu.

The property of stable SQM being low charged led to the proposition to search for strange matter by heavy ion activation [15]. In such an experiment the investigated material is probed by a low energy beam of heavy-ions. If SQM is part of the probed material the normal matter penetrates the Coulomb barrier of the strangelet. The deposited energy is distributed among the quarks and an excited state is created. Part of its energy is spent to regain flavour equilibrium via weak decays [8], while the remaining energy is released in the form of photons. In this picture the emission of nucleons is a negligible decay mode for strangelets with $A > 2000$, because an unlikely high local concentration of energy would be needed to convert deconfined quarks inside the strangelet to gain the configuration of a particle. Similar, pion emission would require an energetic quark near the boundary of the strangelet [16]. It has also been considered that subthreshold pion production and pre-equilibrium nucleon emission, as observed in heavy-ion collisions, have very low cross sections and are negligible for low energy projectiles [17]. As an experimental signature the detection of gamma rays emitted from matter activated by heavy-ion beams would indicate the presence of stable SQM. Results of an experiment using the method of heavy ion activation with the aim to measure the released gammas are reported by the collaboration using the GAMMASPHERE detector [18]. It is detecting gamma rays in an energy band between 20 keV and 20 MeV using germanium and bismuth germanite crystals. Samples of nickel ore, a meteorite and lunar soil have been exposed to a ^{136}Xe beam of 450 MeV which is below the Coulomb barrier of normal matter. Gamma multiplicities and gamma total energies have been measured. If these samples would contain a substantial fraction of SQM one expects either a high gamma multiplicity or a high total gamma energy. None of it was observed leading to an upper limit of strangelet concentration in these material in the order of 10^{-13} to 10^{-16} per ordinary nucleon for strangelet masses between 10^3 and 10^8 amu. These limits are about a factor 100 better compared to earlier measurements mentioned above, which used the technique of Rutherford backscattering [11].

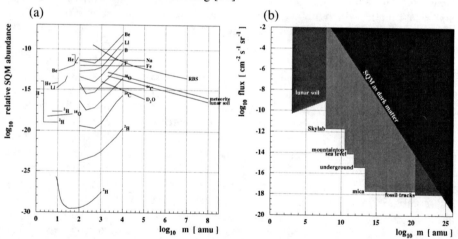

FIGURE 1. Shown are (a) upper limits on the abundance of stable SQM as heavy isotopes. These are determined in various samples as discussed in the text. Fig. (b) shows the deduced limits on the flux of cosmic strange quark nuggets based on various measurements. For the case that dark matter would only consist of SQM its associated maximum flux is also indicated. However, most experimental results exclude this limit by several orders of magnitude over a wide range of assumed SQM masses.

A compilation of achieved limits on the concentration of relative SQM abundance is shown in figure 1(a). Data has been taken from a compilation [19] and several measurements [11, 12, 14, 18]. The concentration limits as found by the experimental investigation allows to set limits on the flux of strange quark nuggets. It is based on the general assumption that galactic cosmic radiation containing stable SQM is absorbed and deposited in the investigated material. The results are shown in figure 1(b).

The region "SQM as dark matter" is based on the assumption that all galactic dark matter is composed of SQM. Experimental results exclude this assumption by several orders of magnitude. The data shown has been taken from a previous compilation [20]. In comparison to these limits an astrophysical assumption can be used to deduce much more stringent exclusions. If no strange star exist and all neutron stars are indeed consisting of nuclear matter, then the cosmic flux of quark nuggets should not be larger than about 10^{-41} cm^{-2} s^{-1} sr^{-1} [21, 22].

SQM IN ASTROPHYSICS?

Different astrophysical aspects of SQM have been proposed. Pulsars as fast spinning compact stars are regarded as valuable objects to test the existence of SQM. For example the time evolution of the rotation of stars as well as the cooling behaviour may reveal the creation and existence of SQM. One striking implication of the hypothesis of stable SQM would be that pulsars, which are conventionally interpreted as rotating neutron stars, almost certainly would be rotating strange stars or strange pulsars. Either these might be created as strange stars or one could think of a conversion process later from nuclear matter to SQM. Under large pressure nuclear matter may be converted to a two flavour quark matter which might get a higher stability through the conversion in SQM. A Coulomb barrier free absorption of neutrons can even enlarge the regions of quark matter [4].

The internal structure of stars can be investigated e. g. by the observation of the spin-down, which is the slow degeneration of rotating frequency in time and sudden deviations from these long on-going process. In a fast rotating star the centrifugal force flattens the star and lowers the effective pressure. Energy is released during the evolution of the star and leads to spin-down with reduced centrifugal force, the star becomes spheric while the internal pressure is rising. This might induce a phase transition where a fast rotating hyperon star is converted to a more packed hybrid star which includes the quark phase. As a consequence the moment of inertia is reduced and a sudden spin-up might take place [23]. As a practical observable the *braking index* has been suggested, which combines the angular velocity Ω and its first and second time derivatives. This braking index would be identical to the intrinsic index of an energy loss mechanism, which is $n = 3$ for magnetic dipole radiation, if the moment of inertia I would be constant. But as soon as this moment changes the rotational frequency and the braking index changes as well. This characteristic value would be observable during about $\frac{1}{100}$ of a typical active pulsar lifetime. Presently about 1000 pulsars are known, thus one could expect that ten of them might signalling anomalous braking indices indicating the growth of a central region of a new phase. Currently only a few braking indices are known without indicating a possible phase transition. The determination of the second time derivative $\ddot{\Omega}$ is difficult. However, it is expected that this difficulty especially occurs during the normal spin-down but should vanish during the spin-up where large accelerations occur. Thus, this fact can be used to deselect candidates. If indeed spin-up will be observed this could indicate a phase transition to metastable SQM.

In contrast to the hybrid stars, stars containing absolute stable SQM are called strange stars, where the energy per baryon is below that of nuclear matter. A signature for strange stars has been determined using their structure compared to neutron stars to develop models about their different cooling behaviour. In contrast to neutron stars an inner crust is missing in strange stars and results in a smaller thermal insulation between the core and its surface. Based on these conditions model predictions of surface temperatures show that young strange stars ($< 30\,\mathrm{y}$) show a significant faster cooling than neutron stars [24]. Measurements of several pulsars done so far are all of ages older than $10^4\,\mathrm{y}$ where the differences between strange and neutron stars vanish. If continued observation of SN 1987A could reveal the temperature of the possibly existing pulsar at its centre, then this is a relevant point to test this cooling scenario and might help to clarify the question about stable SQM in stars.

PRODUCTION OF SQM IN PARTICLE REACTIONS?

The H^0 dibaryon

The H^0 dibaryon is a proposed six quark flavour-singlet dihyperon with strangeness $S = -2$, spin and parity $J^P = 0^+$ [25]. It should have a mass of about $2150\,\mathrm{MeV}/c^2$, contains the quarks uuddss and would be the lightest possible SQM lump. Different reaction schemes have been exploited to study the possible creation of the H^0 baryon. One possibility would be the double associated production scheme, where a strangeness neutral beam hits a nuclear target and one would expect to observe two particles with antistrangeness besides the H^0. An alternative to this production scheme can be a strangeness exchange reaction if one starts with an $S = -1$ incident beam, e. g. K^-, and transferring two units of strangeness to a nuclear system and recoiling the $S = -2$ dibaryon. This environment would be more favourable for the H^0 production, because a strangeness exchange reaction has a considerably larger cross section than the associated-production processes. For an overview see also the contribution "Multi-quark Hadrons and the $S = -2$ Hypernuclei" by Sidney Kahana in this volume.

Among other experiments prior E224 [26] at KEK, has studied the strangeness exchange reaction using an initial K^- beam of $1.66\,\mathrm{GeV}/c$ hitting a scintillating fibre calorimeter. Upper limits for its differential cross section in the forward direction of the K^+ were found to be 0.04 to $0.6\,\mu\mathrm{barn/sr}$ at a 90% confidence level for the H^0 mass range from 1850 to $2215\,\mathrm{MeV}/c^2$

The experiment BNL-AGS-E836 has studied the reaction $^3\mathrm{He}(K^-,K^+)H^0\mathrm{n}$. It uses an initial beam of $1.8\,\mathrm{GeV}/c\ K^-$ on a liquid $^3\mathrm{He}$ target. In the reaction the two protons in $^3\mathrm{He}$ should be converted into an H^0 by the (K^-,K^+) reaction with the neutron assumed to be a spectator. The forward outgoing K^+ are identified in a spectrometer with drift chambers and the help of time-of-flight and Čerenkov counters. The momentum spectrum of these K^+ allows to characterize the reaction. A deeply bound H^0 should manifest itself as a well-separated narrow peak above the region of quasi-free Ξ^- production $K^{-3}He \to K^+\Xi^-pn$ [27, 28]. No evidence for the production of H^0 is seen in the mass interval between 1.85 and 2.2 GeV/c^2. This mass range effectively covers states between two nucleons and two lambdas. The upper limit of the production probability is about a factor ten below a theoretical calculation applicable for masses above 2.05 GeV/c^2. Recently the E885 collaboration [29] which used the same apparatus but a carbon target report on limits for the H^0 production being up to 10 times more significant.

The production of the H^0 dibaryon in relativistic heavy-ion collisions is discussed by [30]. They consider two different possibilities: The production from a hadronic fireball in thermodynamic equilibrium and from a quark-gluon plasma phase. In the latter case they expect an enhancement relative to the rate for a hadron gas. Recently the experiment AGS-E896 searched for H^0 production in gold-gold collisions and tries to identify the decay channels $H^0 \to \Sigma^- + p, \Lambda + p + \pi^-, \Lambda + n$.

Investigations of heavy ion collisions

Heavy-ion experiments allow to test the production of SQM formed in the hot and dense environment of two colliding nuclei. Several types of production models have been suggested to desribe and predict possible strangelet production in heavy ion collisions. In one of these pictures, called coalescence model, an ensemble of quarks, which come as products from the nucleus-nucleus collision form a composite state which fuses to form a strangelet [31]. A second type of production, called thermal models, assume that chemical and thermal equilibrium are achieved prior to final particle production [32]. A further model involves an intermediate quark-gluon plasma (QGP) state, which might be produced in nucleus-nucleus collisions. Strangelets could be regarded as cooled remnants of a QGP and thus their possibly copious production in heavy-ion collisions can be seen as a signature of the quark-gluon plasma. One possibility is that the QGP cools by emitting hadronized π's and K's. The larger amount of up- and down-quarks with respect to the anti-quarks results in a higher chance for a $\bar{s}$-quark to find a u- or d-quark to form a K^+ or K^0 than for a s-quark to form the anti-particle counterparts. Thus, anti-strange quarks are emitted and the remaining part of the QGP is enriched by strange quarks. The emission and expansion is cooling the QGP, leading to bound multi-quark states. If such a conglomerate is stable a strangelet is born. Otherwise, if it is unstable, baryons especially hyperons are left over. The separation of strangeness and anti-strangeness is like a distillation process [33]. On the other hand the quark-gluon plasma might also cool mainly as a result of its expansion as an adiabatic process until the entire QGP reaches the state in which hadronization occurs. In this case there will be little or no strangeness distillation.

Several experimental searches for SQM have been carried out at BNL-AGS and CERN-SPS. Collision systems of Si + Cu, Si + Au, S + W and Pb + Pb have been investigated. The experiment E864 aimed to study heavy ion collisions in an open geometry spectrometer in 11.5 AGeV/c Au + Pb and Au + Pt collisions at BNL-AGS. The total length of this apparatus is about 28 m. This limits any possible detection to lifetimes of the order of 50 ns or longer. Two independent mass measurements, one with a magnet and tracking system and the other with a hadronic calorimeter, allow to identify particles and to reject background powerfully. Achieved limits of the positive strangelet search are in the order of 10^{-8} per 10% most central Au+Pt collision [34]. This data allowed also the collection of light nuclei which have been used to draw restrictions for the production of strangelets in terms of a coalescence model. E864 measures a decrease in the nuclei yield by a factor 48 for every added nucleon [35]. An earlier finding estimates another factor $1/5$ for adding strangeness. From this, together with the sensitivity for low mass strangelets, limits in terms of the baryon number A and the strangeness content S are $A + 0.42 \cdot |S| < 7.2$ at a 90% confidence level [34]. This shows that experimental searches in terms of a pure coalescence model are limited to small strangelets and the production of large strangelets in normal medium is rather unlikely. At negative rigidities the upper limit of strangelets is in the order of 2×10^{-9} per central collision including the extrapolation to full phase space [36]. The apparatus of E864 allows also to search for neutral strangelets. This is done with the help of the combined energy and time measurement of the calorimeter. The lack of tracking information makes the identification and background rejection much more difficult. Null results have been shown [37] and achieved sensitivities are in the order of 10^{-7} per central collision for masses larger than $20 \, \text{GeV}/c^2$.

At CERN-SPS the experiment NA52 searched for SQM. It studied the production of strangelets in $200 \, A\text{GeV}/c$ S + W [38] and $158 \, A\text{GeV}/c$ Pb + Pb [39] collisions. The apparatus consists of a focusing double-bend spectrometer, equipped with time-of-flight scintillator hodoscopes and Čerenkov counters to veto light particles in the trigger and/or further particle identification. No special selection of central events is done, because NA52 recorded all heavy secondary particles resulting from peripheral as well as central collisions at $p_\perp = 0$. Searches at rigidity settings of $\pm 150 \, \text{GeV}/c$ in the sulfur beam reveal limits between 10^{-9} and 10^{-8} per collision [38]. Lead beam data has been also taken with an emphasis at negative polarities where events from several 10^{13} interactions have been recorded. The best limits are achieved for mass to charge ratios between 10 and $30 \, \text{GeV}/c^2$ [40]. Absolute values depend on the assumption of the phase space distribution of the emitted strangelets.

A compilation of limits achieved by several heavy-ion experiments in the search for SQM production is shown in figure 2. Data and values of an assumed phase space distribution (exponential in $p_\perp$, $\langle p_\perp \rangle = 0.5 \ldots 0.7 \sqrt{m} \, \text{GeV}$ and Gaussian in rapidity, $\sigma_y = 0.5$) have been selected to allow a comparison with minimum systematic uncertainty. Nevertheless, it should be noted that a direct comparison of achieved limits remains difficult: on the one hand different colliding systems at different energies have been investigated, on the other hand the selection of interactions vary from minimum bias (E814, E878, E886, NA52) to 10% most central (E864).

The experimental techniques carried out so far limit the search to long-lived candidates. However, the chances for the existence of short-lived metastable candidates are at least from a theoretical point more probable. Therefore, a significant improvement in the detection of short-lived states would be desirable. This might be achieved e. g. by aiming to detect low mass states, which probably decay by the strong interaction and would appear as resonance. Another way would be to study particle correlations which are sensitive to the time dependent charge and spacetime expansion of the system and allow

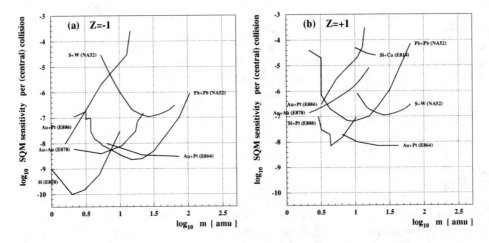

FIGURE 2. Compilation of sensitivity limits (in terms of considered collisions) for the production of long-lived strangelets in heavy-ion collisions. Data is restricted to singly charged objects of (a) negative and (b) positive polarity.

for the measurement of the transient SQM state, even if it decays on strong interaction timescales [41].

FUTURE EXPERIMENTS

Presently the ultra-relativistic heavy-ion collider RHIC at BNL has been started operation and experiments like STAR are aiming at measuring signals which are characteristic of the quark gluon plasma formation [42]. At the Large Hadron Collider (LHC) at CERN it is planned for Pb+Pb periods with beam energies of $7.26 A$TeV, or $\sqrt{s} = 5.5 A$TeV. Here, ALICE, a dedicated detector for heavy ion physics is in preparation. Such detectors are foreseen to explore the mid-rapidity region of the heavy-ion collision, but are not mainly designed for the search of strangelets. Nevertheless, theoretical considerations about the possibility of SQM production in these new kinematic regions of heavy-ion collisions have been expressed. These are in contrast to the common belief that (strange) baryon densities vanish at mid-rapidity, at both RHIC and LHC. [43] claim that this conclusion was premature and argue that fluctuations of the stopping power can provide finite baryochemical potential at mid-rapidity in a small fraction of all events, that fluctuations of the net-baryon and net-strangeness content between different rapidity bins within *one* event can be large and that strange (anti)baryon enhancement can occur due to collective effects such as a chiral phase transition.

A dedicated detector for the search of strangelets at the LHC is in preparation with a new experimental approach by using an unconventional strangelet signature, namely by the energy deposition pattern [44]. This method promises to detect also unstable strangelets. The corresponding specialised detector system CASTOR [45] is an integral part of the CMS experiment. CASTOR will cover the very forward rapidity region. This is assumed to be a preferred region to create a dense quark matter fireball, because it is baryon-rich. This can be understood by the nuclear transparency at LHC energies, where the central rapidity region involves less baryons (however, cf. the former discussion in this section) and those are then shifted towards and found at projectile rapidity. Here now, models of creating a SQM state, e. g. a separation of strangeness from anti-strangeness by a distillation process might be applied. CASTOR is designed to measure the hadronic and photonic contents of an interaction and identify deeply penetrating objects in the forward phase space. It consists of a forward charged particle multiplicity detector and a forward photon multiplicity detector. Both detectors together yield on an event-by-event base the N_γ/N_{ch} ratio and allow to determine possible deviations from normal events. An electromagnetic and hadronic calorimeter is intended to measure the total energy flux of both components into the pseudo-rapidity range $5.6 \leq \eta \leq 7.2$. Longitudinal segmentation allows to identify unusually penetrating components as observed in hadron-rich cosmic-ray events. An azimuthal segmentation allows to determine shower widths.

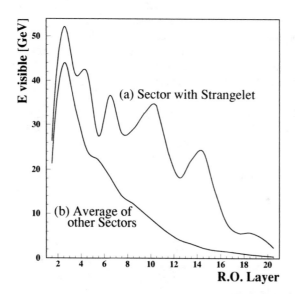

FIGURE 3. Energy deposition versus the readout (R.O.) layers of the CASTOR calorimeter. Curve (a) shows the expected signal of a strangelet and (b) the signal of ordinary particles. While particles usually show a monotonous reduction in energy deposition strangelets are expected to show a much slower attenuation, even several maxima in the energy spectrum could be seen.

Figure 3 shows as an example the comparison of the longitudinal energy spectra of a strangelet and of ordinary particles produced in a central Pb+Pb collision. This picture [45] is based on a simulation of the interaction of a strangelet with the calorimeter material, using the picture described by [46]. It shows similarities to the strongly penetrating cascades detected in the deep chambers of cosmic ray experiments. Their many maxima structure and slow attenuation observed during the shower development could be the result of the successive interaction of the strangelet with the calorimeter absorber nuclei. Stable strangelets would be regarded as wounded nuclei which emit their excitation energy, unstable strangelets would have shorter longitudinal extend, but are still distinguishable from "normal" events. Strangelet evolution, where nucleon emission transforms an unstable strangelet to a stable object is visible as an energy deposition structure containing many maxima.

NO SQM SO FAR ...

Some aspects of the fascinating subject of searches for strangelets as a new form of matter with emphasis on the experimental techniques has been summarized. Various experimental techniques — on the one hand production in particle beams, on the other hand terrestrial searches — are used to identify this form of matter. A basic method for the direct identification is the determination of the mass. Complementary also the presumed low charge of SQM is a stringent characteristic. Both values are determined in various spectrometer experiments with velocity and specific energy loss measurements.

SQM and the discussion about it show a lot of facets. Predictions about production and stability can be found in many variations leading also sometimes to different conclusions. Quantitative values of experimental results usually rely on assumptions which need not to be absolutely true. In despite of these different points of view it is important to investigate the possibilities of new physical spheres and to continue to look for new states of matter.

For example, a direct detection of strangelets in spectrometer experiments carried out so far has a shortcoming to detect long-lived SQM only, while it might be much more valid to probe short-lived states by identifying decay products.

In that sense there is a valid interest for further experimental search of SQM not only in cosmological but also in particle physics aspects. New colliders like RHIC and LHC are and will be playgrounds for new physical regimes and phenomena.

REFERENCES

1. Greiner, C., and Schaffner-Bielich, J., "Physics of Strange Matter," in *Physics of strange matter in Heavy Elements and Related New Phenomena*, edited by R. K. Gupta and W. Greiner, World Scientific, Singapore, 1999.
2. Greiner, C., and Schaffner, J., *Int. J. Mod. Phys. E*, **5**, 239–300 (1996).
3. Bodmer, A. R., *Phys. Rev. D*, **4**, 1601–1606 (1971).
4. Witten, E., *Phys. Rev. D*, **30**, 272–285 (1984).
5. Liu, H.-C., and Shaw, G. L., *Phys. Rev. D*, **30**, 1137–1140 (1984).
6. Chin, S. A., and Kerman, A. K., *Phys. Rev. Lett.*, **43**, 1292–1295 (1979).
7. Koch, P., *Nucl. Phys. B (Proc. Suppl.)*, **24B**, 255–259 (1991).
8. Berger, M. S., and Jaffe, R. L., *Phys. Rev. C*, **35**, 213–225 (1987).
9. Madsen, J., *Phys. Rev. D*, **47**, 5156–5160 (1993).
10. Schaffner-Bielich, J., Greiner, C., Diener, A., and Stöcker, H., *Phys. Rev. C*, **55**, 3038–3046 (1997).
11. Brügger, M., Lützenkirchen, K., Polikanov, S., Herrmann, G., Overbeck, M., Trautmann, N., Breskin, A., Chechik, R., Fraenkel, Z., and Smilansky, U., *Nature*, **337**, 434–436 (1989).

12. Hemmick, T. K., Elmore, D., Gentile, T., Kubik, P. W., Olsen, S. L., Ciampa, D., Nitz, D., Kagan, H., Haas, P., Smith, P. F., McInteer, B. B., and Bigeleisen, J., *Phys. Rev.* D, **41**, 2074–2080 (1990).
13. Smith, P. F., Bennett, J. R. J., Homer, G. J., Lewin, J. D., Walford, H. E., and Smith, W. A., *Nucl. Phys.* B, **206**, 333–348 (1982).
14. Vandegriff, J., Raimann, G., Boyd, R. N., Caffee, M., and Ruiz, B., *Phys. Lett.*, **B365**, 418–422 (1996).
15. Farhi, E., and Jaffe, R. L., *Phys. Rev.* D, **32**, 2452–2455 (1985).
16. Banerjee, B., Glendenning, N. K., and Matsui, T., *Phys. Lett.*, **B127**, 453–457 (1983).
17. Schutz, Y., et al., *Nucl. Phys.* A, **622**, 404–477 (1997).
18. Perillo Isaac, M. C., Chan, Y. D., Clark, R., Deleplanque, M. A., Dragowsky, M. R., Fallon, P., Goldman, I. D., Larimer, R.-M., Lee, I. Y., Macchiavelli, A. O., MacLeod, R. W., Nishiizumi, K., Norman, E. B., Schroeder, L. S., and Stephens, F. S., *Phys. Rev. Lett.*, **81**, 2416–2419 (1998).
19. Blackman, E. G., and Jaffe, R. L., *Nucl. Phys.* B, **324**, 205–214 (1989).
20. Klingenberg, R., *J. Phys. G: Nucl. Part. Phys.*, **25**, R273–R308 (1999).
21. Madsen, J., "Physics and Astrophysics of Strange Quark Matter," in *Hadrons in Dense Matter and Hadrosynthesis: Proc. Eleventh Chris Engelbrecht Summer School (Cape Town, South Africa, 4–13 February 1998)*, edited by J. Cleymans, H. B. Geyer, and F. G. Scholtz, Springer, 1999, vol. 516 of *Lecture Notes in Physics*, pp. 162–203.
22. Madsen, J., *Phys. Rev. Lett.*, **61**, 2909–2912 (1988).
23. Glendenning, N. K., Pei, S., and Weber, F., *Phys. Rev. Lett.*, **79**, 1603–1606 (1997).
24. Schaab, C., Hermann, B., Weber, F., and Weigel, M. K., *J. Phys. G: Nucl. Part. Phys.*, **23**, 2029–2037 (1997).
25. Jaffe, R. L., *Phys. Rev. Lett.*, **38**, 195–198 (1977), (erratum 617).
26. Ahn, J. K., et al., *Phys. Lett.*, **B378**, 53–58 (1996).
27. Aerts, A. T. M., and Dover, C. B., *Phys. Rev.* D, **28**, 450–463 (1983).
28. Aerts, A. T. M., and Dover, C. B., *Phys. Rev. Lett.*, **49**, 1752–1755 (1982).
29. Yamamoto, K., et al., *Phys. Lett.*, **B478**, 401 (2000).
30. Dover, C. B., Koch, P., and May, M., *Phys. Rev.* C, **40**, 115–125 (1989).
31. Baltz, A. J., Dover, C. B., Kahana, S. H., Pang, Y., Schlagel, T. J., and Schnedermann, E., *Phys. Lett.*, **B325**, 7–12 (1994).
32. Braun-Munzinger, P., and Stachel, J., *J. Phys. G: Nucl. Part. Phys.*, **21**, L17–L20 (1995).
33. Greiner, C., Koch, P., and Stöcker, H., *Phys. Rev. Lett.*, **58**, 1825–1828 (1987).
34. Xu, Z., *J. Phys. G: Nucl. Part. Phys.*, **25**, 403–410 (1999).
35. Armstrong, T. A., et al., *Phys. Rev. Lett.*, **83**, 5431–5434 (1999).
36. Nagle, J. L., *Nucl. Phys.* A, **661**, 185c–190c (1999).
37. Munhoz, M. G., *J. Phys. G: Nucl. Part. Phys.*, **25**, 417–422 (1999).
38. Borer, K., Dittus, F., Frei, D., Hugentobler, E., Klingenberg, R., Moser, U., Pretzl, K., Schacher, J., Stoffel, F., Volken, W., Elsener, K., Lohmann, K. D., Baglin, C., Bussière, A., Guillaud, J. P., Appelquist, G., Bohm, C., Hovander, B., Selldèn, B., and Zhang, Q. P., *Phys. Rev. Lett.*, **72**, 1415–1418 (1994).
39. Klingenberg, R., et al., *Nucl. Phys.* A, **610**, 306c–316c (1996).
40. Weber, M., et al., "The NA52 strangelet and particle search in Pb+Pb collisions at 158AGeV/c," in *Proc. 6th Int. Conf. on Strange Quark Matter (SQM2001), Frankfurt am Main, Germany, 24–29 September 2001*, 2002, vol. 28 of *J. Phys. G: Nucl. Part. Phys.*, pp. 1921–1927.
41. Soff, S., Ardouin, D., Spieles, C., Bass, S. A., Stöcker, H., Gourio, D., Schramm, S., Greiner, C., Lednicky, R., Lyuboshitz, V. L., Coffin, J.-P., and Kuhn, C., *J. Phys. G: Nucl. Part. Phys.*, **23**, 2095–2105 (1997).
42. Harris, J. W., et al., *Nucl. Phys.* A, **566**, 277c–286c (1994).
43. Spieles, C., Gerland, L., Stöcker, H., Greiner, C., Kuhn, C., and Coffin, J. P., *Phys. Rev. Lett.*, **76**, 1776–1779 (1996).
44. Angelis, A. L. S., Bartke, J., Gładysz-Diaduś, E., and Włodarczyk, Z., *EPJdirect* C, **9**, 1–18 (2000).
45. Angelis, A. L. S., Bartke, J., Bogolyubsky, M. Y., Filippov, S. N., Gładysz-Diaduś, E., Kharlov, Y. V., Kurepin, A. B., Maevskaya, A. I., Mavromanolakis, G., Panagiotou, A. D., Sadovsky, S. A., Stefanski, P., and Włodarczyk, Z., *Nucl. Phys.* B *(Proc. Suppl.)*, **75A**, 203–205 (1999).
46. Gładysz-Diaduś, E., and Włodarczyk, Z., *J. Phys. G: Nucl. Part. Phys.*, **23**, 2057–2067 (1997).

Multi-Quark Hadrons and S=-2 Hypernuclei

D. E. Kahana* and S. H. Kahana*

*Physics Department, Brookhaven National Laboratory, Upton, NY 11973, USA

Abstract.
The general character of 4-quark (mesonic) and strange 6-quark (baryonic) quark systems is very briefly reviewed a la Jaffe, *i. e.* in the MIT bag, and so far still possibly viable candidates are indicated. Concentration is on $S = -2$ systems. Traditionally, one employs the (K^-, K^+) reaction on a relatively light target and hopes to retain two units of strangeness on a single final state fragment. Alternatively, heavy ion reactions can be used to produce Λ-hyperons copiously and one seeks to observe coalescence of two of these particles into the lightest $S = -2$ nucleus, the H-dibaryon. The complications arising from the presence of a repulsive core in the baryon-baryon interaction on the production of the H are discussed. Also considered is the possible presence in the data from the AGS experiment E906, of slightly heavier $S = -2$ nuclei, in particular $^4_{\Lambda\Lambda}$H.

INTRODUCTION

One very interesting question which arises in our search for examples of quark-gluon matter is the apparent absence, or at least the paucity of examples of elementary hadrons possessing more than three quarks. There is perhaps one good candidate for an exotic meson consisting of two quarks combined with two anti-quarks: *viz* the π_1 1410 thought to be an $I^G(J^{PC} = 1^-(1^{-+})$ state [1], but there is not likely an equally good candidate for a $gg + ggg + \cdots$ state or glueball. The perhaps more distinguishable H-dibaryon [2] has also not yet shown up on its own in any experimental search and it has proved to be equally elusive in theoretical analysis. Surely, however, strange matter must be present at the heart of virtually all gravitationally collapsed objects [3]. Nothing new is offered here with respect to the mesonic possibilities: but the presence of more than a single strange quark in dibaryons and light nuclei is explored in more detail.

During this presentation, we wish to cover two apparently disparate subjects: (1) the production of the H-dibaryon in both elementary and heavy ion induced reactions, and (2) the generation of very light to moderately light $S = -2$ hypernuclei. Both subjects concern $S = -2$ systems and they are quite possibly tied together by the possible presence within finite systems of a hybrid H, possibly constructed both from dibaryon $\Lambda\Lambda$ and from 6-quark bag like $(uuddss)$ components, *viz*:

$$|\Psi\rangle = \alpha|\Lambda\Lambda\rangle + \beta|q^6\rangle \tag{1}$$

with α, β being amplitudes for the two-body and 6-quark components of the hybrid state. The purely Jaffe-like H state [2] corresponds to $\beta = 1$. Our later coalescence calculation for the formation of an H is independent of this parameter. The procedure we follow to estimate the effect of a repulsive core on entry from a doorway $\Lambda\Lambda$ state into the final H is applicable to either the pure bag or hybrid cases.

CP644, *Exotic Clustering: 4th Catania Relativistic Ion Studies*, edited by S. Costa, A. Insolia, and C. Tuvè
© 2002 American Institute of Physics 0-7354-0099-7/02/$19.00

We begin by indicating that the seeming absence of the H in existing searches is perhaps attributable to repulsive (soft-core) forces in the baryon-baryon system, which prevent penetration to short range of a $\Lambda\Lambda$ pair during any mechanism for formation of the dihyperon structure. However, one might anticipate to the contrary, that within a finite nucleus two Λ's could be held together for sufficient time to permit a short range structure to develop.

Since the first of these subjects, H-suppression, has been described elsewhere [4], it will only be briefly dealt with here. To our knowledge all theoretical estimates of production rates [5, 6], irrespective of mechanism, have overlooked the possibility of a repulsive core in the baryon-baryon interaction at short distances. As we show, under reasonable assumptions the core can lead to an appreciable diminution of H yield. We introduce this device in the context of heavy ion collisions where a previous calculation [6] suggested a high formation probability, ~ 0.07 per central Au + Au collision. The AGS experiment E896 [7] is presently analysing some 100 million central Au + Au events and could, in the light of this previous estimate, have provided a definitive search for the H. In Reference [4] we presented an estimate of the extent to which a repulsive core might interfere with this hope.

We put forward first the simplest possible theoretical description of the multi-quark systems; then we consider the, successful or otherwise, experimental searches for such objects. This is followed by a description of the related attempts to produce multi-strange nuclei.

BAG MODEL ANALYSIS OF MULTI-QUARK STATES

Soon after the introduction of the MIT bag model [8], used to consider the normal hadrons, mesons constructed from a single quark-antiquark pair and baryons containing three quarks, Jaffe [2] proposed the insertion into the bag of extra valence quarks could produce more exotic systems. For meson states these additional components could be quarks only $(q\bar{q})^2$ or could be hybrids of quarks and glue $(q\bar{q},g)$. Glueball states, $(gg + ggg + \cdots)$, were already on the table. Jaffe also suggested the existence of the H-Dibaryon which could be described as a 6-quark bag (uu,dd,ss) [2]. Of course, the easiest meson candidates to identify would be the so-called exotics, $i.\ e.$ those with quantum numbers which cannot arise from $(q\bar{q})$ alone. Examples of exotic quantum numbers are states characterised by $J^{PC} = 0^{--}, 0^{+-}, 1^{-+}, 2^{+-}$.

The basic theory covering all these states in the bag model is essentially the same. The total energy for N quarks in a bag of radius R may be written [2]:

$$E(N) = E_Q + E_V + E_0 + E_G \tag{2}$$

with a quark kinetic energy

$$E_Q = \frac{1}{R}\sum_{i=1}^{N}[k(m_iR)^2 + (m_iR)^2], \tag{3}$$

271

where k is a wave number, and

$$E_G = -\frac{g^2}{2}\sum_{i<j}\sum_{a=1}^{8}\int d^3x\,(\bar{B}_i^a\cdot\bar{B}_j^a) = -\frac{\alpha_c}{R}\sum_{i<j}\sum_{a=1}^{8}M(m_iR,m_jR)\,[\bar{\sigma}_i\cdot\bar{\sigma}_j]\,(\lambda_i^a\lambda_j^a)\,. \quad (4)$$

Here $M(m_iR,m_jR)$, is the color magnetic energy of one-gluon exchange from quark to quark. The other terms $E_V = BV$ and $E_0 = -z_0/R$ are the volume bag pressure and the bag zero point energy respectively. The bag parameters suggested by analysis of the normal hadron spectrum involve a rather large gluon exchange coupling $\alpha_c = (g^2/4\pi) = 0.55$, a bag constant $B^{1/4} = 146$ MeV, vacuum energy $z_0/R = 1.84$, and finally a rather large mass for the strange quark, $m_s = 279$ MeV. The matrix element, $M(m_iR,m_jR)$, obtained in the bag for the color-magnetic operators is approximated by a diagonal matrix for strange quarks; $M = M((n_s/N)m_sR,(n_s/N)m_sR)$ with n_s the number of strange quarks.

To proceed, one constructs anti-symmetric wave functions in color, spin and flavour, diagonalises the energy exploiting the existing $SU(6) = SU_c(3) \times SU_s(2)$ symmetry and then minimises the whole with respect to bag radius. The general range found for the masses of $(q\bar{q})^2$ states is from 1460 MeV ($S = 0$) to 2140 ($S = 4$) MeV. A reasonably good candidate for such an exotic is the $\pi_1(1400)$, referred to above [1], identified in π^-N scattering. In particular Thompson et al. [1] find a resonant final state (η,π^-) component to which they assign a mass of $1370 \pm 10^{+50}_{-30}$ MeV and a width of $385 \pm 40^{+65}_{-105}$ MeV. No state of similar status has been isolated for other quark hybrids or glueballs.

The $S = -2$ H-dibaryon, considered more extensively hereafter, is found by Jaffe, through reasoning similar to that above, to be some 80 MeV bound.

H-DIBARYON

This highly symmetric object is in principle more likely to exist than its meson-like compatriots. The H possesses conserved baryon and strangeness numbers, viz $B = 2, S = -2$. If it were a purely hadronic state its wave function would appear as

$$\Psi_H = \sqrt{\frac{1}{8}}|\Psi_{\Lambda\Lambda}\rangle + \sqrt{\frac{3}{8}}|\Psi_{\Sigma\Lambda}\rangle + \sqrt{\frac{4}{8}}|\Psi_{\Xi N}\rangle. \quad (5)$$

More recent estimates of the mass of the H have differed from Jaffe's original estimate and from each other. Indeeed, there is no consensus among theorists that the H is in fact a bound object. There is, however, general agreement on the results of the many searches which have been made for this state: it is not yet found.

Conventional hypernuclear studies [5] exploit the elementary processes $p(K^-,K^+)\Xi^-$, $p(\Xi^-,\Lambda)\Lambda$ The first produces an effective Ξ^- beam, incident on another nucleus in the target and thus generates a di-lambda. The latter pair may or may not form the putative H. The same experimental approach could of course yield doubly strange hypernuclei. An advantageous target for the Ξ^- beam is the deuteron, resulting hopefully in an H plus a monoenergetic neutron. The latter is relatively easily identified.

Another approach, used in the BNL experiment E896 [7] exploits the large numbers of Λ's, some 30 per event, generated in relativistic $Au + Au$ collisions. Here again the results have been so far negative.

One good reason for the lack of success of these experiments lies in the nature of the production of a bound di-lambda state. For two strange baryons to coalesce into a bag they must first penetrate the mutually repulsive core of the potential. Suppression of the yield for a spatially very extended object like the deuteron is minimal. This is not so for the H, which consists of six quarks in a bag comparable in size to that of a single baryon, so that short range repulsion, found in NN interactions and expected to exist for strange baryon-baryon interactions as well, could play a considerable role. H formation from two Λs can be viewed as proceeding in two steps: merging into a broad deuteron-like state followed by barrier penetration into the bag. The overall rate is then proportional to the product of the probability of coalescence [9] with a prefactor giving the penetration probability. Of course there are unknowns, one the $\Lambda\Lambda$ separation a at which the two bags would dissolve into a single bag, the other the nature of the short range forces after dissolution. The first, a, we treated as a parameter; the second we took from the Bonn potential [10], limiting our considerations to the shortest ranged ω and σ components. Since the exchange of a ϕ meson between s quarks very nearly matches that of an ω between ordinary quarks, the Bonn interaction needs little readjustment for strange-strange interaction. Thus we take the short range force, appropriate to the penetration of the core and depicted in Fig. 1, to be of the form

$$V(r) = V_\omega(r) + V_\sigma(r), \tag{6}$$

where

$$V_i(r) = \frac{g_i}{r} \exp(-m_i r). \tag{7}$$

The strong short range σ attraction reduces the effect of the hard core, while the longer range parts of the force are assumed to play a negligible role. The two baryons approach to some outer radius a, in fact a turning point, before being faced with the strong repulsion. The calculation is especially sensitive to this separation a. Although our final results on barrier penetration are consequently somewhat uncertain, it will become apparent that the one thing one cannot do is to ignore them.

Barrier penetration in an effective two body model can the quantified in the transmission coefficient [11] at relative energy E:

$$T(E) = 4\exp(-2\tau), \tag{8}$$

where

$$\tau = \int_a^b dr \sqrt{2m(V(r) - E)}. \tag{9}$$

The chief results, as they relate to coalescence into di-baryons in Au + Au collisions, are presented in Fig. 2, indicating the variation of H-dibaryon yield with a. This latter parameter must not be thought of as an effective hard core radius for the $\Lambda\Lambda$ interaction. The underlying quark-quark forces may be viewed as possessing a repulsive short range component due to the exchange of vector mesons [12]. Even with complete overlap of the parent baryons the average interquark separation for a uniform spatial distribution

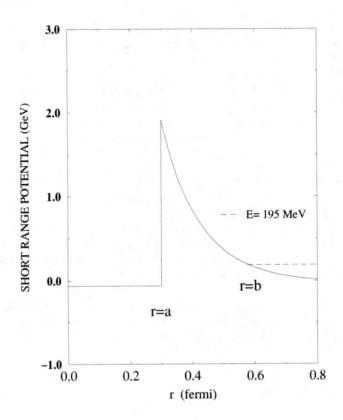

FIGURE 1. The short range ΛΛ potential taken from the σ and ω exchange parts of the Bonn potential. At some separation *a* the two strange baryons are assumed to dissolve into a single bag, represented here by a shallow, attractive potential. Barrier penetration at relative energy *E* from the di-lambda doorway state is represented by the dashed line.

is comparable to the parent radius $R \sim 0.8$ fermis, *i. e.* considerably greater than any conceivable fixed hard core. We have concentrated on baryon centers between 0.25 and 0.40 fermis apart as a reasonable range.

At the largest *a* suppression from the repulsive forces is not inconsiderable, but for the smallest *a*-values observation of the H, should it exist, becomes difficult. Early analysis of the actual experimental setup using the heavy ion simulation ARC [13, 14], suggested a neutral background comparable with the initial estimate of 0.07 *H*'s per central collision. For baryon separations of 0.25 to 0.35 fermis one would have to achieve a tracking sensitivity of 10^{-4} to 10^{-2} relative to background. Even in the worst case scenario one is still left with perhaps a few thousand sample dibaryons, from the very large number of central Au + Au events, but they are immersed in what may prove to be a daunting background.

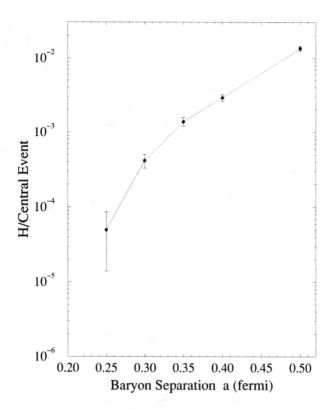

FIGURE 2. Absolute H Production per Central Au+Au collision at 10.6 GeV. The precise separation of Λ bag centers, a, at which a single six quark bag forms is of course not known, but a reasonable value is likely less than 0.3 fm. Even for complete baryon overlap, the average distance between constituent quarks is greater than the conventional nucleon-nucleon hard core radius of 0.4 fm.

We conclude that short range repulsion between strange baryons can profoundly hinder coalescence into objects whose very existence depends on the presence of important bag-like structure. This lesson applies even more to the many H-searches initiated in (K^-, K^+) reactions [5], since these generally involve even lower relative energies E. Unfortunately the very same repulsive forces which made coalescence into a bound state difficult, may also, at the quark level, destroy the existence of any such state.

DOUBLY STRANGE HYPERNUCLEI

There is perhaps one way to circumvent this frustrating barrier to the discovery of the lightest of all possible strangelets. In the event a pair of Λs is attached, through a (K^-, K^+) reaction, to a light nucleus, a hybrid H may form, *i. e.* the H described above containing both dibaryon and six quark bag components. In a light nucleus, for example $_{\Lambda\Lambda}^4$H, the extra nucleons in this four particle system keep the captured hyperons together for some 100 picoseconds, far more than enough time for penetration of the rather modest $1 - 2$ GeV barrier between them. In any case the search for $S = -2$ hypernuclei, the only strangelets we are certain exist, is highly interesting in its own right. An ongoing AGS experiment, E906 [15], focuses on these nuclei and as we indicate may have already provided evidence for their existence [15].

In E906 [15] the (K^-, K^+) reaction on a ^{9}Be target is employed to produce a tagged beam of Ξ^-s. The Ξ^- in turn may convert into a pair of Λs by interaction with a proton, either in the nucleus in which it was produced or by subsequent interaction with another ^{9}Be nucleus. It is in an emulsion experiment at KEK [16] that one such hypernuclei was found with perhaps other examples from Prowse [17] and Danysz et al [18]. Indeed, all three of these candidates for double hypernuclei were interpreted as possessing a $\Lambda\Lambda$ pairing energy $\Delta B_{\Lambda\Lambda} \sim 4.5$ MeV [19] . Such a high value is unexpected from existing, albeit theoretical, knowledge of hyperon-hyperon forces. It was then possible to surmise interesting activity in the $\Lambda\Lambda$ system at short range separation.

Given the K^- beam energy of 1.8 GeV, the tagged Ξ^-s initially possess considerable kinetic energy, ~ 140 MeV, and are more likely than not to escape the nucleus in which they are produced. A guiding principle we employ in qualitatively understanding the broad aspects of the data, is that final states containing the very stable ^{4}He are favoured. This picture is also mindful of an oft used cluster model for ^{9}Be as two α-particles joined together by a weakly bound neutron. The ^{9}Be target is useful in slowing down the Ξ^-, but the observed reactions are for the most part initiated essentially on an α.

Further, in the present experiment which observes $S = -2$ final states by their decay into two momentum-correlated π^-'s, the heaviest $A = 9, 8, 7, \cdots$ systems will decay weakly predominantly through non-mesonic channels. Thus the lightest systems are probably doubly favoured in our data, through both their production and their decay dynamics.

ANALYSIS OF THE E906 DATA

The measurements which contain evidence for possible $\Lambda\Lambda$ hypernuclei are highlighted in Fig. 3. Displayed are the two-dimensional, Dalitz plot for the total sample of correlated two π^- decays recorded in the CDS detector [15], and the projections of this data on the P_L and P_H axes. These denote the low and high momenta for the π_L and π_H pair detected in an event. The circled feature in the 2-D plot is of principle interest, giving rise to the prominent peaks near $P_H = 114$ MeV/c in insert I and a correlated peak at $P_H \sim 104$ MeV/c in II. The experiment is searching for pairs of pion momenta unexplained from previous knowledge of hypernucleon lines. The expected spectrum close to the region of interest in E906 are shown in Fig. 4, with single hypernuclear lines indicated vertically and doubly strange candidates diagonally. The slope in the latter denotes the dependence of the $S = -2$ energy as a function of the important pairing energy $\Delta B_{\Lambda\Lambda}$.

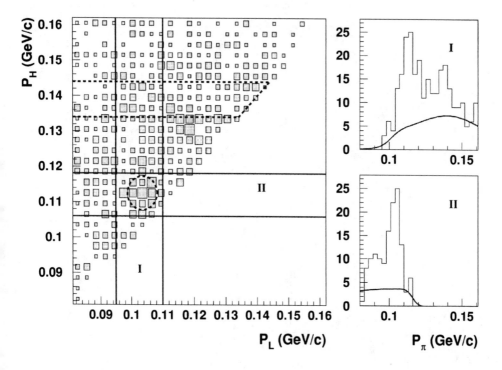

FIGURE 3. E906 Two Pion Data. The Dalitz plot together with projections onto high (inset I) and low (inset II) momentum axes describe the approximately 1000 recorded, correlated, π^- events. The interesting region is circled in the 2-D plot. I contains a broad π_H peak with components from $_\Lambda^3H$ and the two-body decay of $_{\Lambda\Lambda}^4H$. Inset II contains an unexpected narrow π_L peak interpreted as arising from an alternative decay of $_{\Lambda\Lambda}^4H$ involving a resonance in $_\Lambda^3He$.

The higher momentum structure in I near 137 MeV/c is understood as the decay in flight of Ξ^- hyperons, but the width of the lower peak is too broad to be due to a single component line. The very prominent peak in II is completely unexpected. The high momentum π^- peak in I arises in part from the decay of $_\Lambda^3H$ yielding a meson line at 114.3 MeV/c, but considering the experimental resolution of 2.5 MeV, this π_H prominence at 114 MeV must contain more structure.

The projection in II is constructed from a reverse cut $106 MeV/c \leq k^{\pi^-} \leq 120$ MeV/c, *i. e.* under the first compound peak in II, and thus should reveal the low momentum π_L^- correlated with the 114 MeV/c peak in I. The most striking feature in II is the narrow prominence near 104 GeV/c, which cannot for example be accounted for by single Λ decay in a $_\Lambda^3H + \Lambda$ final state, or for that matter by any other known line. Adding even

Expected Signals and Lines

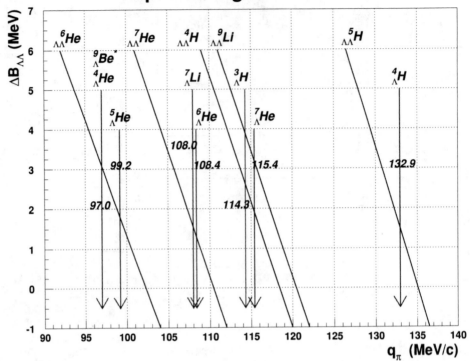

FIGURE 4. Expected $S = -1$ lines together with possible doubly strange hypernuclear energies, the latter given as a function of the pairing energy $\Delta B_{\Lambda\Lambda}$.

a small kinetic energy to the Ξ^- initiating the reaction leading to the final $S = -2$ state would broaden the single Λ decay well beyond that measured for the the dominant peak in this reverse cut. The correlation of this peak in II with the excessively broad dominant peak in Fig. 3 I is strong evidence for the presence of a light double-Λ species in the data.

Fig. 4, [20], indicates where known single Λ hypernuclear lines are expected as well as where $S = -2$ lines are anticipated as a function of the pairing energy $\Delta B_{\Lambda\Lambda}$. Reiterating, the most interesting feature in the present data, centered at $112 - 114$ MeV/c, could only result from the production of $^4_{\Lambda\Lambda}$H and/or $^3_{\Lambda}$H.

$^4_{\Lambda\Lambda}$H has two generic modes of decay:

$$^4_{\Lambda\Lambda}\text{H} \rightarrow {^4_{\Lambda}\text{He}} + \pi_2^-, \tag{10}$$

or

$$\,_{\Lambda\Lambda}^{4}H \rightarrow \,_{\Lambda}^{3}H + p + \pi_1^-. \tag{11}$$

The two body decay in the first mode yields the high energy π_H^- from which the $\,_{\Lambda\Lambda}^{4}H$ $\Delta B_{\Lambda\Lambda}$ can in principle be estimated, and is followed by a three body decay producing the correlated π_L.

The very narrowness of the π_L^- peak in II suggests that the decay mode in Equation 7 is not strictly three body in character. We offer as a candidate a resonance in $\,_{\Lambda}^{4}He$, arrived at by decay from the clearly spatially extended $\,_{\Lambda\Lambda}^{4}H$ and thus not favoured in production from $K^- + {}^4H$. The first, lower, π_1^- momentum from the initial decay would then be sensitive to the sum $E_R + \Delta B_{\Lambda\Lambda}$. Preliminary theoretical calculations [21] indicate that indeed a narrow p-wave resonance can be placed near $E_R = 0.5 - 1.5$ MeV in the p + $\,_{\Lambda}^{3}H$ system , using a potential consistent with the rather extended geometry of $\,_{\Lambda}^{3}H$ and constrained by the known proton binding in the ground state of $\,_{\Lambda}^{4}He$.

COMMENTS AND CONCLUSIONS

There is little to add to the above discussion of multi-quark elementary systems. The search for the H is certainly complicated by our analysis of the mechanism of its formation. The discovery and study of $S = -2$ hypernuclei will, however, continue either at the AGS or at the anticipated Japanese Hadron Facility.

For E906 there were 1040 events in the total two prong sample, some 70 above background in the first feature in Fig. 3 I. The reverse cut in II can be used to isolate the three body decays of $\,_{\Lambda\Lambda}^{4}H$ and from the simulation an estimate of the two body, resonant, decays can also be made. We conclude then a lower limit of $40 + 20$ $\,_{\Lambda\Lambda}^{4}H$ have been produced in these two modes respectively. Unfortunately, the important di-lambda pairing energy is not well determined in the present experiment. A simulation of Fig. 3 I suggests $\Delta B_{\Lambda\Lambda} \sim 1 - 2$ MeV while that in Fig. 3 II involving the combination with E_R is perhaps described by a value closer to $0.5 - 1.0$ MeV. More recent emulsion experiments [22] have definitively identified the species $\,_{\Lambda\Lambda}^{6}He$ and agree with the lower values of $\Delta B_{\Lambda\Lambda}$, finding something less than 1 MeV. The motivation for a hybrid H living within some light nucleus then recedes.

Clearly, better statistics and an improvement in resolution are required to determine the $\Lambda\Lambda$ pairing energy definitively, and more importantly to establish the actual presence of the large numbers of light doubly strange hypernuclei suggested by E906. The principal issues to be settled, presumably in a follow on experiment, are the existence or not of more than one contribution to the broad lower peak in the higher momentum π_L^- spectrum and the very exciting possibility that more than one species of double Λ resides in the data. Significantly higher counting rate should allow cuts to be placed on the Ξ^- kinetic energy and determine the nature of the reaction mechanism producing individual doubly strange species.

ACKNOWLEDGEMENTS

This manuscript has been authored under US DOE contracts and DE-AC02-98CH10886.

REFERENCES

1. D. M. Alde *et al*. Phys. Lett. **B205**, 397 (1986); H. Aoyagi *et al*. Phys. Lett. **B334**, 246 (1993); D. R. Thompson *et al*. Phys. Rev. Lett.**79**,1630 (1997).
2. R. L. Jaffe, Phys. Rev. **D15**, 267(1977); **D15**, 281 (1977); R. L. Jaffe, Phys. Rev. Lett **38**, 195(1977); **38**, (1977) 617(E).
3. S. H. Kahana, J. Cooperstein, and E. Baron, Phys. Lett. **B196**, 259 (1987).
4. D. E. Kahana and S. H. Kahana, Phys. Rev. C **60**, 065206-1 (1999).
5. B. Bassalleck, Nucl. Phys.**A639**, 401 (1998) and included references; J. Beltz *et al*., Phys. Rev. Lett **76**, 3277 (1996); T. Iijima, P. H. D. Dissertation Kyoto University (1995); as well Brookhaven National Laboratory experiments E813, E836, E885, E888, and KEK experiments E224 and E248.
6. A. J. Baltz *et al*., Phys. Lett **B325**,7 (1977).
7. H. Crawford, Nucl. Phys.**A639**,401 (1998).
8. A. Chodos *et al* Phys. Rev **D9** 3471 (1974).
9. D. E. Kahana *et al*., Phys. Rev. C **54**, (1996) 338.
10. R. Machliedt, K. Holinde and Ch. Elster, Physics Reports **149**, 1 (1987).
11. A. Messiah, Quantum Mechanics, Interscience, (New York, 1961).
12. S. Kahana and G. Ripka, Nucl. Phys.**A429**, 462 (1984); Phys. Lett. **B278**,11 (1992).
13. Y. Pang, T. J. Schlagel, and S. H. Kahana, *et al*.. Phys. Rev. Lett **68**, 2743 (1992);
14. E. Judd, Private communication.
15. J. K. Ahn *et al* Phys. Rev. Lett **87** 132504 (2001).
16. S, Aoki *et al*., Phys. Rev. Lett **65**, 1729 (1990); S, Aoki *et al*., Prog. Theor. Phys. **85**, 1287 (1991).
17. D. Prowse, *Phys. Rev. Lett***17**, 782 (1966).
18. M. Danysz *et al*., Nucl. Phys.**49**, 121 (1963).
19. C. B. Dover, D. J. Millener, A. Gal and D. H. Davis, Phys. Rev. C **44**, (1991) 1905.
20. Y. Yamamoto *et al*, Nucl. Phys.**A625**, 107 (1997).
21. D. E. Kahana, S. H. Kahana, and D. J. Millener, Resonances in the Production of S=-2 hypernuclei, Poster Session, INPC2001, Berkeley CA, July 2001.
22. H. Takahashi *et al*., Phys. Rev. Lett. **87** 212502 (2001).

Strangeness Production in Heavy Ion Collisions: What Have We Learnt with the Energy Increase from SPS to RHIC

Grazyna Odyniec

Lawrence Berkeley National Laboratory, Berkeley, CA 94720, USA

Abstract. A review of strange particle production in heavy ion collisions at ultrarelativistic energies is presented. The particle yields and ratios from SPS and RHIC are discussed in view of the newest developments in understanding collision dynamics, and in view of their role in the search for a quark gluon plasma. A strangeness enhancement, most notably observed in CERN Pb-beam results, shows a remarkable two fold global enhancement with a much larger effect seen in the case of multistrange baryons. Hadronic models did fail to explain this pattern. At RHIC energy strangeness assumes a different role, since temperatures are higher and the central rapidity region almost baryon-free. An intriguing question: "Did RHIC change the way we understand strangeness production in heavy ion collisions ?" is discussed.

INTRODUCTION

For the last two decades [1] strange hadrons have been expected to contribute characteristic signals towards the overall understanding of dense hadronic matter and its hypothetical transformation to partonic stage consisting of quarks and gluons (for a review see e.g. [2]).

It has been predicted that nuclear matter will undergo a phase transition into a Quark Gluon Plasma (QGP) at a critical temperature near the rest mass of the pion and at about 10 times the density of normal nuclear matter. These predictions have been supported by lattice gauge calculations, where the presence of a large jump in the energy density for two and three flavor systems at critical temperature, T_c, of 160 MeV has been shown [3, 4, 5] - see Fig.1.

Apparent difficulties in introducing quark mass to the lattice calculations preclude a definite statement on the order of the phase transition. This could be resolved by experiment, i.e. the experimental results may point towards the right order of phase transition (first or second), or perhaps towards a cross-over from one side of the diagram to the other without any new behavior at the cross-over point.

Accordingly, complex heavy ion experimental programs have been launched on both sides of the Atlantic with the goal to investigate the properties of strongly interacting matter at high density and temperature in general, and to confirm theoretical predictions of the formation of a quark gluon plasma phase in particular.

Since 1986, research centers in the US (AGS, top energy $\sqrt{s} = 4.9$, heaviest system Au+Au) and in Europe (CERN, top energy $\sqrt{s} = 17.3$, heaviest system Pb+Pb) have provided a very impressive wealth of experimental information.

CP644, *Exotic Clustering: 4th Catania Relativistic Ion Studies,* edited by S. Costa, A. Insola, and C. Tuvè

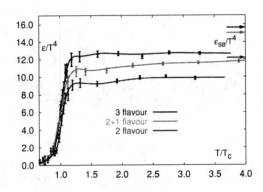

FIGURE 1. QCD lattice calculations with two and three flavor systems from ref [5], see text.

The newest and the most powerful machine, the Relativistic Heavy Ion Collider (RHIC) is located in USA at the Brookhaven National Laboratory on Long Island. It is capable of accelerating a variety of heavy ion beams ranging from protons to gold. The top energy is 250 GeV per beam for protons and 100 GeV per nucleon per beam for gold. RHIC has already completed two run cycles with gold beams of 130 GeV (in 2000) and 200 GeV (in 2001), as well as a first cycle of p+p running at 200 GeV (in 2001). The first results of data analysis from the heavy ion program, although often preliminary, are widely discussed.

In this paper, selected results on strangeness production at CERN SPS ($\sqrt{s}$ = 17, 12 and 9 GeV) and RHIC ($\sqrt{s}$ = 130 and 200 GeV) energies will be reviewed, with main emphasis on their impact for the search for a deconfined quark-gluon plasma state.

RESULTS OF CERN SPS PHASE I PROGRAM AND CHEMICAL EQUILIBRATION

The original idea of strangeness enhancement as a quark-gluon plasma signal was based on the estimate that the strangeness equilibration time in QGP is of the same order ($\approx$ 10 fm/c) as the expected life time of the fireball formed during A+A collisions, while it is much longer (30-40 times) in a hot and baryon-rich hadronic system [6]. This quantitative idea was supported by another, namely that the strange (and anti-strange) quarks are thought to be produced more easily through gluon fusion and hence more abundantly in a deconfined state compared to the production via threshold suppressed inelastic hadronic collisions. Enhanced strangeness production could be seen most easily in the yield of kaons, the most abundantly produced strange particles. And indeed, measurements from the fixed target CERN SPS experiment NA35/NA49 (S+S, S+Ag and Pb+Pb collisions at 158 GeV/c) demonstrate a significant increase in the K/π ratio, approximately by a factor of 2, with respect to pp reactions at the same energy [7, 8] - see Fig.2.

Surprisingly, the strangeness/entropy ratio ($\sim$ K/π, Fig.2) has the same value for a

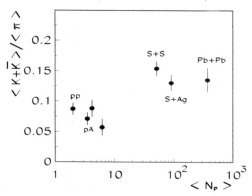

FIGURE 2. Multiplicity ratio $\langle K + \bar{K} \rangle / \langle \pi \rangle$ in full phase space for pp, pA and AA collisions plotted versus the average number of participating nucleons.

loosely bound system (S+S) as for very densely packed one (Pb+Pb). The value for S+Ag collisions is similar to S+S and Pb+Pb. This clearly eliminates the hypothesis of re-scattering and final state interactions being a possible source of the observed strangeness enhancement. While the effect is quite spectacular, it has been argued [9, 10], that a factor 2-3 enhancement can be only considered as an indirect signal for QGP formation.

A much more suitable QGP signal candidate [1] would be the enhanced yield of hyperons (particularly multi-strange anti-hyperons), as the high production thresholds in the various binary hadronic reaction channels [6] preclude the possibility of their abundant formation in a hadronic gas. Experiment WA97 (continued by NA57) was designed to make this measurement, and indeed, the enhancement was observed. Experiment NA49 reports a similar observation. The precise measurements of strange hyperons and their anti-particles reveal a systematic increase with respect to p+Pb (Λ, Ξ, Ω - WA97/NA57 data [11]), and with respect to p+p (Ξ - NA49 data [12]) collisions.

Fig.3 shows particle yields per participant for the five centrality intervals in the NA57 experiment. The two groups (particles with at least one valence quark in common with the nucleon and particles with no valence quarks in common with the nucleon) are kept separate since it is experimentally known that they may exhibit different production mechanisms. The WA97 results (4 centrality classes) are also presented. All yields are shown relative to the WA97 p+Be data. The horizontal line represents the predicted scaling from p+Be to central Pb+Pb with the number of wounded nucleons. All particle yields are enhanced in Pb+Pb collisions and enhancement grows with the strangeness content of the hyperon, up to a factor of about 15 for the $\Omega + \bar{\Omega}$. This is probably the most dramatic effect in the entire CERN SPS heavy ion program, and of course it does not have an explanation within any hadronic gas model. On the other hand, assuming a short lived plasma presence and following simple, statistical coalescence calculations, one concludes that the abundantly produced strange (anti-strange) quarks

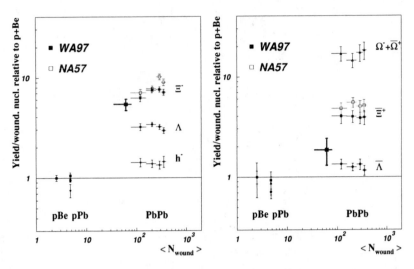

FIGURE 3. Yields per wounded nucleon relative to the p-Be yields as a function of the number of wounded nucleons, for negative particles, Λ and Ξ^- (left) and for $\bar{\Lambda}$, $\bar{\Xi}$ and $\Omega + \bar{\Omega}$ (right).

could be redistributed to combine with the light quarks (anti-quarks) to form the strange baryons (anti-baryons), which may eventually come close to their equilibrium values. Such behavior of nearly chemically equilibrated populations in Pb+Pb collisions at CERN SPS has been demonstrated by comparing data to thermal model [13]. And indeed, the extracted value of the temperature parameter is very close to the critical temperature T_c obtained by thermodynamical lattice QCD calculations.

CERN SPS PHASE II RESULTS

In order to investigate the production of strangeness systematically and in more detail, Phase II of the CERN heavy ion experimental program was set out with lead beams of 40 and 80 GeV/c, to be complimented in the fall of 2002 by additional runs at 20 and 30 GeV/c.

The results obtained so far from top SPS energy suggest some kind of chemical equilibration in Pb+Pb collisions at 158 GeV/c, whereas the lower energy data from AGS at 11 GeV/c (not discussed here, for more information see [14] and references there) was convincingly explained with hadronic rescattering and final state interactions. One would then expect that the energy at which the anomalies in pion and strangeness production set in may be located between these energies, and a dedicated energy scan may allow to identify the point of transition from trivial effects of secondary interactions to the new phenomena observed at CERN SPS experiments.

The first results from the 40 and 80 GeV runs [15] turned out to be very interesting. Comparison of these measurements to data at other energies is shown in figure 4. The K^-/π^- ratio increases steeply in the AGS energy range [16, 17, 18, 19, 20, 21, 22] and

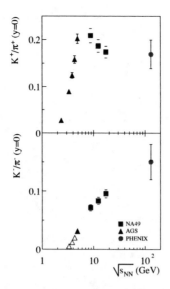

FIGURE 4. The energy dependence of the midrapidity K^+/π^+ and K^-/π^- ratios in central Pb+Pb and Au+Au collisions. The results of NA49 are indicated by squares. Open triangles indicate the A+A results for which a substantial extrapolation was done.

slowly saturates in the SPS [15] region. The K^+/π^+ exhibits a very different behavior: a steep increase at low energies is followed by a rapid turnover around 40 GeV/c into a decreasing trend. The K^+/π^+ ratio for p+p (open symbols) has a rather large error bar, but nevertheless it suggests a monotonic increase with energy.

The difference between the dependence of K^+ and K^- yields on $\sqrt{s}$ can be attributed to their different sensitivity to the baryon density. K^+ (and K^0) carry a dominant fraction of all produced $\bar{s}$ quarks, and therefore in isospin symmetric collisions their yield is nearly proportional to the total strangeness production and only weakly sensitive to the baryon density. Whereas the large fraction of s quarks is carried by hyperons, and therefore the total number of produced anti-kaons (K^- and $\bar{K}^0$) reflects both the strangeness yield and the baryon density.

The energy dependence of K/π ratio has been discussed within a number of models with and without explicitly invoking a QGP phase. Figure 5 shows comparisons with "non plasma" models: statistical hadron gas model, RQMD and UrQMD. The hadron gas model, originally proposed by Hagedorn [23], later developed by many others [24, 25, 26, 27, 28], is probably the simplest and the most intuitive one. It assumes that the hydrodynamical freeze-out creates a hadron gas in equilibrium independently of energy. A recently proposed version [29] was extended to include the energy dependence of thermal parameters (temperature and baryon chemical potential). By construction, the prevailing trend in the data is reproduced by the model, but the decrease of the ratio between 40 and 158 GeV/c is not well described. The hadron-string models, RQMD [30, 31, 32] and UrQMD [33] treat the hadron production in A+A collisions within a

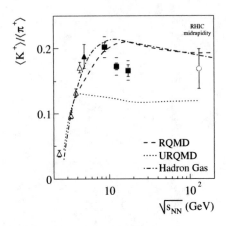

FIGURE 5. The energy dependence of the full phase space $\langle K^+\rangle/\langle\pi^+\rangle$ ratio in central Pb+Pb and Au+Au collisions. The experimental data are compare to model predictions, see text.

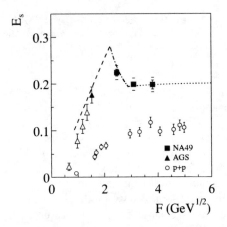

FIGURE 6. The energy dependence of E_s ratio in central Pb+Pb (Au+Au) and p+p collisions. The experimental results are compared with predictions of SMES. Open triangles indicate the A+A results for which a substantial extrapolation was necessary, see text.

string-hadronic framework as a starting point, then extend it with the effects relevant to A+A collisions, like hadronic rescattering and string-string interactions. Both of them, RQMD and UrQMD, like the hadron gas model, fail to describe the decrease of K^+/π^+ at the SPS energy range.

The observed behavior can be, however, understood with the Statistical Model of Early Stage (SMES) [7] - Figure 6 - which explicitly assumes the transient state of deconfined matter in Pb+Pb reactions at energies larger than 40 GeV. The K^+/π^+, replaced in Fig.6 by more general E_s ($E_s = (\langle\Lambda\rangle + \langle K+\bar{K}\rangle)/\langle\pi\rangle$) represents roughly the total strangeness to entropy ratio which in the SMES model is assumed to be preserved from the early stage till freeze-out. It is plotted as a function of the Fermi variable, F which is proportional to $\sqrt{s}$. The dramatic change of the behavior at $F \approx 2$ is attributed to the change of the mass of strangeness carriers at the phase transition point from $m_s \approx 500$ MeV (the kaon mass) to $m_s \approx 170$ MeV (the strange quark mass). Interestingly, the energy dependence of mean pion multiplicity in A+A collisions also shows the change of the behavior approximately at the same value of F [15], which is consistent with the expected increase of the number of degrees of freedom due to phase transition from hadronic gas to quark gluon plasma.

However, the most striking result, an enormously increased multi-strange hyperon production, does not find an explanation with any model so far. The NA57 experiment extended its centrality reach and measured hyperon yields in much less central collisions than WA97 in the first round of CERN experiments. The so called "fifth point" (see Figure 3), corresponding to a peripheral class of the events with approximately 60 wounded nucleons, drops by a factor of 2.6 for $\bar{\Xi}$ yields. Note that the enhancement seems to be saturated above a mean number of wounded nucleons of about 100. For the Ξ^- the yield drops by a factor of 1.3 in the most peripheral collisions. The centrality dependence of the Ξ^- looks different from that of the $\bar{\Xi}$, as the Ξ^- enhancement grows smoothly as a function of the number of wounded nucleons.

The drop in the $\bar{\Xi}$ yield per wounded nucleon, which might indicate the onset of the QGP phase transition, remains a tremendous challenge for the theorists and model builders (phenomenologists). While some attempts to fit WA97 data with the canonical statistical model were initially quite successful e.g. [34, 35], they predicted that the yield of multi-strange hyperons saturates at $N_w \approx 20$ (much too early). The additional constraint arising from the measurement of peripheral collisions ("fifth point") eliminated all previous explanations as they could not reproduce the centrality dependence of the observed hyperon production.

The fact that strangeness in A+A collisions is analyzed with respect to the Wounded Nucleon Model (WNM) imposes an additional complication and precludes a definite conclusion regarding possible QGP formation. The multiple collision mechanism, not understood so far, requires careful experimental studies in order to determine its impact on the final state (including strangeness content) of the collision. Only recently these measurements became available via experiments with p+A collisions at CERN energy (e.g. [36], next section).

LESSON FROM PROTON - NUCLEUS COLLISIONS

A very interesting lesson regarding strangeness enhancement was recently provided by the NA49 analysis of p+A data at the CERN SPS top energy. It has been reported [36] that the change of energy dependence of strangeness/pion ratio is not a unique feature of A+A collisions. It was also observed in pp and pA channels.

Moreover, the pA data clearly demonstrate [36] that the WNM overestimates the number of the participating nucleons in the collisions of protons and heavy nuclei. This calls into question the reliability of using this particular model for analysis of significantly more complex systems like A+A. Therefore, one has to be particularly

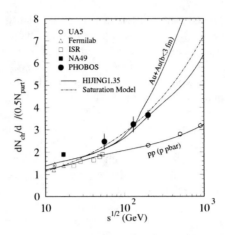

FIGURE 7. Energy dependence of charged particle density in central AA and pp collisions.

careful when analyzing particle yields as a function of the number of wounded nucleons (often used as a measure of the centrality of the collisions) in A+A .

RHIC ENVIRONMENT

The experimental program at CERN SPS has gathered a lot of data from heavy ion collisions establishing the strangeness enhancement effect. It has failed, however, to demonstrate that this is caused by QGP formation, because an effect has also been observed (to lesser degree) in systematic studies of proton-nucleus collisions.

The commissioning of RHIC provided the very first opportunity to explore a higher energy regime, with center of mass energies order of magnitude higher than those previously obtained. Thus, a new environment was expected with larger initial energy density, larger freeze-out volume, longer lifetime, larger freeze-out temperature, smaller baryonic chemical potential, smaller crossing time of two nuclei and, of course, increasing role of quark and gluon degrees of freedom.

Four experiments took data in the 2000/2001 runs: two large multi-purpose experiments, STAR [37] and PHENIX [38], surveying a new energy domain and two smaller ones, BRAHMS [39] and PHOBOS [40], with the specific physics agenda. Gold ions were accelerated during both runs, at 56 and 130 GeV per nucleon in 2000, and 200 GeV per nucleon in 2001. The very first RHIC physics result, the charged particle rapidity density at mid-rapidity has been measured by all four experiments with very good agreement among their results. Figure 7 shows the energy dependence of charged particle density $dN_{ch}/d\eta$, normalized to the number of participating nucleon pairs ($N_{part}/2$) which rises almost logarithmically with $\sqrt{s}$ from AGS up to RHIC energies. This dependence is very different from the $p\bar{p}/pp$ systematics, also shown in Fig.7. At 200 GeV about 65% more particles per pair of participants is produced in central Au+Au collisions than in $p\bar{p}/pp$. For central Au+Au collisions at $\sqrt{s} = 130$ GeV the global average is $dN_{ch}/d\eta = 580 \pm 18$, and it is only about 15% higher at $\sqrt{s} = 200$ GeV [41]. The maximum multiplicity is large, but not large enough to justify the simple idea of a first order phase transition from 3 pionic degrees of freedom to 37 gluonic degrees of freedom.

Most of the models and predictions, made before the data became available, failed to describe the multiplicity. The rare exception, the Hijing model, provides a fairly good description of the measured multiplicities [42].

The transverse energy density, $dE_T/d\eta = 578 + 26 - 39$ GeV, measured by the PHENIX experiment, in the 2% most central collisions [43] combined with the Bjorken formula [44] estimate the initial energy density to be $\varepsilon = 5$ GeV/fm^3, i.e. 60-70% larger than the corresponding value at the top SPS energy. Note that this is a very conservative estimate using formation time $\tau \approx 1$fm/c. Much larger energy density, of the order of 20 GeV/fm^3, would be obtained for shorter formation times advocated by the saturation model [45].

The BRAHMS collaboration data extended significantly the rapidity reach of the dN_{ch}/dy, because their spectrometer can cover a wide range of laboratory angles. The observed dN_{ch}/dy decreases rather fast away from mid-rapidity (e.g. by about 30% at rapidity of 3) [46], restricting the anticipated hypothesis of boost invariant dynamics in nuclear collisions at RHIC energies to the mid-rapidity interval.

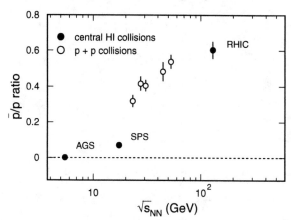

FIGURE 8. Mid-rapidity antiproton to proton ratios measured in central heavy ion collisions (filled symbols). The left end of the abscissa is the $p - \bar{p}$ pair production threshold in p+p.

Other very important information on the initial conditions of the matter created at the early stage of the collision is coded in the baryon number transport (or baryon stopping), which is established presumably very early in the collision. The degree of stopping is expected to affect the overall dynamical evolution of the matter in general, and thermal and/or chemical equilibration, so fundamental for QGP study, in particular. The baryon stopping can be addressed experimentally though analysis of baryon/antibaryon ratio. At RHIC energy the $\bar{p}/p$ ratio is much larger [47] than at SPS (0.07± 10%) or at AGS (0.00025 ±10%; practically zero), but still is significantly smaller than unity over the measured centrality range, indicating an overall excess of protons over antiprotons at the midrapidity. This implies that a certain fraction of the baryon number is transported from the incoming nucleus at beam rapidity to the midrapidity region even in peripheral

Au+Au collisions at $\sqrt{s} = 130$ and 200 GeV. Thus, at this energy the midrapidity region is not yet baryon free. Comparison of the $\bar{p}/p$ ratio at RHIC to lower energies indicates a dramatic increase in the importance of pair production mechanism with the rise of center of mass energy. This is demonstrated in Fig.8, where points for p+p collisions are also shown for reference. The baryon yields have consistently attracted interest not only due to the not yet understood baryon stopping mechanism, but primarily because baryon-antibaryon pair production (from $\bar{p}/p$ to $\bar{\Omega}/\Omega$) is expected to reflect the degree of availability of s and $\bar{s}$ quarks which are suppressed in hadronic matter due to the high mass of strangeness carriers. The sensitivity of course, is expected to be proportional to the strangeness content of the baryon-antibaryon pair. Accurate measurements of the yields of produced baryons and antibaryons (particularly important here are those which contain strange quarks) are available from the STAR experiment (see B.Hippolyte talk in this proceedings).

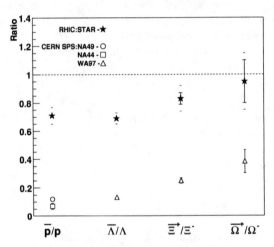

FIGURE 9. Antibaryon to baryon ratios according to their strangeness content for SPS and RHIC data.

Fig. 9 presents antibaryon to baryon ratios according to their strangeness content for SPS (WA97) [14] and RHIC data (STAR) [49, 50]. The observed enhancement of ratios, as strangeness content increases, reflects dramatically the change in the baryon density and confirms the increasing importance of pair production. This sensitivity to baryon density has already been briefly discussed earlier, in the context of K/π ratios at CERN SPS energy. Note, that K^+ and Λ yields are related by associated production mechanism.

Measured particle ratios combining particles of different flavor content and mass allow one to determine thermal model parameters at chemical freeze-out. Data (all four RHIC experiments) analyzed [51] using a statistical canonical ensemble [25] demonstrate rather good agreement with the model. The extracted thermal parameters temperature, $T = 176 \pm 9$ MeV and chemical potential, $\mu_B = 3\,\mu_q = 36 \pm 4$ MeV, show that the nucleus-nucleus collisions at RHIC are characterized by high energy density (high T), and low baryon density (at mid-rapidity). The value of γ_s parameter, called the strangeness saturation factor, obtained in the fit, is consistent with 1 (0.95 ± 0.10) pointing towards chemical equilibrium established at mid-rapidity. Note that particle ratios used for constraining T and μ_B are measured at mid-rapidity, not in full phase space. The mid-rapidity particles most likely reflect a thermal source provided that global equilibrium is approximated by local equilibrium, whereas particle production at high rapidity is expected to have significant content from the initial colliding nuclei. Thus, different statistical descriptions at mid-rapidity and away from mid-rapidity would be appropriate. Due to the large rapidity range at RHIC (≈ 6 units), the contamination from high rapidity sources is expected to be relatively small at mid-rapidity.

Taking the preliminary results of statistical model analysis at face value, one arrives at the conclusion that the initial state is very dense. The energy density is higher by about 30 times than that found in cold nuclear matter. A new territory of ultradense hadronic or pre-hadronic equilibration dynamics has been entered. However, nothing is known about its properties and characteristics.

STRANGENESS AT RHIC

Usually the initial phase of the experimental program, particularly in the new energy domain, brings only information on global characteristics of the collisions. RHIC is not an exception in this respect. For detailed and conclusive analysis of strangeness production at $\sqrt{s} = 130$ and $200\,$GeV, one still needs to wait. However, already available preliminary results seem to be quite interesting and unexpected. In this chapter some of them will be reviewed. A word of caution will be added where appropriate, as they might be revised/replaced with final, large statistics analysis.

As discussed earlier, the ratios of baryons and antibaryons are well described by a thermal model fit. It has been attributed, along the lines of thermal model, to a significant degree of equilibration which sets-in in these nuclear collisions. However, this may very well reflect only the trivial loss of statistical sensitivity as the antibaryon-baryon ratios at RHIC energies (baryon chemical potential ≈ 0) approach unity. The ratios of different mass particles should be free from this problem and may present a more suitable test for the model. Analysis is in progress.

Multi-strange hyperons, due to their low production cross section, will probably not influence the degree of the chemical equilibration in any meaningful way, however they are the only ones to provide information on processes sensitive to the density of strange quarks in the initial state, and perhaps, on some non-equilibrium production mechanisms, if they exist. So, it is not their agreement with thermal model expectations that is important and worth studying, but rather discrepancies from the fits may lead to the most interesting results, and perhaps to the discovery of new phenomena.

Transverse mass spectra have also been measured for strange hadrons. The values of invariant slope parameters extracted from the fits to these spectra are summarized in Fig.10 as a function of particle mass. For comparison, values of slope parameters at SPS are also shown. The behavior at SPS and RHIC is very similar. For each particle species, the RHIC spectrum has an inverse slope about 50 MeV higher than at SPS, which may indicate stronger transverse explosion. It appears that multistrange hadrons deviate from the simple radial flow picture, suggesting that perhaps they do not participate in a common expansion, and decouple rather early from the collision system due to their small hadronic cross section.

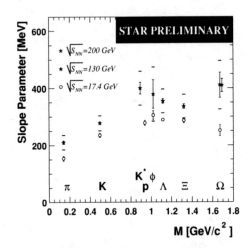

FIGURE 10.   Invariant slope parameter as a function of particle mass.

This observation hints that it might be possible to obtain insights into the very early stages of the collision by studying the flow of strange particles (both: radial and elliptic). The STAR experiment [53, 54] reports first measurements of the azimuthal asymmetry parameter v_2 for strange particles K_s^0, Λ and $\bar{\Lambda}$ for Au+Au at $\sqrt{s} = 130$ GeV. Hydrodynamic model calculations, where collective motion established by a pressure gradient transfers geometrical anisotropy to momentum anisotropy, seems to adequately describe elliptic flow of the strange particles up to a p_t of 2 GeV/c. The v_2 values as a function of p_t from mid-central collisions are higher at each p_t than v_2 from central collisions. The p_t-integrated v_2 as a function of particle mass is also consistent with a hydrodynamical picture. For p_t above 2 GeV/c, however, the observed v_2 seems to saturate whereas hydrodynamical models predict a continued increase with p_t. It has been argued in the framework of pQCD model [55] that the observed shape and numerical value of v_2 above 2-3 GeV/c reflect the energy loss in an early, high parton density, stage of evolution.

The fundamental question regarding the strangeness production differences (or lack of them) at RHIC compare to SPS, and their interpretation remains still open. While it is too early for definite conclusions, preliminary estimates [56] show that the overall strangeness, expressed as the ratio of strange to non-strange particle multiplicity or by the Wroblewski Factor [57] λ_s, defined as $\langle s\bar{s} \rangle / (\langle u\bar{u} \rangle + \langle d\bar{d} \rangle)$[1], does not change as a function of $\sqrt{s}$ between SPS and RHIC.

The solid and dotted lines in Fig.11 indicate different versions of statistical model

---

[1] quantities in angular brackets refer to the number of newly formed quark-antiquark pairs, i.e. excludes all quarks that were present in the target and projectile

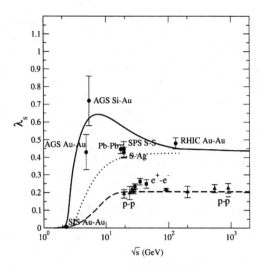

FIGURE 11. The Wroblewski Factor λ_s as a function of $\sqrt{s}$ - see text for details.

calculations for A+A, whereas the dashed line represents a statistical model for elementary collisions. There are two distinct differences in the behavior of λ_s in elementary and heavy ion collisions: (1) strangeness content is smaller by a factor of two in elementary collisions compared to nucleus-nucleus, and (2) there is no maximum in λ_s dependence on $\sqrt{s}$ in a compilation of pp, $p\bar{p}$, and e^+e^-. Note, that the SPS data are well below the solid line (complete equilibrium case) in Fig.11. The new CERN SPS data at 40 and 80 GeV/c [7] demonstrate the presence of this discrepancy very clearly.

SUMMARY

Fifteen years of heavy ion physics at the CERN SPS have resulted in several suggestive results regarding formation of the partonic state, but no clear discovery was made.

Early RHIC data have allowed for significant progress in mapping out the soft physics regime. The global conditions are indeed very different from the ones at SPS. However, the amount of strangeness produced is almost constant with growing energy. The enhancement in AA as compared to pp is preserved from the lower energy, however the pair production mechanism seems to dominate at RHIC and K^+/π^+ and K^-/π^- ratios seem to be approaching the same value as the baryon density goes to zero at mid-rapidity.

For the first time in the history of relativistic nuclear physics, the hydrodynamical calculations, which assume local equilibrium, agree with central collision Au+Au data in the low p_t range (less than 2 GeV/c), suggesting that thermalization might be obtained early in the collision.

It is clearly too early to draw definite conclusions, or to make firm statements about how data fit theoretical expectations. To a large extent the first RHIC results have not changed the picture that has emerged from the analysis of earlier AA collisions.

A complete picture of heavy ion collision dynamics at high energies requires the analysis of complementary information gained at both CERN and RHIC. It will be extended in the future by LHC findings. Despite the new aspects of the picture emerging from the analyzed RHIC collisions at $\sqrt{s} = 130$ and 200 GeV, the highly excited matter produced at RHIC energies is not quite yet at the baryon chemical potential $\mu_B = 0$. The opportunity to discover fascinating and challenging new science in nuclear collisions is still ahead.

ACKNOWLEDGMENTS

This work was supported by the Director, Office of Energy Research, Office of High Energy and Nuclear Physics, Division of Nuclear Physics of the US Department of Energy under Contract DE-AC03-76SF00098.

REFERENCES

1. Muller, B., and Rafelski, J., *Nucl.Phys.Lett.*, **48** (1982).
2. Odyniec, G., *Nucl.Phys.A*, **638**, 135 (1998).
3. Karsch, F. (2001), hep-lat/0106019.
4. Karsch, F. (2001), hep-ph/0103314.
5. Karsch, F., *Nucl.Phys.A*, **698**, 198 (2002).
6. Koch, P., Muller, B., and Rafelski, J., *Phys.Rep.*, **142**, 167 (1986).
7. Gazdzicki, M., et al., *Nucl.Phys.A*, **498**, 375c (1989).
8. Bamberger, A., et al., *Nucl.Phys.A*, **498**, 133c (1989).
9. Kapusta, J., and Mekjian, A., *Phys.Rev.D*, **33**, 1304 (1986).
10. Matsui, T., Svetitsky, B., and McLerran, L., *Phys.Rev.D*, **34**, 2047 (1991).
11. Fanebust, K., et al., *J.Phys.G*, **28**, 1607 (2002).
12. Appelshauser, H., et al., *Phys.Lett.B*, **444**, 523 (1998).
13. Becattini, F., et al., *Phys.Rev.C*, **64**, 024901 (2001).
14. Odyniec, G., editor, J. Phys. G, 2001, proc. of Strangeness 2000 Conf.
15. Afanasiev, S., et al. (2002), nucl-ex/0205002.
16. Ahle, L., et al., *Phys.Rev.C*, **57**, 466 (1998).
17. Ahle, L., et al., *Phys.Rev.C*, **58**, 3523 (1998).
18. Ahle, L., et al., *Phys.Rev.C*, **60**, 044904 (1999).
19. Ahle, L., et al., *Phys.Lett.B*, **476**, 1 (2000).
20. Ahle, L., et al., *Phys.Lett.B*, **490**, 53 (2000).
21. Barrette, J., et al., *Phys.Rev.C*, **62**, 024901 (2000).
22. Pelte, D., et al., *Z.Phys.A*, **357**, 215 (1997).
23. Hagedorn, R. (1994), cERN-TH-7190-94.
24. Cleymans, J., and Satz, H., *Z.Phys.C*, **57**, 135 (1993).
25. Sollfrank, J., et al., *Z.Phys.C*, **61**, 659 (1994).
26. Braun-Munzinger, P., et al., *Phys.Lett.B*, **365**, 1 (1996).
27. Yen, G., et al., *Phys.Rev.C*, **56**, 2210 (1997).
28. Yen, G., et al., *Phys.Rev.C*, **59**, 2788 (1999).
29. J.Cleymans, and Redlich, K., *Z.Phys.C*, **60**, 054908 (1999).
30. Sorge, H., et al., *Nucl.Phys.A*, **498**, 567c (1989).
31. Sorge, H., *Phys.Rev.C*, **52**, 3291 (1995).
32. Wang, F., et al., *Phys.Rev.C*, **61**, 064904 (2000).
33. Bass, S., et al., *Prog.Part.Nucl.Phys*, **41**, 255 (1998).
34. S.Hamiel, et al. (2000), hep-ph/0006024.

35. Yacoob, S., Canonical Strangeness Conservation in the Hadron Gas Model of Relativistic Heavy Ion Collisions (2002), 2002, Master Thesis, University of Cape Town, South Africa.
36. Kadija, K., et al., *J.Phys.G*, **28**, 1675 (2002).
37. (2000), URL http://www.star.bnl.gov, web page for STAR experiment.
38. (2000), URL http://www.phenix.bnl.gov, web page for PHENIX experiment.
39. (2000), URL http://www.rhic.bnl.gov/brahms, web page for BRAHMS experiment.
40. (2000), URL http://www.phobos.bnl.gov, web page for PHOBOS experiment.
41. Back, B., et al., *Phys.Rev.Lett.*, **88**, 022302 (2002).
42. Wang, X., and Gyulassy, M., *Phys.Rev.Lett.*, **86**, 3496 (2001).
43. Adcox, K., et al., *Phys.Rev.Lett*, **87**, 052301 (2001).
44. Bjorken, J., *Phys.Rev.D*, **27**, 140 (1983).
45. Karzeev, D., and Nardi, M., *Phys.Lett.B*, **507**, 121 (2001).
46. Bearden, I., et al., *Phys.Lett.B*, **523**, 227 (2001).
47. Adler, C., et al., *Phys.Rev.Lett*, **86**, 4776 (2001).
48. VanBuren, G., et al., *Nucl.Phys.A* (2002), in print.
49. Adler, C., et al. (2002), nucl-ex/0203016.
50. Xu, N., and Kaneta, M., *Nucl.Phys.A*, **698**, 306c (2002).
51. Kaneta, M., et al., *Nucl.Phys.A* (2002), in print.
52. Suire, C., et al., *Nucl.Phys.A* (2002), in print.
53. Kunde, G., et al., *Nucl.Phys.A* (2002), in print.
54. Filimonov, K., et al., *Nucl.Phys.A* (2002), in print.
55. Gyulassy, M., et al., *Phys.Rev.Lett*, **86**, 2537 (2001).
56. Braun-Munzinger, P., et al., *Nucl.Phys.A*, **697**, 902 (2002).
57. Wroblewski, A., *Acta Physica Polonica B*, **16**, 379 (1985).

Strange Baryons in the STAR Experiment

B. Hippolyte for the STAR Collaboration

Institut de Recherches Subatomiques, Strasbourg, FRANCE

Abstract. Preliminary results from the STAR experiment are presented here and complement the studies related to the hyperon production at mid-rapidity in central Au+Au collisions for the RHIC energy of $\sqrt{s_{NN}} = 130\ GeV$. The identification of the triply strange Ω^- via the invariant mass reconstruction of the decay products in the (Λ, K^-) channel allows us to calculate transverse mass spectra for this particle and its anti-particle $\overline{\Omega}^+$. We thus report an inverse slope of $\sim 410\ MeV$ and mid-rapidity yields of 0.32 ± 0.09 (stat.) and 0.34 ± 0.09 (stat.) for the Ω and $\overline{\Omega}^+$ respectively. Furthermore, a study of the (Λ, p, π) and (Λ, Λ) decay channels has been performed in order to search for the predicted H-dibaryon in STAR. Due to the statistics reached during the first year of data taking, it is difficult to draw any conclusion. However we expect it to be possible to measure some limits on its production in the near future.

INTRODUCTION

The primary goal for performing relativistic heavy ion collisions is the creation and the study of a quark-gluon plasma. The new state of matter governed by quark and gluonic degrees of freedom, should be achieved under extreme conditions such as high temperature [1]. Among the signatures of this parton-deconfined phase, the strangeness enhancement was one of the earliest predicted [2]. Due to an increasing mass threshold, a hierarchy in this enhancement should be observed from light to heavy baryons. The measurements of the strange baryon production [3, 4, 5] in central Au+Au collisions at $\sqrt{s_{NN}} = 130\ GeV$ have been performed in STAR [6] utilizing its full azimuthal coverage and large acceptance. We will mainly focus on the Ω reconstruction since we report here for the first time preliminary results of its yields and inverse slope parameters. Then we will show how we can extend the identification method to the search for some of the hypothetical H-dibaryon decay modes. Such a strangelet (a bag of six quarks: uuddss) is supposed to exist according to early predictions [7, 8] and, if stable against strong decay, a possible decay mode to look for in STAR would be the $H^0 \to \Lambda p \pi^-$ channel. In the case that the mass of the H^0 lies above the $\Lambda\Lambda$ threshold, or if only the hadronic counterpart (a bound state of two Λs) exists, looking for the $H^0 \to \Lambda\Lambda$ decay channel is also conceivable [9] with STAR. Preliminary invariant mass distributions of the aforementioned decay channels are presented although higher statistics are needed to complete such analyses.

CP644, *Exotic Clustering: 4th Catania Relativistic Ion Studies*, edited by S. Costa, A. Insolia, and C. Tuvè
© 2002 American Institute of Physics 0-7354-0099-7/02/$19.00

THE STAR EXPERIMENT

For the first year of data taking, the main tracking detector of the STAR experiment consisted of a large time projection chamber (TPC) located inside a solenoidal magnetic field. Two hadronic zero degree calorimeters (ZDC) [10] upstream, along the beam axis, were used to detect spectator neutrons. A central trigger barrel scintillator (CTB) surrounding the TPC measured the particle multiplicity. A trigger based on a coincidence in the ZDCs as well as a minimum signal in the CTB was used to record around $500,000$ events corresponding to approximatively the 14% of the cross-section with the smallest impact parameters. An off-line analysis based on the primary track multiplicity was used to estimate centralities more precisely [11].

STRANGE BARYON RECONSTRUCTION IN THE STAR TPC

Strange and Multi-Strange Baryons

The momenta of charged particles are determined from the tracks left in the TPC which are curved by a uniform magnetic field of $0.25\ T$. Particle identification at mid-rapidity is obtained using the mean energy loss method (dE/dx) in the fiducial volume of the TPC (i.e. ± 1.5 units of pseudorapidity).

A strange baryon which decays via weak interaction is identified if its charged daughters are reconstructed in the TPC and if the resulting vertex topology can be distinguished from the primary vertex of the collision (the typical decay length required is a few centimeters). Therefore a Λ is reconstructed via the charged decay $\Lambda \to p\pi^-$ which has a 64% branching ratio. Positive and negative track pairs are extrapolated back towards the primary vertex and are required to approximately intersect to form a so-called $V0$ [3]. The same topology technique is used for the corresponding anti-particle and the method is extended to identify the multi-strange baryons by combination of a $V0$ and another extrapolated track: the charged decays $\Xi \to \Lambda\pi^-$ and $\Omega \to \Lambda K^-$ (and the corresponding anti-particles) have a branching ratio of 100% and 68% respectively. Since these particles have different characteristics (e.g. lifetime), different geometric and kinematic selections are applied at both reconstruction and analysis levels. To reduce further the combinatoric background resulting from close but uncorrelated tracks, an identification of daughter tracks based on energy loss is used as mentioned above. The resulting invariant mass distributions exhibit clear signals as shown in Figure 1.

Omega yields and inverse slope parameter

The sample of events used to extract the transverse mass spectra of the Ω particle contains $\sim 500,000$ collisions corresponding to those of the 14% most central events where the primary vertex of each event has a position along the beam axis within $\pm 100\ cm$ of the center of the TPC.

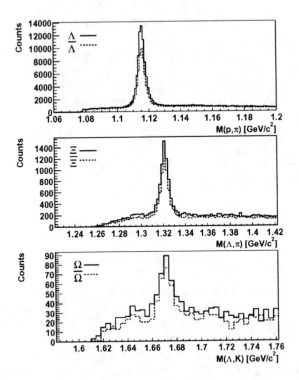

FIGURE 1. Invariant mass distributions for Λ (upper panel), Ξ (middle panel) and Ω (lower panel). Solid and dashed lines are used for respectively particles and anti-particles.

The extraction of the raw yields of Ω^- and $\overline{\Omega}^+$ is done by summing the entries within $\pm 15~MeV$ around the expected mass and subtracting an estimate of the background which is determined by simply counting the entries on either side of the signal region. The invariant mass distributions are histogrammed in four bins of transverse mass, $M_t = \sqrt{p_t^2 + M_\Omega^2}$. The corrections for acceptance and efficiency of the detector are obtained by generating Monte Carlo (MC) Ω^- and using a GEANT simulation of the detector. A simulation of the response of the TPC then creates the charge clusters for the corresponding MC tracks. These particles are then embedded into real events, and the standard reconstruction chain run, which does not distinguish between real and MC information. A correction factor is subsequently calculated as a function of M_t in a restricted window of rapidity where the total efficiency is flat with respect to this variable (i.e. $|y| < 0.75$). The yields are corrected independently for the four M_t bins in the integrated rapidity window. The transverse mass spectra are fit with a simple exponential, Equation 1, to extract both the invariant yield (dN/dy) and the inverse slope parameter (T, in MeV).

$$\frac{1}{2\pi M_t}\frac{d^2N}{dM_t dy} = \frac{dN/dy}{2\pi T(M_\Omega + T)}e^{-\left[\frac{(M_t - M_\Omega)}{T}\right]} \tag{1}$$

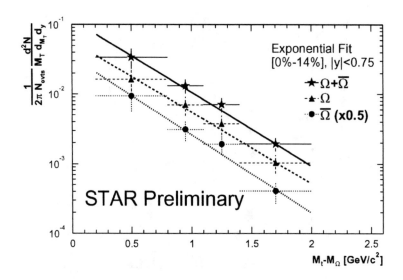

FIGURE 2. Transverse mass distributions for the $\Omega^- + \overline{\Omega}^+$, Ω^- and $\overline{\Omega}^+$. The different lines correspond to the exponential fits used to extract yields and inverse slope parameters. Only statistical errors are shown. A factor of 0.5 has been used to scale the $\overline{\Omega}^+$ for clarity.

TABLE 1. Yields and inverse slope parameters resulting from exponential fits. The errors are statistical only.

Particle	$T\,(MeV)$	dN/dy	χ^2/ddf
$\Omega^- + \overline{\Omega}^+$	412 ± 44	0.64 ± 0.14	$1.044/2$
Ω	424 ± 65	0.32 ± 0.09	$0.693/2$
$\overline{\Omega}^+$	387 ± 52	0.34 ± 0.09	$1.166/2$

Figure 2 shows the corrected spectra for Ω^-, $\overline{\Omega}^+$ and the sum of both particles. The measured range in M_t covers $\sim 65\%$ of the spectra. The results of the related fits are presented in Table 1. The inverse slope parameters are compatible with each other within statistical errors. Systematic errors are estimated to be $\sim 20\%$.

The combined $(\Omega^- + \overline{\Omega}^+)$ inverse slope parameter is shown in Figure 3 along with those of other particles measured by the STAR experiment at $\sqrt{s_{NN}} = 130\ GeV$ and SPS experiments at $\sqrt{s_{NN}} = 17.2\ GeV$ to provide a more extensive picture. We observe the following: the value of Ω slope parameter is i) higher than that of from SPS and ii) it does not follow a collective flow pattern which is shown by pion, kaon, proton. i) indicates that flow at RHIC is stronger than at SPS and ii) indicates that observed Ωs are produced earlier than other hadrons. It is also interesting to note that the Ω, and the other strange particles, seem to again confirm the two group classification proposed in [12].

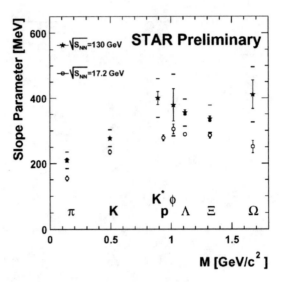

FIGURE 3. Slope parameter as a function of particle mass. The stars correspond to STAR measurements for Au+Au collisions at $\sqrt{s_{NN}} = 130\,GeV$ (error bars are statistic whereas dashes take into account the statistic plus systematic errors), the open circles are for SPS measurements at $\sqrt{s_{NN}} = 17.2\,GeV$ (only statistic errors are represented).

SEARCH FOR DI-BARYONS: THE H-DIBARYON

The dataset and the selection for central events are the same as the ones used for the Ω analysis. Although four different decay modes may be experimentally doable, only two of them are currently under investigation. The $H^0 \rightarrow \Sigma^- p \rightarrow n\pi^- p$ weak decay mode is difficult to search for, due to the relatively[1] short-lived Σ^-. Also, since we will see that there is already a lack of statistics for observing the decay $H^0 \rightarrow \Lambda\Lambda$, the other strong decay mode, $H^0 \rightarrow \Xi^- p$, has not been examined. However this study has been shown to be feasible with STAR if provided statistics are sufficient[13].

$H^0 \rightarrow \Lambda p\pi^-$ *decay mode.* Compared to the charged decay channel of the Ω, this possible weak decay channel of the H^0 is only slightly more difficult since one has to identify 4 daughter tracks instead of 3. Preliminary simulations as well as the main steps of the reconstruction of this mode have already been reported [14].

---

[1] Its identification would require both that the H^0 reaches the TPC and that the Σ^- decay length is sufficiently long enough to identify it unambiguously from the π^- decay daughter.

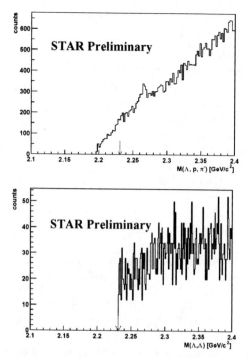

FIGURE 4. Invariant mass distributions for the $H^0 \to \Lambda p \pi^-$ weak decay mode (upper panel) and for the $H^0 \to \Lambda\Lambda$ strong decay mode (lower panel).

$H^0 \to \Lambda\Lambda$ *decay mode.* For this analysis, the invariant mass of all possible Λ candidate pairs within an event are computed. The resulting distribution thus contains a possible signal superimposed upon a combinatorial background. It is desirable to firstly reduce the background by selecting purified Λ samples with optimized cuts [15]. A mixed event technique is then applied to subtract the combinatoric background: the invariant mass distribution of Λ pairs where the two Λ are from different events provides a reliable estimation of the uncorrelated background.

The invariant mass spectra resulting from these two analyses are shown in Figure 4. In the upper panel, no clear signal is seen in the significant region i.e. below the $\Lambda\Lambda$ mass threshold represented by the arrow (2231 MeV/c^2). It is notable that no selection based on the H^0 candidate decay length has been applied. Thus, the background peak at $\sim 2265\ MeV/c^2$ for this mode needs to be investigated and is not likely to correspond to the $\Xi^- p$ decay channel. In the lower panel, one can see that due to insufficient statistics[2], no conlusion can be drawn.

---

[2] The Λ reconstruction rate is approximately 0.7 per event, leading to a very low number of events containing two candidate Λs.

CONCLUSION

We have presented preliminary results related to the Ω production at mid-rapidity for the 14% most central Au+Au collisions at the RHIC energy of $\sqrt{s_{NN}} = 130\ GeV$. We reported the yields of the Ω^- and $\overline{\Omega}^+$ particles and the corresponding inverse slope parameters. The overall inverse slope parameter is shown to be higher than that at the SPS. Omega particles exhibit a transverse mass spectra which deviates significantly from a simple flow picture including all particles. The high efficiency of STAR to identify strange and multi-strange baryons was naturally taken advantage of to undertake a preliminary search for strange dibaryons with no conclusive results. The increased statistics of the subsequent runs at RHIC may help to probe the existence of such objects.

ACKNOWLEDGMENTS

We wish to thank the RHIC Operations Group and the RHIC Computing Facility at Brookhaven National Laboratory, and the National Energy Research Scientific Computing Center at Lawrence Berkeley National Laboratory for their support. This work was supported by the Division of Nuclear Physics and the Division of High Energy Physics of the Office of Science of the U.S. Department of Energy, the United States National Science Foundation, the Bundesministerium fuer Bildung und Forschung of Germany, the Institut National de la Physique Nucléaire et de la Physique des Particules of France, the United Kingdom Engineering and Physical Sciences Research Council, Fundacao de Amparo a Pesquisa do Estado de Sao Paulo, Brazil, the Russian Ministry of Science and Technology and the Ministry of Education of China and the National Natural Science Foundation of China.

REFERENCES

1. Blaizot, J.-P., Nucl. Phys. A, **661**, 3-12 (1999).
2. Rafelski, J., and Müller, B., Phys. Rev. Lett., **48**, 1066-1069 (1982).
3. Adler, C. *et al.* (STAR Collaboration), Phys. Rev. Lett., **89**, 092301 (2002).
4. Lamont, M.A.C. (STAR Collaboration), J. Phys. G: Nucl. Part. Phys., **28**, 1721-1728 (2002).
5. Castillo, J. (STAR Collaboration), J. Phys. G: Nucl. Part. Phys., **28**, 1987-1991 (2002).
6. Wieman, H. *et al.*, IEEE Trans. Nucl. Sci., **44**, 671 (1997).
7. Jaffe, R.L., Phys. Rev. Lett., **38**, 195-198 (1977).
8. Donoghue, J.F., Golowich, E. and Holstein, B.R., Phys. Rev. D, **34**, 3434-3443 (1986).
9. Schaffner-Bielich, J., Mattiello, R. and Sorge, H., Phys. Rev. Lett., **84**, 4305-4308 (2000).
10. Adler, C. *et al.*, Nucl. Inst. Meth. A, **470**, 488-499 (2001).
11. Ackermann, K.H. *et al.* (STAR Collaboration), Phys. Rev. Lett. **86**, 402-407 (2001).
12. van Hecke, H., Sorge, H. and Xu, N., Phys. Rev. Lett. **81**, 5764 (1998).
13. Paganis, S.D. *et al.*, Phys. Rev. C, **62**, 024906 (2000).
14. Kuhn, C. *et al.*, J. Phys. G: Nucl. Part. Phys., **28**, 1707-1714 (2002).
15. Lamont, M.A.C., PhD Thesis, University of Birmingham (2002).

Strange Baryon Production from the NA57 Experiment at the CERN SPS

F.Riggi for the NA57 Collaboration

F.Antinori[l], A.Badalà[g], R.Barbera[g], A.Belogianni[a], A.Bhasin[e],
I.J.Bloodworth[e], G.E.Bruno[b], S.A.Bull[e], R.Caliandro[b], M.Campbell[h],
W.Carena[h], N.Carrer[h], R.F.Clarke[e], A.Dainese[l], A.P.de Haas[s], P.C.de
Rijke[s], D.Di Bari[b], S.Di Liberto[o], R.Divia[h], D.Elia[b], D.Evans[e],
K.Fanebust[c], F.Fayazzadeh[k], J.Fedorisin[j], G.A.Feofilov[q], R.A.Fini[b],
J.Ftáčnik[f], B.Ghidini[b], G.Grella[p], H.Helstrup[d], M.Henriquez[k],
A.K.Holme[k], A.Jacholkowski[b], G.T.Jones[e], P.Jovanovic[e], A.Jusko[i],
R.Kamermans[s], J.B.Kinson[e], K.Knudson[h], A.A.Kolojvari[q], V.Kondratiev[q],
I.Králik[i], A.Kravcakova[j], P.Kuijer[s], V.Lenti[b], R.Lietava[f], G.Løvhøiden[k],
M.Lupták[i], V.Manzari[b], G.Martinska[j], M.A.Mazzoni[o], F.Meddi[o],
A.Michalon[r], M.Morando[l], D.Muigg[s], E.Nappi[b], F.Navach[b], P.I.Norman[e],
A.Palmeri[g], G.S.Pappalardo[g], B.Pastirčák[i], J.Pisut[f], N.Pisutova[f], F.Posa[b],
E.Quercigh[l], F.Riggi[g], D.Röhrich[c], G.Romano[p], K. Šafařík[h], L. Šándor[i],
E.Schillings[s], G.Segato[l], M.Senè[m], R.Senè[m], W.Snoeys[h], F.Soramel[l],
M.Spyropoulou-Stassinaki[a], P.Staroba[n], T.A.Toulina[q], R.Turrisi[l],
T.S.Tveter[k], J.Urbán[j], F.F.Valiev[q], A. van den Brink[s], P. van den Ven[s],
P.Vande Vyvre[h], N. van Eijndhoven[s], J. Van Hunen[h], A.Vascotto[h], T.Vik[k],
O.Villalobos Baillie[e], L.Vinogradov[q], T.Virgili[p], M.F.Votruba[e],
J.Vrláková[j] and P.Závada[n]

[a] Physics Department, Unversity of Athens, Athens, Greece
[b] Dipartimento IA di Fisica dell'Università e del Politecnico di Bari and INFN, Bari, Italy
[c] Fysik Institutt, Universitetet i Bergen, Bergen, Norway
[d] Høgskolen i Bergen, Bergen, Norway
[e] University of Birmingham, Birmingham, UK
[f] Comenius University, Bratislava, Slovakia
[g] University of Catania and INFN, Catania, Italy
[h] CERN, European Laboratory for Particle Physics, Geneva, Switzerland
[i] Institute of Experimental Physics, Slovak Academy of Science, Košice, Slovakia
[j] P.J. Šafařík University, Košice, Slovakia
[k] Fysik Institutt, Universitetet i Olso, Oslo, Norway
[l] University of Padua and INFN, Padua, Italy
[m] Collège de France, Paris, France
[n] Institute of Physics, Prague, Czech Republic

CP644, Exotic Clustering: 4th Catania Relativistic Ion Studies, edited by S. Costa, A. Insolia, and C. Tuvè
© 2002 American Institute of Physics 0-7354-0099-7/02/$19.00

[o] *University "La Sapienza" and INFN, Rome, Italy*
[p] *Dipartimento di Scienze Fisiche "E.R.Caianiello" dell'Università and INFN, Salerno, Italy*
[q] *State University of St.Petersburg, St.Petersburg, Russia*
[r] *Institut de Recherche Subatomique, IN2P3/ULP, Strasbourg, France*
[s] *Utrecht University and NIKHEF, Utrecht, The Netherlands*

Abstract. The production of strange baryons and anti-baryons in PbPb and pBe collisions has been studied by the NA57 experiment at the CERN SPS, extending down the centrality range covered by the previous WA97 experiment, and collecting data at two different energies. The enhanced production of such particles in central PbPb collisions with respect to pBe has been experimentally confirmed. This paper discusses results on Ξ and Λ hyperon production obtained at 40 and 158 A GeV/c.

INTRODUCTION

The investigation of the behavior of nuclear matter under extreme conditions of energy density, temperature and pressure has received a strong boost, especially in the last decade, with the advent of accelerator facilities allowing the study of ultra-relativistic heavy-ion collisions.

Experiments carried out at the CERN SPS with Pb projectiles at beam momentum of 158 A GeV/c showed that in central PbPb collisions a strongly interacting nuclear state is indeed produced, which has an energy density larger by at least one order of magnitude than that of normal nuclear matter. A deconfinement phase transition is predicted under such conditions by the Quantum ChromoDynamics (QCD). Several possibilities of observing specific signatures originating from the Quark Gluon Plasma (QGP) deconfined phase, have been discussed in the past.

One of such signals, originally proposed by Rafelski and Müller[1], is the strangeness enhancement. This effect is expected to be larger for particles with more than one strange quark, such as the cascade baryons Ξ^- and Ω^- (and their antiparticles).

Such a pattern of strangeness enhancement has been observed by the WA97 Collaboration, which has studied[2] strange baryon and antibaryon production in PbPb and pBe collisions at 158 A GeV/c. The NA57 experiment (successor of WA97) was designed to understand how the strangeness enhancement depends on the collision centrality down to small centrality regions, on the bombarding energy and on the system size. The NA57 centrality range was lowered to about 50 wounded nucleons (N_{wound}), i.e. the nucleons which are found within the geometrical overlap of the two colliding nuclei[3] (the corresponding limit for WA97 was about 100). The experiment has collected data both at 158 and 40 A GeV/c, corresponding to a factor of 2 in $\sqrt{s}$, while WA97 collected data at 158 A GeV/c only.

This paper reports some recent results on the production of Λ, $\bar{\Lambda}$, Ξ^- and $\bar{\Xi}^+$ in PbPb and pBe collisions at SPS energies. Results on strange baryon production from the NA57 experiment have also been reported in recent Conferences[4].

THE NA57 EXPERIMENT

The experimental setup of the NA57 experiment is shown in Fig.1. The main tracking device is a compact telescope made by 13 silicon pixel planes with a sensitive area of about 5 x 5 cm^2. The telescope is placed 60 cm downstream of the target position (30 cm at 40 A GeV/c) and is inclined with respect to the beam in such a way as to cover the central rapidity region. Charged particles are tracked in a magnetic field of 1.4 T, provided by the GOLIATH magnet.

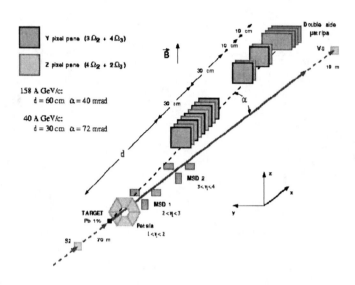

FIGURE 1. Experimental lay-out of the NA57 experiment at the CERN SPS.

Two stations of silicon strip detectors (MSD1, MSD2), covering the pseudorapidity interval $2 < \eta < 4$, are used to measure the charged particle multiplicity and to determine the collision centrality for PbPb collisions. A set of scintillators (Petals), placed 10 cm downstream of the target ($1 < \eta < 2$) is used to trigger on the centrality of the collision. In case of PbPb collisions at 158 A GeV/c, the most central 60 % of the total inelastic cross section has been selected.

A more detailed description of the experimental lay-out can be found in Ref[5]. The main differences with respect to the previous experiment, WA97, are the use of a different beam line with a smaller spot size, a new telescope layout, fully based on silicon pixel detectors, a new magnet (GOLIATH), a new data acquisition system, and a revised set of software programs for detector alignment, track finding and event reconstruction.

Strange and multistrange baryons are identified by the topology of their weak decays into charged particles in the final state (Table 1). Particles are detected in a kinematical window $\Delta y = \pm 0.5$ around midrapidity, with a minimum transverse momentum of about 0.3 GeV/c, depending on the detected particle. Fig.2 shows the invariant mass spectra of Ξ^- and $\bar{\Xi}^+$ (left), together with their acceptance window (right).

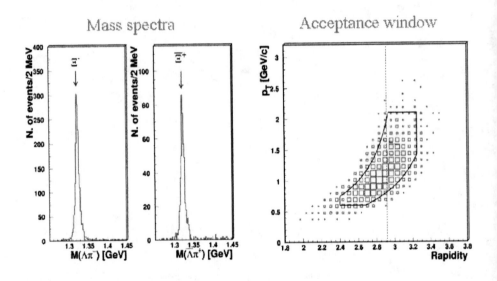

FIGURE 2. Lambda-pion invariant mass spectra (left) and the corresponding acceptance window (right), for PbPb collisions at 158 A GeV/c.

To correct for acceptance and efficiency losses, each individual detected particle is weighted by generating Monte Carlo particles with the same rapidity and transverse momentum as the real particles and tracing them through the experimental set-up. Generated hits are then embedded into real events to account for background. These events are finally processed by the same analysis chain used for real data.

TABLE 1. Strange particles decay parameters.

Particle	Decay	B.R. (%)	cτ (cm)
Λ ($\bar{\Lambda}$)	$p\pi^-$ ($\bar{p}\,\pi^+$)	63.9	7.89
Ξ^- ($\bar{\Xi}^+$)	$\Lambda\pi^-$ ($\bar{\Lambda}\pi^+$)	99.9	4.91
Ω^- (Ω^+)	ΛK^- ($\bar{\Lambda}K^+$)	67.8	2.46

DATA ANALYSIS AND RESULTS

Table 2 reports a summary of the data sets measured so far by the NA57 experiment.

TABLE 2. Data sets measured by NA57.

System	Beam Momentum	Sample size	Data taking	Event reconstruction
PbPb	158 A GeV/c	230 Mevents	1998	Completed
pBe	40 A GeV/c	60 Mevents	1999	Completed
PbPb	40 A GeV/c	290 Mevents	1999	Completed
PbPb	158 A GeV/c	230 Mevents	2000	Completed
pBe	40 A GeV/c	100 Mevents	2001	Ongoing

Collision centrality

The collision centrality is estimated by the number of wounded nucleons, according to the method described in Ref [6]. Events are divided into five centrality classes (0 to IV), class 0 being the most peripheral one ($<N_{wound}> \cong 50$). Classes I to IV correspond approximately to the four centrality classes of WA97. Fig.3 (left) shows the distribution of the charged particle multiplicity in the range $2 < \eta < 4$, for PbPb collisions. The boundary of the five centrality classes are also shown in the figure. The distribution of the number of wounded nucleons for the five centrality classes (0 to IV) is shown in fig.3 (right).

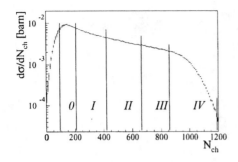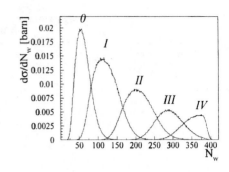

FIGURE 3. (Left) Charged particle multiplicity distribution in the range $2 < \eta < 4$ measured by NA57, together with the five centrality classes used in the analysis. (Right) Distribution of the number of wounded nucleons for the five centrality classes.

Particle ratios

Particle production ratios provide information on the chemical equilibrium in a heavy-ion collision. Antibaryons to baryons ratios in particular give information on the baryon density. Table 3 lists the present status of antibaryon to baryon ratios at 40 and 158 A GeV/c, extracted from the analyzed sets of data. Results are still preliminary. For the Ξ's in pBe at 40 A GeV/c only a part of the statistics has been analyzed so far.

A significant decrease, by a factor 6 for Λ's and a factor 3 for Ξ's, is observed for antibaryons to baryons ratios when going from 158 to 40 A GeV/c for PbPb collisions. The same reduction is observed for Λ's in the pBe case. This behavior suggests a larger baryon density at lower bombarding energy.

TABLE 3. Particle ratios in NA57. The values of the particle ratios in pBe are from WA97.

Particle ratio	PbPb @ 158 A GeV/c	PbPb @ 40 A GeV/c	pBe @ 158 A GeV/c	pBe @ 40 A GeV/c
$\bar{\Lambda}/\Lambda$	0.146 ± 0.005	0.023 ± 0.001	0.332 ± 0.008	0.059 ± 0.007
$\bar{\Xi}^+/\Xi^-$	0.270 ± 0.020	0.080 ± 0.025	0.45 ± 0.07	---

308

Transverse mass spectra

Transverse momentum distributions $d^2N/dm_T dy$ have been extracted from the data for Λ, $\bar{\Lambda}$, Ξ^- and $\bar{\Xi}^+$ produced in PbPb collisions at 158 A GeV/c. The following parametrization

$$\frac{d^2N}{dm_T dy} = Am_T \exp(-m_T/T)$$

has been used to evaluate the inverse slope T via a log likelihood fit. Fig. 4 shows the results for Λ and $\bar{\Lambda}$. The values of the inverse slopes (see Table 4) are found to be compatible for particles and their antiparticles; they are also consistent with those measured by WA97. In case of Λ's the transverse mass spectra have been analyzed for different centrality bins, going from class 0 (the most peripheral one) to class IV (the most central one). A slight increase of the inverse slopes with the collision centrality is observed for Λ's, whereas no significant variation, within the present statistics, is evidenced for Ξ's.

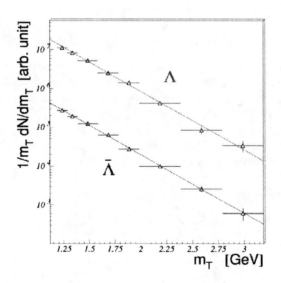

FIGURE 4. Transverse mass spectra of Λ's and their antiparticles.

TABLE 4. Inverse slopes (MeV) for the five centrality classes (0-IV). PbPb at 158 A GeV/c (preliminary).

Particle	0 - IV	0	I	II	III	IV
Λ	284 ± 6	258 ± 19	261 ± 11	276 ± 11	313 ± 13	310 ± 15
$\bar{\Lambda}$	287 ± 6	274 ± 18	258 ± 10	279 ± 10	307 ± 13	309 ± 14
Ξ^-	303 ± 11					
$\bar{\Xi}^+$	321 ± 23					

Particle Yields

Particle yields are obtained by integration of previous equation over y and extrapolating to full m_T:

$$Yields = \int_m^\infty dm_T \int_{y_{CM}-0.5}^{y_{CM}+0.5} dy \frac{d^2N}{dm_T dy}$$

Fig. 5 shows the yields per wounded nucleon, rescaled to the pBe values, for Λ, $\bar{\Lambda}$, Ξ^- and $\bar{\Xi}^+$. Yields are evaluated separately for the five classes, as discussed before. NA57 results confirm that the strange particle yields per participant in PbPb collisions are enhanced with respect to the proton-nucleus case. By comparison to the previous WA97 data, yields are found to be larger by up to 20-30%. Extensive checks of the data analysis procedure have been carried out. The results of these checks give us confidence in the recent NA57 results. Most of the differences in the yields are mainly due to underestimation of acceptance corrections due to the transversal beam spread at target position in case of the WA97 runs.

The data indicate a general increase in the yields per wounded nucleon with the centrality of the event. The $\bar{\Xi}^+$ yield per wounded nucleon increases by a factor 2.6 (3.5 σ).

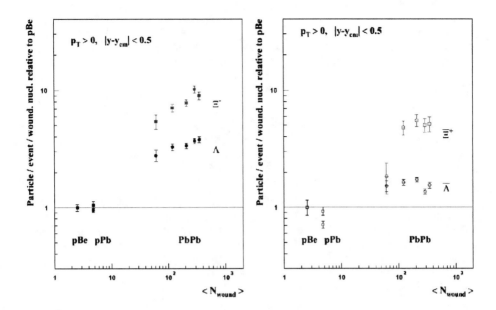

FIGURE 5. Yields per wounded nucleon relative to pBe, at 158 A GeV/c, as a function of the number of wounded nucleons.

CONCLUSIONS AND OUTLOOK

Preliminary results from NA57 experiment in PbPb and pBe collisions at 40 and 158 A GeV/c have been reported. Antibaryon to baryon ratios exhibit a reduction when going from 158 to 40 A GeV/c. The Ξ^-, $\overline{\Xi}^+$ and Λ yields per wounded nucleon, normalized to the pBe yields show a significant decrease for the most peripheral bin, up to a factor 2.6 for $\overline{\Xi}^+$, which could be indicative of the onset of the QGP phase transition.

The data analysis is still in progress to extract the Ω^- and $\bar{\Omega}^+$ yields from the PbPb data sample at 158 A GeV/c and to double the overall statistics, and to extract the absolute values of the yields and enhancements for the data sample at 40 A GeV/c.

REFERENCES

1. Rafelski, J., and Müller, B., *Phys. Rev. Letters* **48**, 1066 (1982).
2. Fini, R.A., et al., *J.Phys. G: Nucl. Physics* **27**, 375 (2001); Antinori, F., et al., *Nucl. Physics* **A661**, 130c (1999).
3. Wong, C.Y., *Introduction to High-Energy Heavy-Ion Collisions*, Singapore, World Scientific Publishing, 1994, p.251, and references therein.
4. Carrer, N., et al., *Nucl. Physics* **A698**, 118c(2002); Fanebust, K., et al., Int. Conf. On Strangeness in Quark Matter, Frankfurt, Germany, September 2001; Bruno, G., et al., Rencontres de Moriond, March 16-23, 2002
5. Manzari, V. et al., *J.Phys. G: Nucl.Physics* **27**, 383 (2001).
6. Carrer, N., et al., *J.Phys.G: Nucl. Physics* **27**, 391 (2001).

Quark Dynamics in phase space

S. Terranova* and A. Bonasera*

*Laboratori Nazionali del Sud, Via S. Sofia 44, I-95123 Catania, Italy

Abstract. We propose a dynamical model to reproduce Equation of State of Nuclear matter starting from quark degrees of freedom with color at zero temperature and finite density. We solve molecular dynamics with constraint and we use as input a phenomenological potential: the Richardson's potential. At low densities the quarks condensate into approximately color white cluster (nucleon). At high densities, for quark masses (u,d) of 0.18 GeV we find some evidence for a second order phase transition to Quark Gluon Plasma.

One of the most challenging open problems in theoretical nuclear physics is how to obtain the well known nuclear properties starting from the quark degrees of freedom [1]. In addition, an interesting question is how the nuclear matter composed of nucleons (which are by themselves composite three-quark objects) dissolve into quark matter. Such transition may occur in high-energy heavy ion collisions [2] and in the central core of neutron stars [3].

Starting from quark degrees of freedom, some dynamical approaches have been proposed in [4, 5, 6] based on the Vlasov equation [7, 8], and/or molecular dynamics type approach. Of course, in such approaches it is important to get quark clusterization and the correct properties of nuclear matter (NM) at the ground state (gs) baryon density $\rho_0 = 0.15$ fm^{-3}. However the property of ground state nuclear matter, together with the high density phenomena, is not sufficiently studied from the view point of quark degrees of freedom.

It is the purpose of this work to show that taking into account the color properties of the quarks and their confinement in the low density limit we obtain some basic features of NM and we can pose initial conditions for the simulation of relativistic ions collisions to produce the QGP.

The color degrees of freedom of quarks are taken into account through the Gell-Mann matrices and their dynamics is solved classically, in phase space, through the evolution of the distribution function.

The exact (classical) one-body distribution function $f(\mathbf{r}, \mathbf{p}, t)$ satisfies the equation:

$$\partial_t f + \frac{\mathbf{p}}{E} \cdot \vec{\nabla}_{\mathbf{r}} f - \vec{\nabla}_{\mathbf{r}} U \cdot \vec{\nabla}_{\mathbf{p}} f = 0, \tag{1}$$

where $E = \sqrt{p^2 + m_q^2}$ is the energy, m_q is the (u,d) quark mass and $U = U(\mathbf{r}) = \sum_{ij} V(\mathbf{r}_i, \mathbf{r}_j)$ is the exact potential, with $V(\mathbf{r}_i, \mathbf{r}_j)$ Richardson's potential:

$$V(\mathbf{r}_{i,j}) = 3 \sum_{a=1}^{8} \frac{\lambda_i^a}{2} \frac{\lambda_j^a}{2} \left[\frac{8\pi}{33 - 2n_f} \Lambda \left(\Lambda r_{ij} - \frac{f(\Lambda r_{ij})}{\Lambda r_{ij}} \right) + \frac{8\pi}{9} \alpha_s \frac{\langle \sigma_{qi} \sigma_{qj} \rangle}{m_{qi} m_{qj}} \delta(r_{ij}) \right], \tag{2}$$

CP644, *Exotic Clustering: 4th Catania Relativistic Ion Studies*, edited by S. Costa, A. Insolia, and C. Tuvè
© 2002 American Institute of Physics 0-7354-0099-7/02/$19.00

and[9]

$$f(t) = 1 - 4 \int \frac{dq}{q} \frac{e^{-qt}}{[\ln(q^2 - 1)]^2 + \pi^2}. \tag{3}$$

λ^a are the Gell-Mann matrices. We fix the number of flavors $n_f = 2$ and the parameter $\Lambda = 0.25$ GeV. Here we assume the potential to be dependent on the relative coordinates only. The first term is the linear term, that has part in the confinement, the second term is the Coulomb term and the last term is the chromomagnetic term (ct) which depends on the relative spins orientation of the quarks through the Pauli matrices σ_{qi} [10]. Since in this work we will be dealing with infinite matter, the ct can be neglected, which is very important for reproducing the hadron masses in the vacuum.

When the particles are embedded in a dense medium such as in nuclear matter, the potential becomes screened in a similar fashion as ions and electrons in condensed matter. This is the Debye screening. The Debye screening is explicitly obtained by introducing a cut-off of the color potential, which in our case is a free parameter. When quark distances is greater then the cut-off the interaction is equal to zero.

Numerically the equation (1) is solved by writing the one body distribution function through the delta function :

$$f(\mathbf{r}, \mathbf{p}, t) = \sum_{i=1}^{Q} \delta(\mathbf{r} - \mathbf{r}_i(t)) \delta(\mathbf{p} - \mathbf{p}_i(t)), \tag{4}$$

where $Q = q + \bar{q}$ is the total number of quarks (q) and antiquarks $(\bar{q})$ (in this work $\bar{q} = 0$). Inserting this expression in exact equation we get the Hamilton equations:

$$\frac{d\mathbf{r}_i}{dt} = \frac{\mathbf{p}_i}{E_i}, \tag{5}$$

$$\frac{d\mathbf{p}_i}{dt} = -\vec{\nabla}_{\mathbf{r}_{ij}} U(\mathbf{r}_{ij}). \tag{6}$$

Hence we must solve these equations of motion for our system of quarks. Initially we distribute randomly the quarks in a box of side L in coordinate space and in a sphere of radius P_f in momentum space. P_f is the Fermi momentum estimated in a simple Fermi gas model by imposing that a cell in phase space of size $h = 2\pi\hbar$ can accommodate at most g_q identical quarks of different spins, flavors and colors. A simple estimate gives the following relation between the quark density, ρ_q, and the Fermi momentum:

$$\rho_q = \frac{g_q}{6\pi^2} p_f^3 \tag{7}$$

Where $g_q = n_f \times n_c \times n_s$ is the degeneracy number, n_c is the number of colors (for quarks and antiquarks three different colors are used: red, green and blue) hence $n_c = 3$, $n_s = 2$ is the number of spins[11].

To describe the Fermionic nature of the system we impose that average distribution function for each particle is less than 1 or equal to 1 ($\bar{f}_i \leq 1$). We create many events

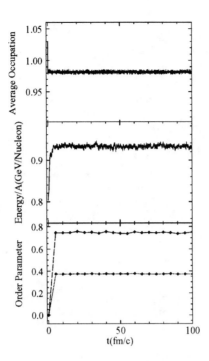

FIGURE 1. Time dependence of Average Occupation (top), Energy for Nucleon (middle) Order Parameter (bottom up) and Reduced Order Parameter (bottom down) for $\rho = 0.15\text{fm}^{-3}$

and take the average over all events in each cell on the phase space. At each time step we control the value of average distribution function and consequently we change momentum of particles multiplying the momenta for a quantity ξ: $P_i = P_i \times \xi$, ξ is greater then 1 or less then 1 if $\bar{f}_i$ is greater or less then 1 respectively. This is the constraint.

With this procedure the total energy, the average occupation and the order parameter after a given time will reach stationary values. We can see this in Fig. 1 in a typical event with $\rho = 0.15\text{fm}^{-3}$. In the top part of figure we display the time dependence of the average occupation. The value of distribution function is greater than 1 when we initially distribute randomly the quarks in the box and later it becomes nearly 1 when it saturates. We can see the same feature for the energy per nucleon (middle) and the order parameter (bottom). The order parameter is a important quantity in order to check the order of phase transition, it is defined through the Gell-Mann matrices as[12]:

$$M_c = \frac{1}{N} \sum_{i=1}^{N} \sum_{a=3,8} \lambda_j^a \lambda_k^a + \lambda_i^a \lambda_j^a + \lambda_i^a \lambda_k^a = M_{cr} + \frac{1}{N} \sum_{a=3,8} \lambda_j^a \lambda_k^a + \lambda_i^a \lambda_k^a, \tag{8}$$

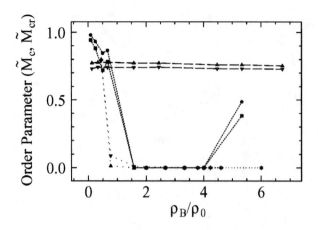

FIGURE 2. Normalized Order Parameter and Reduced Order Parameter vs density (divided by a normal density ρ_0) for quark masses $m_u = 5$ MeV and $m_d = 10$ MeV and the cut-off 3 fm. The long dashed line is obtained omitting the linear term and the dotted line is without the Coulomb term. The triangles up refer to order parameter while triangles down refer to reduced order parameter. For dotted dense line we have linear and Coulomb terms (circles for order parameter, squares for reduced order parameter).

where j and k are the two quarks closest to the quark i. M_{cr} is the reduced order parameter which gives the color of the particle j closest to particular quark i.

We normalize order parameter in this way:

$$\tilde{M}_c = \frac{2}{9}[M_c + 3], \tag{9}$$

$$\tilde{M}_{cr} = \frac{2}{3}[M_{cr} + 1] \tag{10}$$

From the properties of the Gell-Mann matrices it is easy to derive the following result for the order parameter: if the three closest quarks have different colors then $\tilde{M}_c = 1$ and $\tilde{M}_{cr} = 1$, in this case we have isolated white nucleons. This case is recovered in the calculation at small densities, then the system is locally invariant for rotation in color space. If three closest quarks have two different colors $\tilde{M}_c = \tilde{M}_{cr} = \frac{2}{3}$, i.e. the color of closest particle to quark i is randomly chosen. This is the case of Quark Gluon Plasma. If three closest quark have the same color $\tilde{M}_c = \tilde{M}_{cr} = 0$ we have a condition that we call EXOTIC COLOR CLUSTERING. The corresponding potential energy is very large and repulsive.

In Fig. 2 we plot the normalized order parameter and the reduced order parameter versus density (divided by a normal density ρ_0) for quark masses $m_u = 5$ MeV and $m_d = 10$ MeV and cut-off 3 fm.

We have distinguished three different cases: the triangles refer to two calculation where the linear term is equal to zero (long dashed line) and where the Coulomb term is turned off (dotted line). The circles and squared refer to calculation where we have

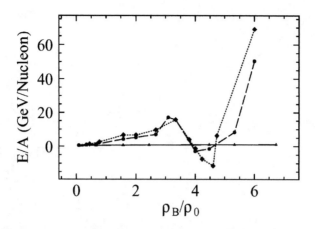

FIGURE 3. Energy per nucleon vs density (divided by a normal density ρ_0) for the quark masses $m_u = 5$ MeV and $m_d = 10$ MeV and the cut-off 3 fm for three different cases: Coulomb term only (solid line), linear term only (dotted line), and Coulomb plus linear terms (dashed line).

both terms: linear and Coulomb (dotted dense line). For small densities the quarks are condensed in cluster of three and the net color is zero: the system is locally white. The normalized order parameter and reduced order parameter are equal to 1 and the closest particle to quark i always has a different color (isolated white nucleon) $\tilde{M}_c = \tilde{M}_{cr} = 1$. At very high densities, the quark are not in cluster but randomly distributed, $\tilde{M}_c = \tilde{M}_{cr} = \frac{2}{3}$, and we have the QGP. At density about $2 \sim 4$ times the normal density $\tilde{M}_c = \tilde{M}_{cr} = 0$, and we have the exotic color clustering where three closest quarks have the same color. At about 5 times the normal nuclear matter density we can see a first order phase transition.

In order to see better the phase transition we plot in Fig. 3 the energy per nucleon versus density in the same conditions: $m_u = 5$ MeV, $m_d = 10$ MeV, cut-off equal 3 fm. The triangles refers to results obtained by considering Coulomb term only (solid line), the square by considering the linear term only (dotted line) and in full circles we have both terms: Coulomb and linear (dashed line). At about 5 times ρ_0 we can see a discontinuity of energy per nucleon indicating the first order phase transition. We have repeated the same calculations with different conditions $m_u = 180$ MeV, $m_d = 180$ MeV, cut-off equal 1.26 fm. In Fig. 4 we plot normalized order parameter and reduced order parameter (left figure) and energy per nucleon (right figure) versus density (divided by ρ_0). The displayed values of $\tilde{M}_c$ and $\tilde{M}_{cr}$ are always positive, i.e. it never happens that three equal quark color states are on average in the same region in r space. Hence for different quark masses and cut-off we don't have the exotic color clustering. However at very high densities (almost 50 times the normal density), in the figure relative of energy per nucleon we can see some fluctuations, indicating a second order phase transition.

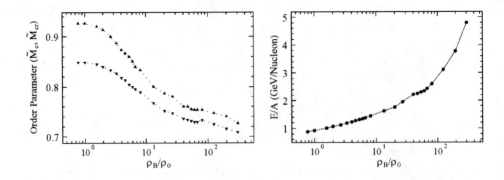

FIGURE 4. In the left figure we plot normalized order parameter (triangles up) and reduced order parameter (triangles down) while in the right figure we have energy per nucleon vs density (divided by ρ_0) for the quark masses $m_u = m_d = 180$ MeV and the cut-off equal 1.26 fm

SUMMARY

In summary we have studied Equation of State of quarks with color and Pauli principle at $T = 0$ and $\rho_B \neq 0$ by microscopic dynamical simulation. We have found exotic color clustering at high densities and small quark masses and first order phase transition. For larger quark masses we don't have exotic color clustering, however we have a second order phase transition to QGP at high densities. We would like to extend our approach to RHIC including $q\bar{q}$ creation and annihilation.

REFERENCES

1. B. Povh, K. Rith, C. Scholz, F. Zetsche, *Particles and Nuclei: an introduction to the physical concepts*, Springer, Berlin, 1995.
2. For example, Proceedings of *Quark Matter '97*, ed. T. Hatsuda et al., Nucl. Phys. **A638**, 1 (1998).
3. G. Baym, Nucl. Phys. **A590**, 233 (1995).
4. A. Bonasera, Phys. Rev. **C60**, 65212 (1999); Phys. Rev. **C62**, 52202(R) (2000); Nucl. Phys. **A681**, 64c (2001).
5. T. Maruyama and T. Hatsuda, Phys. Rev. **C61**, 62201(R) (2000).
6. S. Loh, C. Greiner and U. Mosel, Phys. Lett. **B404**,238(1997), and references therein.
7. A. Bonasera, F. Gulminelli and J. Molitoris, Phys. Rep. **243**,1(1994); G. Bertsch, S. Dasgupta, Phys. Rep.**160**,189 (1988), and references therein.
8. E. M. Lifshitz and L. P. Pitaevskii, *Physical Kinetics*, Pergamon Press 1991.
9. J. L. Richardson, Phys. Lett. **82B** (1979) 272.
10. A. Bonasera, nucl-th/9905025, nucl-th/9908036, Phys. Rev. **C60** (1999).
11. C. Y. Wong, *Introduction to High-Energy Heavy ion Collisions*, World Scientific Co., Singapore,1994; L. P. Csernai *Introduction to Relativistic Heavy ion Collisions*, John Wiley and Sons, New York, 1994.
12. A. Bonasera, Phys. Rev. **C62**, 0522XX(R) (2000).

Magnetic properties and EoS of spin polarized isospin asymmetric nuclear matter

I. Vidaña and I. Bombaci

Dipartimento di Fisica "E. Fermi", Università di Pisa & INFN Sezione di Pisa, Via Buonarroti 2, I-56127 Pisa, Italy

Abstract. Bulk and single-particle properties of spin polarized isospin asymmetric nuclear matter are studied within the framework of the Brueckner–Hartree–Fock approximation. The single-particle potentials of neutrons and protons with spin up and down are determined for several values of the neutron and proton spin polarizations and the asymmetry parameter. An analytic parametrization of the total energy per particle as a function of these parameters is constructed, and employed to compute the magnetic susceptibility of nuclear matter. The results show no indication of a ferromagnetic transition at any density for any asymmetry of nuclear matter.

INTRODUCTION

The study of the magnetic properties of dense matter is of considerable interest in connection with the physics of pulsars. These objects are believed to be rapidly rotating neutron stars endowed with strong magnetic fields of $\sim 10^{12} - 10^{13}$ G [1,2]. There is, however, no general consensus about the origin of such a strong magnetic field. The field could be either a fossil remnant from the one of the progenitor star, or it could be generated after the neutron star's formation by some kind of dynamo process or electric currents flowing in the highly conductive neutron star material [3]. There are strong theoretical and observational arguments which indicate that the magnetic field decays during the "life" of a neutron star. Studies on magnetic field evolution of neutron stars by population codes [4] give a field decay time of $\sim 10^7 - 10^9$ yr and a residue magnetic field of $\sim 10^8$ G. If this residual field is assumed to be permanent and not generated by some dynamo process, it could be produced by a spontaneous ferromagnetic transition in the dense stellar core. Several authors have studied the possible existence of a ferromagnetic core in the liquid interior of neutron stars [5-18].

In this work we study, within the framework of the Brueckner–Hartree–Fock (BHF) approximation, the bulk and single-particle properties of spin polarized isospin asymmetric nuclear matter. We use the nucleon-nucleon part of the recent realistic baryon-baryon interaction (model NSC97e) of the Nijmegen group to describe the bare nucleon-nucleon interaction [19]. We study the dependence of the single-particle potentials and the total energy per particle on the neutron and proton spin polarizations and the asymmetry parameter. Further, we calculate the magnetic susceptibility for several values of the asymmetry parameter and explore the possibility of a ferromagnetic transition in the high density region relevant for neutron stars.

CP644, *Exotic Clustering: 4th Catania Relativistic Ion Studies*, edited by S. Costa, A. Insolia, and C. Tuvè
© 2002 American Institute of Physics 0-7354-0099-7/02/$19.00

THEORETICAL BACKGROUND

Our calculations are based on the BHF approximation extended to the case in which it is assumed that nuclear matter is arbitrarily asymmetric both in the isospin and spin degrees of freedom (i.e., $\rho_{n\uparrow} \neq \rho_{n\downarrow} \neq \rho_{p\uparrow} \neq \rho_{p\downarrow}$). Therefore, our many-body scheme starts by constructing all the nucleon-nucleon G matrices which describe in an effective way the interaction between two nucleons (nn, np, pn and pp) for each one of the spin combinations ($\uparrow\uparrow, \uparrow\downarrow, \downarrow\uparrow$ and $\downarrow\downarrow$). The G matrices can be obtained by solving the well known Bethe–Goldstone equation, which schematically reads

$$\langle N_1 N_2 | G(\omega) | N_3 N_4 \rangle = \langle N_1 N_2 | V | N_3 N_4 \rangle + \sum_{ij} \langle N_1 N_2 | V | N_i N_j \rangle \frac{Q_{N_i N_j}}{\omega - E_{N_i} - E_{N_j} + i\eta} \langle N_i N_j | G(\omega) | N_3 N_4 \rangle , \quad (1)$$

where N_k indicates the isospin and spin projections ($n\uparrow(\downarrow), p\uparrow(\downarrow)$) of the two nucleons in the initial, intermediate and final states, and their corresponding quantum numbers, V is the bare nucleon-nucleon interaction, $Q_{N_i N_j}$ is the Pauli operator which allows only intermediate states compatible with the Pauli principle, and ω is the so-called starting energy. Note that the G matrices are obtained from a coupled channel equation.

The single–particle energy of a nucleon with momentum k and spin projection $\sigma = \uparrow(\downarrow)$ is given by $E_N = \frac{\hbar^2 k^2}{2M_N} + U_N(k)$, where the single–particle potential $U_N(k)$ represents the mean field "felt" by the nucleon due to its interaction with the other nucleons of the system, given in the BHF approximation by

$$U_N(k) = \sum_{N'} U_{NN'}(k) = \sum_{N'} \sum_{k' \leq k_F^{N'}} \mathrm{Re}\langle NN' | G(\omega = E_N + E_{N'}) | NN' \rangle_{\mathscr{A}} , \quad (2)$$

where a sum over the Fermi seas of neutrons and protons with spin up and down is performed and the matrix elements are properly antisymmetrized. Once a self-consistent solution of Eqs. (1) and (2) is obtained, the total energy per particle is easily calculated

$$\frac{E}{A} = \frac{1}{A} \sum_N \sum_{k \leq k_F^N} \left(\frac{\hbar^2 k^2}{2M_N} + \frac{1}{2} U_N(k) \right) \equiv \frac{T}{A} + \frac{V}{A} . \quad (3)$$

This quantity is a function of $\rho_{n\uparrow}, \rho_{n\downarrow}, \rho_{p\uparrow}$ and $\rho_{p\downarrow}$ or, equivalently, of the total density $\rho = \rho_n + \rho_p = \rho_{n\uparrow} + \rho_{n\downarrow} + \rho_{p\uparrow} + \rho_{p\downarrow}$, the asymmetry parameter $\beta = (\rho_n - \rho_p)/\rho$, and the neutron and proton spin polarizations $S_n = \frac{\rho_{n\uparrow} - \rho_{n\downarrow}}{\rho_n}$, $S_p = \frac{\rho_{p\uparrow} - \rho_{p\downarrow}}{\rho_p}$.

The magnetic susceptibility χ of a system characterizes the response of this system to a magnetic field and gives a measure of the energy required to produce a net spin alignment in the direction of the field. In the case of nuclear matter it is defined by the 2×2 matrix

$$\frac{1}{\chi} = \begin{pmatrix} 1/\chi_{nn} & 1/\chi_{np} \\ 1/\chi_{pn} & 1/\chi_{pp} \end{pmatrix} , \quad \text{with} \quad \frac{1}{\chi_{ij}} = \frac{\rho}{\mu_i \rho_i \mu_j \rho_j} \left(\frac{\partial^2(E/A)}{\partial S_i \partial S_j} \right)_{S_i = S_j = 0} . \quad (4)$$

It is convenient to study the magnetic susceptibility of the system in terms of the ratio $\det(1/\chi)/\det(1/\chi_F)$, where χ_F is the magnetic susceptibility of the two component free

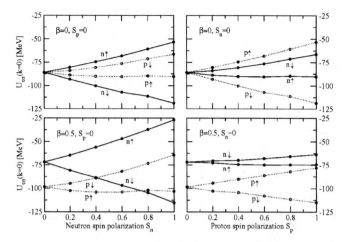

FIGURE 1. Neutron and proton single-particle potentials at $k = 0$ and $\rho = 0.17$ fm^{-3} as a function of neutron (left panels) and proton (right panels) spin polarizations.

Fermi gas. Writing the total energy per particle as a sum of the kinetic, T/A, and the potential, V/A, energy contributions, the ratio between both determinants can be written as

$$\frac{\det(1/\chi)}{\det(1/\chi_F)} = \left(1 + \frac{\frac{\partial^2(V/A)}{\partial S_n^2}}{\frac{\partial^2(T/A)}{\partial S_n^2}}\right)\left(1 + \frac{\frac{\partial^2(V/A)}{\partial S_p^2}}{\frac{\partial^2(T/A)}{\partial S_p^2}}\right) - \left(\frac{\frac{\partial^2(V/A)}{\partial S_n \partial S_p}}{\frac{\partial^2(T/A)}{\partial S_n^2}}\right)\left(\frac{\frac{\partial^2(V/A)}{\partial S_p \partial S_n}}{\frac{\partial^2(T/A)}{\partial S_p^2}}\right). \quad (5)$$

Stability of matter against spin fluctuations is guaranteed if $\det(1/\chi)/\det(1/\chi_F) > 0$. A change of sign of the ratio indicates the onset of a ferromagnetic phase in the system.

RESULTS

In Fig. 1 we show the value at $k = 0$ of the single-particle potentials $U_{n\uparrow}, U_{n\downarrow}, U_{p\uparrow}$ and $U_{p\downarrow}$ at $\rho = 0.17$ fm^{-3} as a function of neutron (left panels) and proton (right panels) spin polarizations for two values of the asymmetry parameter: $\beta = 0$ (upper panels) and $\beta = 0.5$ (lower panels). As it can be seen, the single-particle potentials split off when a partial polarization is assumed. This splitting can be mainly ascribed to two reasons: (i) the change in the number of pairs which the nucleon under consideration can form with the remaining nucleons of the system as nuclear matter is spin polarized, and (ii) the spin dependence of the nucleon-nucleon G-matrix in the spin polarized nuclear medium. It can be seen also that the variation is almost linear and symmetric with deviations from this behavior at higher values of the asymmetry and spin polarizations.

The total energy per particle of neutron, asymmetric, and symmetric matter is shown on the left, middle, and right panels of Fig. 2. In the three panels solid lines show results fon nonpolarized matter, whereas those for totally polarized matter are reported

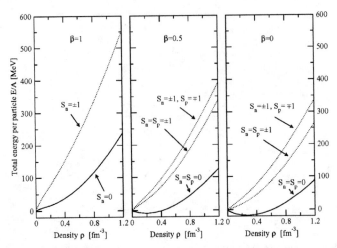

FIGURE 2. Total energy per particle as a function of the density for nonpolarized (solid lines) and totally polarized (dotted lines) matter.

by dotted lines. Note that in totally polarized matter asymmetric and symmetric matter, we have distinguished two possible orientations of the neutron and proton spins: that in which neutron and proton spins are orientated along the same direction (i.e., $S_n = S_p = \pm 1$), and that in which neutron and proton spins are orientated along opposite directions (i.e., $S - n = \pm 1, S_p = \mp 1$). As can be seen from the figure, totally polarized matter has always more energy than nonpolarized matter in all the density range explored for any value of the asymmetry parameter. Note also that the case in which neutron and proton spins are parallelly orientated has less energy than the case in which they have an antiparallel orientation. All these features can be understood, firstly, in terms of the kinetic energy contribution, which is larger in the totally polarized case than in the nonpolarized one, simply due to the fermionic nature of nucleons. Secondly, in terms of the spin singlet and triplet channel contributions to the potential energy which are different in the totally polarized and nonpolarized cases (see ref. [17]). An interesting conclusion which can be inferred from these results is that a spontaneous transition to a ferromagnetic state is not to be expected. If such a transition would exist, a crossing of the energies of the totally polarized and nonpolarized cases would be observed in neutron, asymmetric or symmetric matter at some density, indicating that the ground state of the system would be ferromagnetic from that density on. As can be seen in the figure, there is no sign of such a crossing and, on the contrary, a spin polarized state becomes less favorable as the density increases.

Performing a phase space analysis of the single-particle potentials, it is possible to infer the dependence of the total energy per particle on the asymmetry and spin polarizations in a very simple way:

$$\frac{E}{A}(\rho,\beta,S_n,S_p) = E_0(\rho) + E_1(\rho)\beta^2 + E_2(\rho)(1+\beta)^2 S_n^2$$
$$+ E_2(\rho)(1-\beta)^2 S_p^2 + E_3(\rho)(1-\beta^2)S_n S_p \,. \tag{6}$$

322

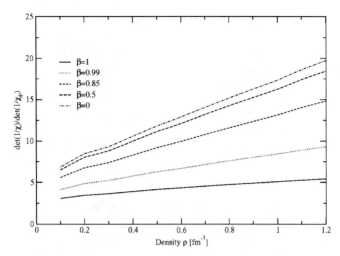

FIGURE 3. Ratio between the determinats of the matrices $1/\chi$ and $1/\chi_F$ as a function of the density for several values of the asymmetry parameter.

It is clear that the determination of the coefficients is not unique. We have chosen them in order to reproduce with a good quality the results of the BHF calculation for small values of β, S_n and S_p (see ref. [17] for a more detailed discussion).

The ratio $\det(1/\chi)/\det(1/\chi_F)$ can be evaluated in a very simple analytic way from Eq. (5) if the parametrization of Eq. (6) is assumed. The result is shown in Fig. 3 as a function of the density for several values of the asymmetry parameter β from neutron to symmetric matter. According to the criteria for the appearance of a ferromagnetic instability, such an instability should appear when $\det(1/\chi)/\det(1/\chi_F) = 0$. Nevertheless, it can be seen from the figure that $\det(1/\chi)/\det(1/\chi_F)$ increases monotonically with density, and a decrease is not to be expected even at higher densities for any value of the asymmetry parameter. Therefore, it can be inferred from these results that there is no sign at any density of a possible ferromagnetic phase transition for any asymmetry of nuclear matter. We note here, that our results confirm the reduction of about a factor 3 of the magnetic susceptibility of neutron matter with respect to its Fermi gas value found recently by Fantoni *et al.* [18].

CONCLUSIONS

We have studied bulk and single-particle properties of spin polarized isospin asymmetric nuclear matter within the Brueckner–Hartree–Fock approximation. We have found that the potentials exhibit an almost linear and symmetric variation as a function of the asymmetry and spin polarizations.

We have found that in the range of densities considered (up to $\sim 7\rho_0$) totally polarized matter has always more energy than nonpolarized matter for any asymmetry. We have also found that matter in which neutron and proton spins are antiparallel has higher energy than matter in which all the spins are aligned along the same direction.

We have constructed an analytic parametrization of the total energy per particle as a function of the asymmetry parameter and spin polarizations. Employing this parametrization we have determined the magnetic susceptibility of nuclear matter in terms of the ratio $\det(1/\chi)/\det(1/\chi_F)$. We have found that this quantity increases monotonically with density, from which it can be inferred that a phase transition to a ferromagnetic state is not to be expected in nuclear matter at any density for any asymmetry. Finally, our results confirm the reduction of about a factor 3 of the magnetic susceptibility of neutron matter with respect to its Fermi gas value found recently by Fantoni *et al.* [18].

ACKNOWLEDGMENTS

The authors are very grateful to professors A. Fabrocini, A. Polls, A. Ramos and S. Rosati for useful discussions and comments.

REFERENCES

1. Pacini, F., *Nature* **216**, 567 (1967).
2. Gold, T., *Nature* **218**, 731 (1968).
3. Ostriker, J. P. and Gunn, J. E., *Astrophys. Jour.* **157**, 1395 (1969).
4. Colpi, M., Possenti, A., Popov, S., and Pizzolato, F., "Spin and magnetism in old neutron stars" in *Physics of the neutron star interiors*, edited by D. Blaschle, N. K. Glendenning and A. Sedriakan. Lecture Notes in Physics **578**, Springer Verlag (2001).
5. Brownell, D. H. and Callaway, J., *Nuovo Cimento* **60B**, 169 (1969).
6. Rice, M. J., *Phys. Lett.* **29A**, 637 (1969).
7. Silverstein, S. D., *Phys. Rev. Lett.* **23**, 139 (1969).
8. Østgaard, E., *Nucl. Phys. A* **154**, 202 (1970).
9. Clark, J. W., *Phys. Rev. Lett.* **23**, 1463 (1969).
10. Pearson, J. M. and Saunier, G., *Phys. Rev. Lett.* **24**, 325 (1970).
11. Pandharipande, V. R., Garde, V. K., and Srivastava, J. K. *Phys. Lett.* **38B**, 485 (1972).
12. Bäckmann S. O. and Källman, C. G., *Phys. Lett.* **43B**, 263 (1973).
13. Jackson, A. D., Krotscheck, E., Meltzer, D. E., and Smith, R. A., *Nucl. Phys. A* **386**, 125 (1982).
14. Vidaurre, A., Navarro, J., and Bernabéu, J., *Astron. Astrophys.* **135**, 361 (1984).
15. Marcos, S., Niembro, R., Quelle, M. L., and Navarro, J., *Phys. Lett. B* **271**, 277 (1991).
16. Vidaña, I., Polls, A., and Ramos, A., *Phys. Rev. C* **65**, 035804 (2002).
17. Vidaña, I. and Bombaci, I. *Phys. Rev. C*, in press; nucl-th/0203061.
18. Fantoni, S., Sarsa, A., and Schmidt, E., *Phys. Rev. Lett.* **87**, 18110 (2001).
19. Stoks, V. G. J. and Rijken, Th. A., *Phys. Rev. C* **59**, 3009 (1999).

Strangeness and Antimatter

Superheavy Nuclei and Clusters of Hyper- and Antimatter

Walter Greiner

Institut für Theoretische Physik, J.W. Goethe-Universität,D-60054 Frankfurt, Germany

Abstract. The extension of the periodic system into various new areas is investigated. Experiments for the synthesis of superheavy elements and the predictions of magic numbers are reviewed. Furtheron, investigations on hypernuclei and the possible production of antimatter–clusters in heavy-ion collisions are reported. Various versions of the meson field theory serve as effective field theories at the basis of modern nuclear structure and suggest structure in the vacuum which might be important for the production of hyper– and antimatter.

INTRODUCTION

There are fundamental questions in science, like e. g. "how did life emerge" or "how does our brain work" and others. However, the most fundamental of those questions is "how did the world originate?". The material world has to exist before life and thinking can develop. Of particular importance are the substances themselves, i. e. the particles the elements are made of (baryons, mesons, quarks, gluons), i. e. elementary matter. The vacuum and its structure is closely related to that. On this we want to report, beginning with the discussion of modern issues in nuclear physics.

The elements existing in nature are ordered according to their atomic (chemical) properties in the **periodic system** which was developped by Mendeleev and Meyer. The heaviest element of natural origin is Uranium. Its nucleus is composed of $Z = 92$ protons and a certain number of neutrons ($N = 128 - 150$). They are called the different Uranium isotopes. The transuranium elements reach from Neptunium ($Z = 93$) via Californium ($Z = 98$) and Fermium ($Z = 100$) up to Lawrencium ($Z = 103$). The heavier the elements are, the larger are their radii and their number of protons. Thus, the Coulomb repulsion in their interior increases, and they undergo fission. In other words: the transuranium elements become more instable as they get bigger.

In the late sixties the dream of the superheavy elements arose. Theoretical nuclear physicists around S.G. Nilsson (Lund) and from the Frankfurt school[1] predicted that so-called closed proton and neutron shells should counteract the repelling Coulomb forces. Atomic nuclei with these special **"magic" proton and neutron numbers** and their neighbours could again be rather stable. These magic proton (Z) and neutron (N) numbers were thought to be $Z = 114$ and $N = 184$ or 196. Typical predictions of their life times varied between seconds and many thousand years. Fig.1 summarizes the expectations at the time. One can see the islands of superheavy elements around $Z = 114$, $N = 184$ and 196, respectively, and the one around $Z = 164$, $N = 318$.

CP644, *Exotic Clustering: 4th Catania Relativistic Ion Studies*, edited by S. Costa, A. Insolia, and C. Tuvè
© 2002 American Institute of Physics 0-7354-0099-7/02/$19.00

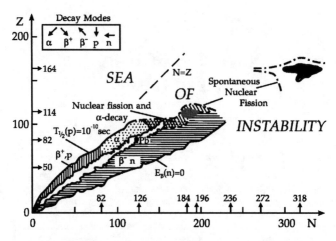

FIGURE 1. The periodic system of elements as conceived by the Frankfurt school in the late sixties. The islands of superheavy elements ($Z = 114$, $N = 184$, 196 and $Z = 164$, $N = 318$) are shown as dark hatched areas.

COLD VALLEYS IN THE POTENTIAL

The important question was how to produce these superheavy nuclei. There were many attempts, but only little progress was made. It was not until the middle of the seventies that the Frankfurt school of theoretical physics together with foreign guests (R.K. Gupta (India), A. Sandulescu (Romania))[1] theoretically understood and substantiated the concept of bombarding of double magic lead nuclei with suitable projectiles, which had been proposed intuitively by the russian nuclear physicist Y. Oganessian[1]. The two-center shell model, which is essential for the description of fission, fusion and nuclear molecules, was developed in 1969-1972 by W. Greiner and his students U. Mosel and J. Maruhn[1]. It showed that the shell structure of the two final fragments was visible far beyond the barrier into the fusioning nucleus. The collective potential energy surfaces of heavy nuclei, as they were calculated in the framework of the two-center shell model, exhibit pronounced valleys, such that these valleys provide promising doorways to the fusion of superheavy nuclei for certain projectile-target combinations (Fig. 2). If projectile and target approach each other through those **"cold" valleys**, they get only minimally excited and the barrier which has to be overcome (fusion barrier) is lowest (as compared to neighbouring projectile-target combinations). In this way the correct projectile- and target-combinations for fusion were predicted. Indeed, Gott-fried Münzenberg and Sigurd Hofmann and their group at GSI [1] have followed this approach. With the help of the SHIP mass-separator and the position sensitive detectors, which were especially developped by them, they produced the pre-superheavy elements $Z = 106$, 107, . . . 112, each of them with the theoretically predicted projectile-target combinations, and only with these. Everything else failed. This is an impressing success, which crowned the laborious construction work of many years. The before last example of this success, was the discovery of element 112 and its long α-decay chain. Very

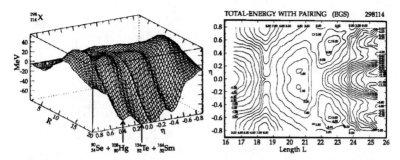

FIGURE 2. The collective potential energy surface of ^{184}114, calculated within the two center shell model by J. Maruhn et al., shows clearly the cold valleys which reach up to the barrier and beyond. Here R is the distance between the fragments and $\eta = \dfrac{A_1 - A_2}{A_1 + A_2}$ denotes the mass asymmetry: $\eta = 0$ corresponds to a symmetric, $\eta = \pm 1$ to an extremely asymmetric division of the nucleus into projectile and target. If projectile and target approach through a cold valley, they do not "constantly slide off" as it would be the case if they approach along the slopes at the sides of the valley. Constant sliding causes heating, so that the compound nucleus heats up and gets unstable. In the cold valley, on the other hand, the created heat is minimized.

recently the Dubna–Livermore–group produced two isotopes of $Z = 114$ element by bombarding ^{244}Pu with ^{48}Ca. Also this is a cold–valley reaction (in this case due to the combination of a spherical and a deformed nucleus), as predicted by Gupta, Sandulescu and Greiner in 1977. There exist also cold valleys for which both fragments are deformed [1], but these have yet not been verified experimentally.

SHELL STRUCTURE IN THE SUPERHEAVY REGION

Studies of the shell structure of superheavy elements in the framework of the meson field theory and the Skyrme-Hartree-Fock approach have recently shown that the magic shells in the superheavy region are very isotope dependent [3]. Additionally, there is a strong dependency on the parameterset and the model. Some forces hardly show any shell structure, while others predict the magic numbers $Z = 114, 120, 126$. Using the heaviest known gg-nucleus Hassium $^{264}_{154}$108 as a criterium to find the best parameter sets in each model, it turns out that PL-40 and SkI4 produce best its binding energy. These two forces though make conflicting predictions for the magic number in the superheavy region: SkI4 predicts $Z = 114, 120$ and PL-40 $Z = 120$. Most interesting, $Z = 120$ **as magic proton number seems to be as probable as** $Z = 114$. Deformed calculations within the two models [3] reveal again different predictions: Though both parametrizations predict $N = 162$ as the deformed neutron shell closure, the deformed proton shell closures are $Z = 108$ (SkI4) and $Z = 104$ (PL-40) (see Fig. 3). Calculations of the potential energy surfaces [3] show single humped barriers, their heights and widths strongly depending on the predicited magic number. Furtheron, recent investigations in a chirally symmetric mean–field theory (see also below) result also in the prediction of these two magic numbers [9, 10]. The corresponding magic neutron numbers are

329

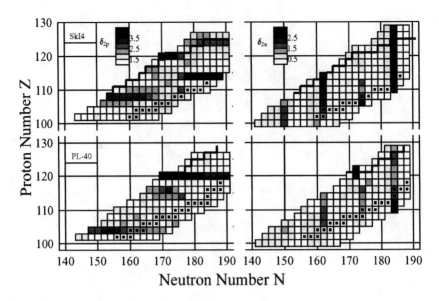

FIGURE 3. Grey scale plots of proton gaps (left column) and neutron gaps (right column) in the N-Z plane for deformed calculations with the forces SkI4 and PL-40. Besides the spherical shell closures one can see the deformed shell closures for protons at $Z = 104$ (PL-40) and $Z = 108$ (SkI4) and the ones for neutron at $N = 162$ for both forces.

predicted to be $N = 172$ and - as it seems to a lesser extend - $N = 184$. Thus, this region provides an open field of research.

The charge distribution of the $Z = 120, N = 184$ nucleus indicates a hollow inside. This leads us to suggest that it might be essentially a fullerene consisting of 60 α-particles and one binding neutron per alpha.

The "cold valleys" in the collective potential energy surface are basic for understanding this exciting area of nuclear physics! It is a master example for understanding the **structure of elementary matter**, which is so important for other fields, especially astrophysics, but even more so for enriching our "Weltbild", i.e. the status of our understanding of the world around us.

EXTENSION OF THE PERIODIC SYSTEM INTO THE SECTIONS OF HYPER– AND ANTIMATTER

Nuclei that are found in nature consist of nucleons (protons and neutrons) which themselves are made of u (up) and d (down) quarks. However, there also exist s (strange) quarks and even heavier flavors, called charm, bottom, top. The latter has just recently been discovered. Let us stick to the s quarks. They are found in the 'strange' relatives of the nucleons, the so-called hyperons ($\Lambda, \Sigma, \Xi, \Omega$). The Λ-particle, e. g., consists of one u, d and s quark, the Ξ-particle even of an u and two s quarks, while the Ω (sss) contains

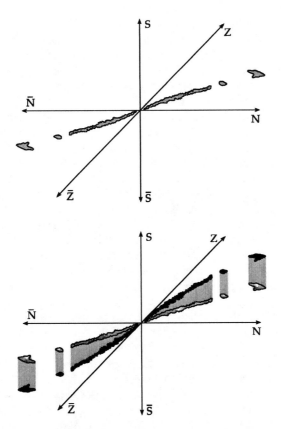

FIGURE 4. The extension of the periodic system into the sectors of strangeness (S, $\bar{S}$) and antimatter ($\bar{Z}$, $\bar{N}$). The stable valley winds out of the known proton (Z) and neutron (N) plane into the S and $\bar{S}$ sector, respectively. The same can be observed for the antimatter sector. In the upper part of the figure only the stable valley in the usual proton (Z) and neutron (N) plane is plotted, however, extended into the sector of antiprotons and antineutrons. In the second part of the figure it has been indicated, how the stable valley winds out of the Z-N-plane into the strangeness sector.

strange quarks only.

If such a hyperon is taken up by a nucleus, a **hyper-nucleus** is created. Hyper-nuclei with one hyperon have been known for 20 years now, and were extensively studied by B. Povh (Heidelberg)[1]. Several years ago, Carsten Greiner, Jürgen Schaffner and Horst Stöcker[1] theoretically investigated nuclei with many hyperons, **hypermatter**, and found that the binding energy per baryon of strange matter is in many cases even higher than that of ordinary matter (composed only of u and d quarks). This leads to the idea of extending the periodic system of elements in the direction of strangeness.

One can also ask for the possibility of building atomic nuclei out of **antimatter**, that means searching e. g. for anti-helium, anti-carbon, anti-oxygen. Fig. 4 depicts this idea. Due to the charge conjugation symmetry antinuclei should have the same magic numbers and the same spectra as ordinary nuclei. However, as soon as they get in touch with ordinary matter, they annihilate with it and the system explodes.

Now the important question arises how these strange matter and antimatter clusters can be produced. First, one thinks of collisions of heavy nuclei, e. g. lead on lead, at high energies (energy per nucleon ≥ 200 GeV). Calculations with the URQMD-model of the Frankfurt school show that through **nuclear shock waves** [1, 6] nuclear matter gets compressed to 5–10 times of its usual value, $\rho_0 \approx 0.17$ fm^3, and heated up to temperatures of $kT \approx 200$ MeV. As a consequence about 10000 pions, 100 Λ's, 40 Σ's and Ξ's and about as many antiprotons and many other particles are created in a single collision. It seems conceivable that it is possible in such a scenario for some Λ's to get captured by a nuclear cluster. This happens indeed rather frequently for one or two Λ-particles; however, more of them get built into nuclei with rapidly decreasing probability only. This is due to the low probability for finding the right conditions for such a capture in the phase space of the particles: the numerous particles travel with every possible momenta (velocities) in all directions. The chances for hyperons and antibaryons to meet gets rapidly worse with increasing number. In order to produce multi-Λ-nuclei and antimatter nuclei, one has to look for a different source.

In the framework of meson field theory the energy spectrum of baryons has a peculiar structure, depicted in Fig. 5. It consists of an upper and a lower continuum, as it is known from the electrons (see e. g. [5]). Of special interest in the case of the baryon spectrum is the potential well, built of the scalar and the vector potential, which rises from the lower continuum. It is known since P.A.M. Dirac (1930) that the negative energy states of the lower continuum have to be occupied by particles (electrons or, in our case, baryons). Otherwise our world would be unstable, because the "ordinary" particles are found in the upper states which can decay through the emission of photons into lower lying states. However, if the "underworld" is occupied, the Pauli-principle will prevent this decay. Holes in the occupied "underworld" (Dirac sea) are antiparticles.

The occupied states of this underworld including up to 40000 occupied bound states of the lower potential well represent the **vacuum**. The peculiarity of this strongly correlated vacuum structure in the region of atomic nuclei is that — depending on the size of the nucleus — more than 20000 up to 40000 (occupied) bound nucleon states contribute to this polarization effect. Obviously, we are dealing here with a **highly correlated vacuum**. A pronounced shell structure can be recognized [7]. Holes in these states have to be interpreted as bound antinucleons (antiprotons, antineutrons). If the primary nuclear density rises due to compression, the lower well increases while the upper decreases and soon is converted into a repulsive barrier. This compression of nuclear matter can only be carried out in relativistic nucleus-nucleus collision with the help of shock waves, which have been proposed by the Frankfurt school (see W. Scheid et al. in [1]) and which have since then been confirmed extensively (for references see e. g. [8]). These **nuclear shock waves** are accompanied by heating of the nuclear matter. Indeed, density and temperature are intimately coupled in terms of the hydrodynamic Rankine-Hugoniot-equations. Heating as well as the violent dynamics cause the creation of many holes in the very deep (measured from $-M_B c^2$) vacuum well. These numerous bound holes resemble antimatter clusters which are bound in the medium; their wave functions have large overlap with antimatter clusters. When the primary matter density decreases during the expansion stage of the heavy ion collision, the potential wells, in particular the lower one, disappear.

332

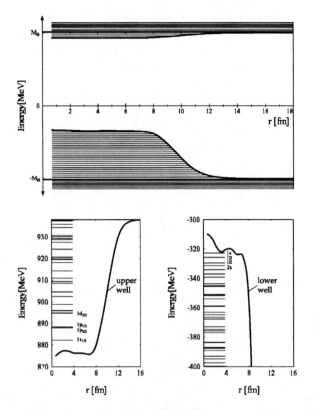

FIGURE 5. Baryon spectrum in a nucleus. Below the positive energy continuum exists the potential well of real nucleons. It has a depth of 50-60 MeV and shows the correct shell structure. The shell model of nuclei is realized here. However, from the negative continuum another potential well arises, in which about 40000 bound particles are found, belonging to the vacuum. A part of the shell structure of the upper well and the lower (vacuum) well is depicted in the lower figures.

The bound antinucleons are then pulled down into the (lower) continuum. In this way antimatter clusters may be set free. Of course, a large part of the antimatter will annihilate on ordinary matter present in the course of the expansion. However, it is important that this mechanism for the production of antimatter clusters out of the highly correlated vacuum does not proceed via the phase space. The required coalescence of many particles in phase space suppresses the production of clusters, while it is favoured by the direct production out of the highly correlated vacuum. In a certain sense, the highly correlated vacuum is a kind of cluster vacuum (vacuum with cluster structure). The shell structure of the vacuum levels (see Fig. 5) supports this latter suggestion. Fig. 6 illustrates this idea.

The mechanism is similar for the production of multi-hyper nuclei (Λ, Σ, Ξ, Ω). Meson field theory predicts also for the Λ energy spectrum at finite primary nucleon density the existence of upper and lower wells. The lower well belongs to the vacuum and is fully occupied by Λ's.

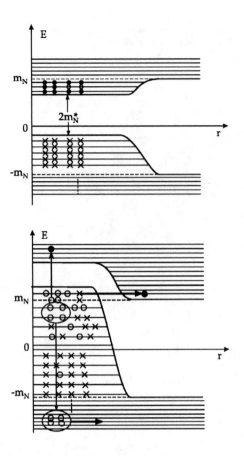

FIGURE 6.
Due to the high temperature and the violent dynamics, many bound holes (antinucleon clusters) are created in the highly correlated vacuum, which can be set free during the expansion stage into the lower continuum. In this way, antimatter clusters can be produced directly from the vacuum. The horizontal arrow in the lower part of the figure denotes the spontaneous creation of baryon-antibaryon pairs, while the antibaryons occupy bound states in the lower potential well. Such a situation, where the lower potential well reaches into the upper continuum, is called supercritical. Four of the bound holes states (bound antinucleons) are encircled to illustrate a "quasi-antihelium"
formed. It may be set free (driven into the lower continuum) by the violent nuclear dynamics.

Dynamics and temperature then induce transitions ($\Lambda\bar{\Lambda}$ creation) and deposit many Λ's in the upper well. These numerous bound Λ's are sitting close to the primary baryons: in a certain sense a giant multi-Λ hypernucleus has been created. When the system disintegrates (expansion stage) the Λ's distribute over the nucleon clusters (which are most abundant in peripheral collisions). In this way multi-Λ hypernuclei can be formed.

Of course this vision has to be worked out and probably refined in many respects. This means much more and thorough investigation in the future. It is particularly important to gain more experimental information on the properties of the lower well by (e, e' p) or (e, e' p p') and also ($\bar{p}_c p_b$, $p_c \bar{p}_b$) reactions at high energy ($\bar{p}_c$ denotes an incident antiproton from the continuum, p_b is a proton in a bound state; for the reaction products the situation is just the opposite). Also the reaction (p, p' d), (p, p' ^{3}He), (p, p' ^{4}He) and others of similar type need to be investigated in this context. The systematic studies of antiproton scattering on nuclei can contribute to clarify these questions. Various effective theories, e. g. of Walecka-type on the one side and theories with chiral invariance on the other side, have been constructed to describe dense strongly interacting matter [9]. It is important to note that they seem to give different strengths of the potential wells and also different dependence on the baryon density.

According to chirally symmetric meson field theories the antimatter-cluster-production and multi-hypermatter-cluster production out of the highly correlated vacuum takes place at considerably higher heavy ion energies as compared to the predictions of the Dürr-Teller-Walecka-type meson field theories. This in itself is a most interesting, quasi-fundamental question to be clarified. In the future, the question of the nucleonic substructure (form factors, quarks, gluons) and its influence on the highly correlated vacuum structure has to be studied. The nucleons are possibly strongly modified in the correlated vacuum: the Δ resonance correlations are probably important. Is this highly correlated vacuum state, especially during the compression, a preliminary stage to the quark-gluon cluster plasma? To which extent is it similar or perhaps even identical with it?

CONCLUDING REMARKS – OUTLOOK

The extension of the periodic system into the sectors of hypermatter (strangeness) and antimatter is of general and astrophysical importance. Indeed, microseconds after the big bang the new dimensions of the periodic system, we have touched upon, certainly have been populated in the course of the baryo- and nucleo-genesis. In the early history of the universe, even higher dimensional extensions (charm, bottom, top) may play a role, which we did not pursue here. It is an open question, how the depopulation (the decay) of these sectors influences the structure and composition of our world today. Our conception of the world will certainly gain a lot through the clarification of these questions.

REFERENCES

1. W. Greiner, Int. J. Modern Physics E, Vol. 5, No. 1, (1995) 1
2. R. K. Gupta, G. Münzenberg and W. Greiner, J. Phys. G: Nucl. Part. Phys. 23 (1997) L13,
    V. Ninov, K. E. Gregorich, W. Loveland, A. Ghiorso, D. C. Hoffman, D. M. Lee, H. Nitsche, W. J. Swiatecki, U. W. Kirbach, C. A. Laue, J. L. Adams, J. B. Patin, D. A. Shaughnessy, D. A. Strellis and P. A. Wilk, preprint
3. K. Rutz, M. Bender, T. Bürvenich, T. Schilling, P.-G. Reinhard, J.A. Maruhn, W. Greiner, Phys. Rev. C 56 (1997) 238,
    T. Bürvenich, K. Rutz, M. Bender, P.-G. Reinhard, J. A. Maruhn, W. Greiner, EPJ A 3 (1998) 139-147,
    M. Bender, K. Rutz, P.-G. Reinhard, J. A. Maruhn, W. Greiner, Phys. Rev. C 58 (1998) 2126-2132.

4. B. Fricke and W. Greiner, Physics Lett *30*B (1969) 317
    B. Fricke, W. Greiner, J.T. Waber, Theor. Chim. Acta (Berlin) *21* (1971) 235
5. W. Greiner, B. Müller, J. Rafelski, QED of Strong Fields, Springer Verlag, Heidelberg (1985). For a more recent review see W. Greiner, J. Reinhardt, *Supercritical Fields in Heavy–Ion Physics*, Proceedings of the 15th Advanced ICFA Beam Dynamics Workshop on Quantum Aspects of Beam Physics, World Scientific (1998)
6. H. Stöcker, W. Greiner and W. Scheid Z. Phys. A 286 (1978) 121
7. I. Mishustin, L.M. Satarov, J. Schaffner, H. Stöcker and W.Greiner
    Journal of Physics G (Nuclear and Particle Physics) *19* (1993) 1303,
    P.K. Panda, S.K. Patra, J. Reinhardt, J. Maruhn, H. Stöcker, W. Greiner, Int. J. Mod. Phys. E 6 (1997) 307,
    N. Auerbach, A. S. Goldhaber, M. B. Johnson, L. D. Miller and A. Picklesimer, Phys. Lett. B182 (1986) 221
8. H. Stöcker and W. Greiner, Phys. Rep. 137 (1986) 279.
9. a) Brian D. Serot, John Dirk Walecka, Advances in Nuclear Physics **Volume 16**, Plenum Press, New-York - London
    b) J. Theis, G. Graebner, G. Buchwald, J. Maruhn, W. Greiner, H. Stöcker and J. Polonyi, Phys. Rev. D 28 (1983) 2286,
    c) S. Klimt, M. Lutz, W. Weise, Phys. Lett. B*249* (1990) 386
    d) I. N. Mishustin, Proc. Int. Conference "Structure of Vacuum and Elementary Matter" (South Africa, Wilderness, March 9-15, 1996), p. 499, World Scientific, Singapour
    e) P. Papazoglou, D. Zschiesche, S. Schramm, J. Schaffner–Bielich, H. Stöcker, W. Greiner,Phys. Rev. C **59** (1999) 411-427
    f) I. N. Mishustin, L. M. Satarov,H. Stöcker, W. Greiner, Phys. Rev. C **62** (2000) 034901
10. P. Papazoglou, Ph.D. thesis, University of Frankfurt, 1998; Ch. Beckmann, P. Papazoglou, D. Zschiesche, S. Schramm, H. Stöcker, and W. Greiner, Phys. Rev C **65**, 024301 (2002) .

Importance of multi-mesonic fusion processes on (strange) antibaryon production

C. GREINER

Institut für Theoretische Physik, Universität Giessen, Heinrich-Buff-Ring 16, D-35392 Giessen, Germany

Abstract. Sufficiently fast chemical equilibration of (strange) antibaryons in an environment of nucleons, pions and kaons during the course of a relativistic heavy ion collision can be understood by a 'clustering' of mesons to buildt up baryon-antibaryon pairs. This multi-mesonic (fusion-type) process has to exist in medium due to the principle of detailed balance. Novel numerical calculations for a dynamical setup are presented. They show that - at maximum SPS energies - yields of each antihyperon specie are obtained which are consistent with chemical saturated populations of $T \approx 150 - 160$ MeV, in line with popular chemical freeze-out parameters extracted from thermal model analyses.

BRIEF OVERVIEW ON ANTIHYPERON PRODUCTION

Strangeness enhancement has been predicted a long time ago as a potential probe to find clear evidence for the temporary existence of a quark gluon plasma (QGP) in relativistic heavy ion collisions. A strong experimental effort has been made since and is still made for measuring strange particle abundancies in experiments at Brookhaven and at CERN (for a short recent review see [1]). In particular because of high production thresholds in binary hadronic reaction channels antihyperons had been advocated as the appropriate QGP candidates [2]. Indeed, a satisfactory picture of nearly chemically saturated populations of antihyperons has been experimentally demonstrated over the last years with the Pb+Pb experiments NA49 and WA97 at CERN-SPS. For this statement, of course, a quantitative, theoretical analysis by employing a thermal (or 'statistical') model has to be invoked by fitting the thermodynamical parameters to the set of individual (strange) hadronic abundancies [3].

On the other hand, already since the first measurements with the lighter ions have been undertaken in the early nineties, the theoretical description of the antibaryon production within hadronic transport schemes in comparison to these data faced some severe difficulties. Some phenomenological motivated attempts to explain a more abundant production of antihyperons within a hadronic transport description [4] had been proposed like the appearance of color ropes, the fusion of strings, the percolation of strings, or the formation of high-dense hadronic clusters. The underlying mechanisms, however, have to be considered exotic and to some extent ad hoc, their purpose lies mainly to create (much) more antibaryon in the very early intial stage of the reaction (compared to simple rescaled p+p collisions). To some extent indeed the philosophy behind these mechanisms was to model a precursor of initial QGP formation within a hadronic trans-

CP644, *Exotic Clustering: 4th Catania Relativistic Ion Studies,* edited by S. Costa, A. Insolia, and C. Tuvè
© 2002 American Institute of Physics 0-7354-0099-7/02/$19.00

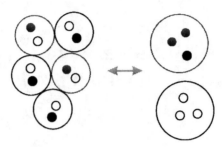

FIGURE 1. Schematic picture for the multi-mesonic fusion-like reaction $3\pi + 2K \leftrightarrow \bar{\Xi} + N$.

port scheme. On the other hand, in most of the transport calculations a dramatic role of subsequent antibaryon annihilation is observed, which, in return, has to be more than counterbalanced by these more exotic, initially occuring mechanisms. Large annihilation rates result if the free-space cross section is employed. The situation seems even more paradox with respect to the fact that the chemical description within thermal models works indeed amazingly well for the antihyperons and being, of course, completely nondependent on the (large) magnitude of the annihilation cross section. For all of this reasons the theoretical and dynamical understanding of the production of (strange) antibaryons has remained a delicate and challenging task [5].

As we will demonstrate, a correct incorporation of the baryonic annihilation channels had actually not been done consistently. We have conjectured recently that a sufficiently fast redistributions of strange and light quarks into (strange) baryon-antibaryon pairs should be achieved by multi-mesonic fusion-type reactions of the type

$$n_1\pi + n_2 K \leftrightarrow \bar{Y} + p \tag{1}$$

occuring in a moderately dense hadronic system [6] (illustrated in Fig. 1). The beauty of this argument lies in the fact that (at least) these special kind of multi-hadronic reactions have to be present because of the fundamental principle of detailed balance. As the annihilation of antihyperons on baryons is of dramatic relevance, the multi-mesonic (fusion-like) 'back-reactions' involving n_1 pions and n_2 kaons, where n_2 counts the number of anti-strange quarks within the antihyperon $\bar{Y}$), must, in principle, be taken care of in a dynamical simulation. These reactions are then the most dominant source of production. This crucial fact had been overseen in all of the aforementioned treatments. The underlying reasoning was first raised by Rapp and Shuryak who described the maintenance of nearly perfect chemical equilibrium of antiprotons together with pions and nucleons during the very late stage of the expanding fireball before the particles loose contact [7].

The crucial input, though plausible, is now to assume that the annihilation cross sections for any strange or nonstrange antibaryon on any baryon are approximately the same order as for $N\bar{p}$ at the same relative momenta, i.e. $\sigma_{B\bar{Y}\to n\pi+n_y K} \approx \sigma_{N\bar{p}\to n\pi}$, being in the range of 50-100 mb for characteristic and moderately low momenta occuring in an expanding hadronic fireball. The equilibration timescale $(\Gamma_{\bar{Y}})^{(-1)} \sim 1/(\sigma_{B\bar{Y}} v_{B\bar{Y}} \rho_B)$ is to a good approximation proportional to the inverse of the density of baryons and their resonances. Adopting an initial density of 1–2 times normal nuclear matter density ρ_0 for the initial and thermalized hadronic fireball, the antihyperons do equilibrate on a timescale of 1–3 fm/c well within the expansion timescale of the late hadronic fireball. Hence, fast chemical equilibration of the antihyperon abundancies is guaranteed by detailed balance with respect to the strong annihilation, the final yields then being independent of the actual size of the (large) annihilation cross section, solving the aforementioned paradox.

To be quantitative, (some) novel results by solving rate calculations for a dynamical setup are presented in the following. Before turning to their discussion, we will review briefly on some general ideas of the baryon-antibaryon annihilation process, in order to strengthen the one main assumption concerning the general size of its cross section, and on how to come to general master equations.

ANTIBARYON ANNIHILATION AND ITS EFFECTIVE DESCRIPTION BY A MASTER EQUATION

Antinucleon-nucleon annihilation is the strongest of all strong interaction processes. The strong annihilation of a nucleon and an antinucleon can be thought quantum mechanically as a complete absorbtive scattering process given approximately by the black disk formula $\sigma_{abs} = \pi(R + \lambda)^2$ or by a more sophisticated boundary condition description [8]. For $p + \bar{p}$-annihilation one finds for the 'black disk' radius $R = 1.07$ fm, for which then this decription reproduces very accurately the total inelastic cross section as a function of the beam momentum [8] and also its steep increase and diverging behaviour at low momenta. Hence, the quantum mechanical interpretation of the annihilation (being exotherm) is a picture of complete absorption. One is now tempted to generalize this simple and intuitive picture for all baryon-antibaryon annihilation processes. The only 'free' parameter which then can change is the radius R. Again, as it basically reflects the radius of the proton in the case of $p\bar{p}$-annihilation, one would expect that this canonical value of 1 fm is of general validity for all the annihilation processes. Thus it is plausible to assume that the cross section for annihilation between any baryon and antibaryon should be rather the same for the same relative momenta. Indeed, there does exist old data on the total $\bar{\Lambda} + p$ cross section [9] with reasonably large value, although the $\bar{\Lambda}$ momenta was exceeding 4 GeV/c in all data taken. (These data are not conclusive to really extrapolate dwn to lower momenta, though they give a right indication: The authors of [9] reported an extrapolated formula of $\sigma_{abs} = 47(\pm10)p_\Lambda^{-1/2}$ mb GeV$^{-1/2}$, which is not fully analogue to the black disk formula, but has already a rather correct shape with a sizeable magnitude.)

A more microscopic and quite popular picture, which can describe quantitatively the complicated final states of individual outgoing mesonic channels, is the concept of two meson doorway states [10] given by

$$\bar{B} + B \rightarrow M_1 + M_2 \rightarrow n_1\pi + n_2 K \, . \tag{2}$$

The main assumption is that the annihilation occurs exclusively via two mesons with nearest threshold dominance. M_1 and M_2 can be rather highlying resonances and/or hybrid states which come close with their individual threshold. The final decay of these two mesons into the various channels of multiple pions and kaons is then treated statistically and microcanonically. It turns out that such a description describes rather nicely reproduces the available data and can also be generalized straightforwardly for antihyperon annihilation [10]. There do exist a lot of other phenomenological and microscopic descriptions like quark models or flux tube models, which try to describe quantitatively the complicated process of the annihilation in more dynamical terms (for a review see [11]).

How can one incorporate such complicated processes into a description of transport dynamics? It is clear that as the very microscopy of the annihilation processes is not fully understood, one has to abandon any detailed local modelling, but has to turn to an effective coarse grained description which is guided by physical principles. Following the concepts of relativistic kinetic theory, the microscopic starting point is a Boltzmann-type equation of the form

$$
\partial_t f_{\bar{Y}} + \frac{\mathbf{p}}{E_{\bar{Y}}} \nabla f_{\bar{Y}} = \sum_{\{n_1\};B} \frac{1}{2E_{\bar{Y}}} \int \frac{d^3 p_B}{(2\pi)^3 2E_B} \prod_{\{n_1,n_2\}} \int \frac{d^3 p_i}{(2\pi)^3 2E_i} (2\pi)^4 \delta^4(p_{\bar{Y}} + p_B - \sum_{\{n_1,n_2\}} p_i)
$$

$$
|\langle\langle n_1, n_2 | T | \bar{Y}B \rangle\rangle|^2 \left\{ (-) f_{\bar{Y}} f_B \prod_{\{n_1,n_2\}} (1 + f_i) + \prod_{\{n_1,n_2\}} f_i (1 - f_{\bar{Y}})(1 - f_B) \right\},
$$

where

$$
\sigma_{\bar{Y}B}^{\{n_1\}} \equiv \frac{1}{'Flux'} \sum_{\{n_1\}} \prod_{\{n_1,n_2\}} \int \frac{d^3 p_i}{(2\pi)^3 2E_i} (2\pi)^4 \delta^4 \left(p_{\bar{Y}} + p_B - \sum_{\{n_1,n_2\}} p_i \right) |\langle\langle n_1, n_2 | T | \bar{Y}B \rangle\rangle|^2
$$

corresponds to the total annihilation cross section into the various possible multiple pion states. This form of the collision exactly incorporates the principle of detailed balance. The static fixed point of this equation (together with the usual binary kinetic processes) are thermal and chemical saturated distributions. Without the back reaction channels the equations would only have vanishing distributons as stable fixed points, which is, of course, unacceptable. Hence, the back reactions have to be considered as very basic and important ingredient of the transport description when trying to implement the baryon-antibaryon annihilation processes.

Furthermore it is obvious that the back reactions will guarantee that the (strange) antibaryons become chemical saturated with the pions, kaons and nucleons on a very short timescale. To see this more explicit, and also for the further numerical treatment, one can bring the above Boltzmann equation into a more intuitive form of a master or rate equation. Assuming $v_{rel} \sigma_{\bar{Y}B}(\sqrt{s})$ to be roughly constant, which is actually a good approximation for the $p\bar{p}$-annihilation, or, invoking a standard description of further effective coarse graining by using thermally averaged cross sections and distributions, and furthermore taking the distributions in the Boltzmann approximantion, the following master equation for the respectively considered antihyperon density is obtained

$$
\frac{d}{dt} \rho_{\bar{Y}} = -\langle\langle \sigma_{\bar{Y}B} v_{\bar{Y}B} \rangle\rangle \left\{ \rho_{\bar{Y}} \rho_B - \sum_{\{n_1\}} \hat{M}_{(n_1,n_2)}(T,\mu_B,\mu_s)(\rho_\pi)^{n_1}(\rho_K)^{n_2} \right\}, \tag{3}
$$

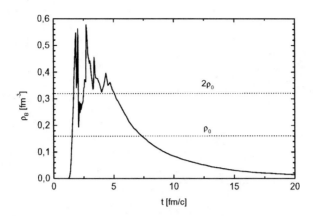

FIGURE 2. Time evolution of the (average) net baryon density for midrapidity $|\Delta Y| \leq 1$ and central Pb+Pb-collision at 160 AGeV obtained within a dynamical transport simulation. Here the amount of baryon number residing still in string-like excitations is explicitly discarded. String-like excitations have disappeared after about 3-4 fm/c, so that from this time on a pure hadronic fireball develops and expands. Its initial net baryon-density starts slightly above $\rho_B = 2\rho_0$.

where the 'back-reactions' of several effectively clustering pions and kaons are incorporated in the 'mass-law' factor

$$\hat{M}_{(n_1,n_2)}(T,\mu_B,\mu_s) = \frac{\rho_{\bar{Y}}^{eq.} \rho_B^{eq.}}{(\rho_\pi^{eq.})^{n_1}(\rho_K^{eq.})^{n_2}} \, p_{n_1} \, .$$

Here p_{n_1} (which will generally depend on the thermodynamical parameters) states the relative probability of the reaction (1) to decay into a specific number n_1 of pions and ρ_B denotes the total number density of baryonic particles. As is well-known, the mass-law factor $\hat{M}$ depends only on the temperature and the baryon and strange quark chemical potentials. $\Gamma_{\bar{Y}} \equiv \langle\langle \sigma_{\bar{Y}N} v_{\bar{Y}N}\rangle\rangle \rho_B$ gives the effective annihilation rate of the respective antihyperon specie on a baryon.

For a further manipulation one has to make assumptions for the various abundancies occuring in the master equation. Nonequilibrium inelastic hadronic reactions can explain to a good extent the overall strangeness production seen experimentally: The major amount of the produced kaons at SPS-energies can be understood in terms of still early and energetic non-equilibrium interactions [12]. In Fig. 2 the time evolution of netbaryon density at midrapidiy obtained within a dynamical transport simulation [13] is depicted. All strangeness is being produced still when string-like excitations are governing the dynamics. In the later hadronic stage the number of strange quarks stays more or less constant and can only be redistributed among the various hadrons [14]. Also the pions and nucleons do stay more or less at thermal equilibrium. Refering to the master equation (3), one can then take the pions, baryons and kaons to stay approximately in thermal equilibrium throughout the later hadronic evolution of the collision, the later being modelled to be an isentropic expansion with fixed total entropy content being specified

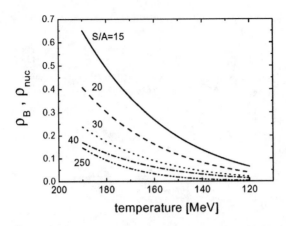

FIGURE 3. The netbaryon density ρ_B of a chemically fully equilibrated thermal hadronic resonance gas as function of decreasing temperature for the situation of an isentropic expansion, i.e. for various constant entropy/baryon ratios. For the value $S/A = 250$ (a situation expected at RHIC) the total baryon number density ρ_{nuc} is plotted instead.

via the entropy per baryon ratio S/A (compare with Fig. 3). (3) has the intuitive form

$$\frac{d}{dt}\rho_{\bar{Y}} = -\Gamma_{\bar{Y}}\left\{\rho_{\bar{Y}} - \rho_{\bar{Y}}^{eq}\right\}.\tag{4}$$

The production rate per unit volume $dN_{\bar{Y}}/dtdV$ is given by the non-vanishing, though small value $Gamma_{\bar{Y}}\rho_{\bar{Y}}^{eq}$. Still this rate is enough to populate the antihyperons to their (very small) equilibrium value.

For the calculations solving this master equation (4) one has to employ an 'effective' volume $V(t)$ in order to extrapolate from Fig. 2 to Fig. 3, i.e. simulating the global characteristic of the expansion and the dilution of the baryon density and thus the annihilation rate $\Gamma_{\bar{Y}}(t)$. The 'effective' (global or at midrapidity) volume $V(t)$ is parametrized as function of time by longitudinal Bjorken expansion and including a transversal expansion either with a linear profile

$$V_{eff,lin}(t \geq t_0) = \pi\,(ct)\,\left(R_0 + v_{lin}(t-t_0)\right)^2\tag{5}$$

(taking v_{lin} as an appropriate parameter to simulate slower or faster expansion) or with an accelerating radial flow

$$V_{eff,acc}(t \geq t_0) = \pi\,(ct)\,\left(R_0 + v_0(t-t_0) + 0.5a_0(t-t_0)^2\right)^2\tag{6}$$

with $R_0 = 6.5\,fm$, $v_0 = 0.15\,c$ and $a_0 = 0.05\,c^2/fm$ for the later modeling.

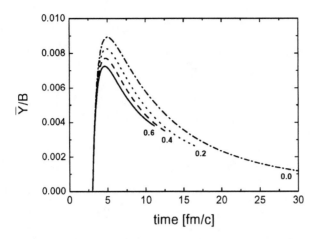

FIGURE 4. The anti-Λ to baryon number ratio $N_{\bar{\Lambda}}/N_B(t)$ as a function of time for various velocity parameters v_{lin} for the transverse expansion. The entropy per baryon is taken as $S/A = 30$, $t_0 = 3$ fm/c and $T_0 = 190$ MeV.

RESULTS AND IMPLICATIONS

At starting time t_0 an initial temperature T_0 is chosen. (T_0 is set to 190 MeV for the SPS and 150*MeV* for the AGS situation, while the initial energy densities are then about 1 GeV/fm^3.) From (5) or (6) together with the constraint of conserved entropy the temperature and the chemical potentials do follow as function of time and thus also $\rho_B(t)$ as well as $\rho_{\bar{Y}}^{eq}(t)$ within the hadronic resonance gas. As a *minimal* assumption the initial abundancy of antihyperons is set to zero. Equation (4), taking into account the volume dilution, is solved for each specie. For the thermally averaged cross section we take a simple constant value of $\langle \sigma_{ann} v \rangle := \sigma_0 \equiv 40$ mb.

In Fig. 4 the number of $\bar{\Lambda}$s (normalized to the conserved net baryon number) as a function of time is depicted. The entropy per baryon is chosen as $S/A = 30$ being characteristic to global ('4π') SPS results [15]. Here we have chosen the ansatz (5)for the expansion of the volume being linear in time in the transverse direction. The parameter v_{lin} is varied to simulate slow or fast expansion of the late hadronic fireball. The general characteristics is that first the antihyperons are dramatically being populated, and then in the very late expansion some more are still being annihilated, depending on how fast the expansion goes. A rapid expansion gives a higher yield, which can increase the final yield by a factor of 2 to 3. However, the typical expansion behaviour obtained from simulations or extracted from the analysis of transverse momentum slopes of individual hadrons (pions and protons) is that at the late stages the transverse expansion velocity shoul be about 0.5 c. In the following we stay to the second ansatz (6) which extrapolates from an initially slower to a later faster expansion.

In Fig. 5 the number of $\bar{\Lambda}$s as a function of time is depicted, where now the cross

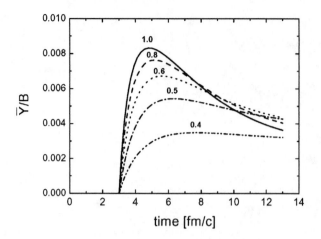

FIGURE 5. The anti-Λ to baryon number ratio $N_{\bar{\Lambda}}/N_B(t)$ as a function of time for various implemented annihilation cross section $\sigma_{eff} \equiv \lambda\,\sigma_0$. The entropy per baryon is taken as $S/A = 30$, $t_0 = 3$ fm/c and $T_0 = 190$ MeV.

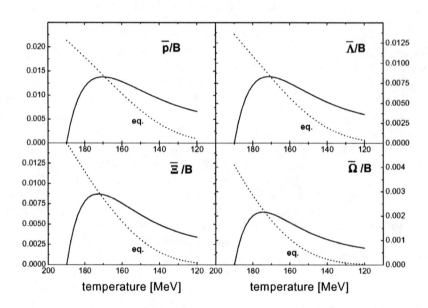

FIGURE 6. The antihyperon to baryon number ratio $N_{\bar{Y}}/N_B(T)$ and $N_{\bar{Y}}^{eq.}/N_B(T)$ (dotted line) as a function of the decreasing temperature. Parameters are the same as in Fig. 5.

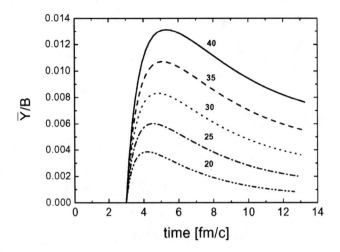

FIGURE 7. $N_{\bar{\Lambda}}/N_B(t)$ as a function of time for various entropy content described via the entropy per baryon ratio ($S/A = 20 - 40$). Other parameters are as in Fig. 5.

section employed is varied by a constant factor, i.e. $\sigma_{eff} \equiv \lambda \sigma_0$. The results are rather robust against a variation by a factor of 2 in the cross section. Typically (for $\lambda = 1$) about more than 5 times in number of antihyperons are created during the evolution compared to the final number freezing out, thus reflecting the fast back ('annihilation') and forth ('creation') processes at work dictated by detailed balance.

In Fig. 6 the number of antihyperons of each specie are now shown as a function of the decreasing temperature $T(t)$ of the hadronic system. For a direct comparison the instantaneous equilibrium abundancy $N_{\bar{Y}}^{eq} \cdot (T(t), \mu_B(t), \mu_s(t))/N_B$ is also given. As noted above, after a fast initial population, the individual yields of the antihyperons do overshoot their respective equilibrium number and then do finally saturate at some slightly smaller value. Moreover, one notices that the yields effectively do saturate at a number which can be compared to an equivalent equilibrium number at a temperature parameter around $T_{eff} \approx 150 - 160$ MeV, being strikingly close to the ones obtained within the various thermal analyses [3].

In Fig. 7 the number of anti-Λs as a function of time is given for various entropy per baryon ratios. One notices that the final value in the yield significantly depends on the entropy content, or, in other words, on the baryochemical potential. We note that the results at midrapidity from WA97 can best be reproduced by employing an entropy to baryon ratio $S/A = 40$. Indeed, at midrapidity one qualitatively expects a higher entropy content due to the larger pion to baryon ratio as compared to full '4π' data over all rapidities. At this point it will also be very interesting to compare our semi-quantitative calculations with the new results from NA49 on the $\bar{\Lambda}$-yield at lower SPS energies of 80 AGeV and 40 AGeV with lower entropy contents, respectively. As the presented results are very sensitive on the entropy content, one first needs a clean analysis to obtain a

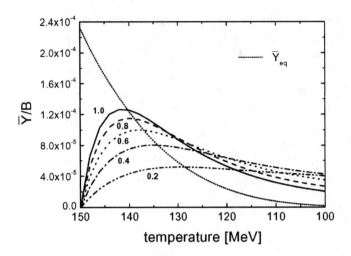

FIGURE 8. $N_{\bar{\Lambda}}/N_B(T)$ and $N_{\bar{\Lambda}}^{eq}/N_B(T)$ as a function of decreasing temperature for a characteristic AGS situation with an entropy content of $S/A = 12$ for various implemented annihiation cross section $\sigma_{eff} \equiv \lambda\,\sigma_0 \cdot t_0 = 5$ fm/c and $T_0 = 150$ MeV.

rather accurate S/A number from the measured pion and proton abundancies.

There is also a clear hint at AGS energies of enhanced anti-Λ production [16]. For most central collisions more anti-Λs are found compared to anti-protons, which is quite puzzling as within a thermal model analysis this ratio is found not to be larger than 1. This enhanced ratio of anti-Λs compared to anti-protons at AGS energies one can understand in a way that one assumes that their annihilation cross section on baryons is just slightly smaller than for the antiprotons. In Fig. 8 a similar study like that of Fig. 5 is shown for a characteristic situation at AGS. For smaller, yet not too small effective cross sections the final yield can here be enlarged by a factor of 2 compared to the case with a 'full' crossection, as the final reabsorption is not as effective. But, of course, this idea is speculation at present. Also, we remark that the $\bar{\Lambda}$s effectively do saturate at an equivalent equilibrium number at a temperature parameter around $T_{eff} \approx 120 - 130$ MeV. Unfortunately, there are no data for $\bar{\Xi}$ at AGS. Again the new NA49 data at lower energies are worthwhile to pursue. A detailed measurement of all antihyperons represents also an excellent opportunity for future heavy ion facilities at an energy upgraded GSI.

To summarize, multi-mesonic production of antihyperons is a consequence of detailed balance and, as the annihilation rate is large, it is by far the most dominant source in a hadronic gas. This is a remarkable observation, as it clearly demonstrates the importance of hadronic multi-particle channels, occuring frequently enough in a (moderately) dense *hadronic* environment in order to populate and chemically saturate the rare antibaryons. In order to be more competitive for a direct comparison with various experimental findings, new strategies have to be developed to describe for such multi-particle interactions within present day transport codes. A significant first step forward was very recently made; first results concerning the production of anti-protons at AGS and SPS energies are quite impressive [17]. Another strategy could be to exploit microscopically the concept of two meson doorway states (2) and their sequential decay by standard binary scattering processes.

REFERENCES

1. R. Stock, *arXiv:hep-ph/0204032*.
2. P. Koch, B. Müller and J. Rafelski, *Phys. Rep.* **142**, 167 (1986); J. Rafelski, *Phys. Lett.* B **262**, 333 (1991).
3. P. Braun-Munzinger, I. Heppe and J. Stachel, *Phys. Lett.* B **465**, 1 (1999); F. Becattini, J. Cleymans, A. Keränen, E. Suhonen and K. Redlich, PRC **64**, 024901 (2001).
4. H. Sorge et al, *Phys. Lett.* B **289**, 6 (1992); H. Sorge, *Z. Phys.* C **67**, 479 (1995); *Phys. Rev.* C **52**, 3291 (1995); K. Werner and J. Aichelin, *Phys. Lett.* B **300**, 158 (1993) and *Phys. Lett.* B **308**, 372 (1993); N. Armesto, M.A. Braun, E.G. Ferreiro and C. Pajares, *Phys. Lett.* B **344**, 301 (1995); E.G. Ferreiro and C. Pajares, *Z. Phys.* C **73**, 309 (1997); M. Bleicher et al, *Phys. Lett.* B **485**, 133 (2000).
5. H. Sorge, *Nucl. Phys.* A **630**, 522c (1998).
6. C. Greiner and S. Leupold, *J. Phys.* G **27**, L95 (2001); C. Greiner, *arXiv:nucl-th/0011026*; *Nucl. Phys.* A **698**, 591 (2002).
7. R. Rapp and E. Shuryak, *Phys. Rev. Lett.* **86**, 2980 (2001).
8. J. Vandermeulen, PRC **33**, 1101 (1986).
9. F. Eisele et al, *Phys. Lett.* B **60**, 297 (1976); G.J. Wang, G. Welke, R. Bellwied and C. Pruneau, *arXiv:nucl-th/9807036*.
10. J. Vandermeulen, ZPC **37**, 563 (1988); J. Cugnon and J. Vandermeulen, PRC **39**, 181 (1989).
11. C. Dover, T. Gutsche, M. Maruyama and A. Faessler, *Prog. Part. Nucl. Phys.* 29, 87 (1992).
12. J. Geiss, W. Cassing and C. Greiner, *Nucl. Phys.* A **644**, 107 (1998).
13. W. Cassing, private communication.
14. C. Greiner, *J. Phys.* G **28**, 1631 (2002).
15. J. Cleymans, B. Kämpfer and S. Wheaton, *arXiv:nucl-th/0110035*.
16. L. Ahle et al, (*E-802* Collaboration), *Phys. Rev. Lett.* **81**, 2650 (1998); B. Back et al, (*E-917* Collaboration), *Phys. Rev. Lett.* **87**, 242301 (2001).
17. W. Cassing, *Nucl. Phys.* A **700**, 618 (2002).

Stability of strange quark matter: model dependence

W. M. Alberico* and C. Ratti*

*Dipartimento di Fisica Teorica, Università di Torino and INFN, Sezione di Torino, Via P. Giuria 1, 10125 Torino, Italy

Abstract. The minimum energy per baryon number of strange quark matter is studied, as a function of the strangeness fraction, in the MIT bag model and in two different versions of the Color Dielectric Model: a comparison is made with the hyperon masses having the same strangeness fraction, and coherently calculated within both models. Calculations are carried out in mean field approximation, with one gluon exchange corrections. The results allow to discuss the model dependence of the stability of strangelets: they can be stable in the MIT bag model and in the double minimum version of the Color Dielectric Model, while the single minimum version of the Color Dielectric Model excludes this possibility.

INTRODUCTION

The production of strange quark matter and/or hypermatter in central heavy ion collisions has been suggested long ago, either in the form of multi-hypernuclear objects (strange hadronic matter) [1, 2, 3], or strangelets (strange multiquark droplets) [4, 5, 6, 7, 8]. The formation of the latter would be rather appealing, since it would be an unambiguous signature that a deconfined, strangeness rich state of quark gluon plasma has been created during the reaction. The investigation of the strangelet stability is therefore of primary importance for their detection in heavy ion experiments.

The idea is that, even if no strangeness is present in the initial state of the collision, and no *net* strangeness is expected after the reaction, nevertheless a large number of $s\bar{s}$ pairs can be produced in a single central event; the antiquarks $\bar{s}$ are then able to rapidly combine with the abundantly available u and d quarks to form antikaons that immediately leave the fireball region, which becomes strangeness rich matter. The hadronization process is then of fundamental importance: the copious formation of strange particles cannot be considered as a reliable signature of QGP formation, since kaons and hyperons can be produced in hadronic reactions as well [9]. If, on the contrary, after the formation of the deconfined plasma, this strangeness rich matter could coalesce into colorless multiquark states, the so-called strangelets, this would be an unambiguous signature of QGP formation; this process might be favoured by a rapid QGP cooling due to the prompt anti-kaon (and also pion) emission from the surface of the fireball.

Up to now, the stability of strangelets has been only investigated within the MIT bag model [10], also including $\mathscr{O}(\alpha_s)$ corrections to the properties of bulk strange matter: according to this pioneering work, heavy, slightly positively charged, strangelets could be more stable than ordinary nuclei. A detailed calculation of strangelet properties within

CP644, *Exotic Clustering: 4th Catania Relativistic Ion Studies*, edited by S. Costa, A. Insolia, and C. Tuvè the MIT bag model, including shell effects and all the hadronic decay channels has been performed by J. Schaffner *et al.* [11]: a valley of stability clearly appears for $A_B = 5 \div 16$ with charge fraction Z/A between 0 and -0.5. On the other hand, strangelets having a larger mass should be positively charged according to the results of Ref. [12].

In ref. [13] the authors confront the predictions about the stability of strangelets within the MIT bag model and the Color Dielectric Model (CDM): the equilibrium energy of the strange matter is compared with the masses of hyperons having the same strangeness fraction, and coherently calculated within both models. The main goal is to find out whether and to which extent the stability of strange matter and/or strangelets depends on the model employed to describe the confined system of quarks. The present contribution is largely based on the results of ref. [13], with a special focus on the model dependence of the strangelets stability. We consider homogeneous quark matter made up of u, d and s quarks, without imposing chemical equilibrium on the density of the strange quarks. Rather, we assume that there exists in the system a definite strange fraction $R_s = \rho_s/\rho$, ρ being the total baryon density of quarks and ρ_s the baryonic density of strange quarks. This is coherent with the hypothesis that, during a high energy collision between heavy ions, this state of matter, if formed at all, can only survive for a very short time, so that it has no time to reach β equilibrium; hence the minimal energy per baryon number can be studied as a function of the strange fraction R_s. We also consider the effect induced by the introduction of perturbative gluons in both models. Since electromagnetic interaction has been neglected, the minimum of the energy corresponds to an equal number of u and d quarks.

We consider, for simplicity, an infinite and homogeneous system, but strangelets are indeed finite objects, and therefore one should remember that the energy of the infinite system appears to be a lower limit with respect to the envelop of strangelet energies versus strangeness fraction: the latter was nicely illustrated by Schaffner *et al.* [11] calculating the strangelet masses within the MIT bag model with shell mode filling. We simply recall that surface effects, which we do not consider, would increase the energy curves of bulk matter, typically of 50-100 MeV: hence, if hyperons should turn out to be more stable than strange matter, then this would exclude also the stability of strangelets. If, on the contrary, strange matter is more stable, then this provides only an indication in favour of stable strangelets unless the mass gap between the two states is large enough.

STRANGELETS IN THE MIT BAG MODEL

MIT bag without gluons

We consider first the simplest version of the MIT bag, not including one gluon exchange corrections; therefore the model has only two parameters: the vacuum pressure B and the strange quark mass m_s. In order to discuss the various possible scenarios, we have used a wide range for both parameters:

$$
\begin{aligned}
B &= 60, 100, 150 \text{ MeV/fm}^3; \\
m_s &= 100, 200, 300 \text{ MeV}.
\end{aligned}
\tag{1}
$$

The single flavor contribution to the energy density of the system is given by:

$$\varepsilon_f = 6 \int \frac{d\vec{k}}{(2\pi)^3} E_f(k) \, \theta \left(k_{F_f} - k\right), \tag{2}$$

where $E_f(k) = \sqrt{k^2 + m_f^2}$ and k_{F_f} is the Fermi momentum of flavor f. It can be analytically expressed, for u, d (massless) and s quarks, respectively:

$$\varepsilon_{u,d} = \frac{3}{(2\pi)^2} k_{F_{u,d}}^4, \tag{3}$$

$$\varepsilon_s = \frac{3}{8\pi^2} \left[m_s^4 \ln \left(\frac{m_s}{k_{F_s} + \sqrt{k_{F_s}^2 + m_s^2}} \right) + k_{F_s} \sqrt{k_{F_s}^2 + m_s^2} \left(2k_{F_s}^2 + m_s^2\right) \right].$$

The total energy density of our system turns then out to be:

$$\varepsilon_{tot} = 2\varepsilon_{u,d} + \varepsilon_s + B. \tag{4}$$

The dependence of the above formula on R_s and ρ can be easily found by recalling the following relations for the various Fermi momenta:

$$\rho_s = R_s \rho \tag{5}$$

$$k_{F_s} = \left(3\pi^2 \rho_s\right)^{1/3}$$

$$k_{F_{u,d}} = \left(\frac{3\pi^2}{2} \rho (1 - R_s)\right)^{1/3},$$

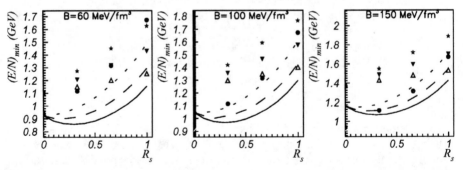

FIGURE 1. Minimal energy per baryon number in the MIT bag model, as a function of the strangeness fraction $R_s = \rho_s/\rho$, for various values of the model parameters. The continuous line corresponds to $m_s = 100$ MeV, the dashed line to $m_s = 200$ MeV and the dotted line to $m_s = 300$ MeV. Full circles correspond to experimental masses, the other points to the masses evaluated in the model, with $m_s = 100$ MeV (open triangles), $m_s = 200$ MeV (full triangles), $m_s = 300$ MeV (stars), respectively.

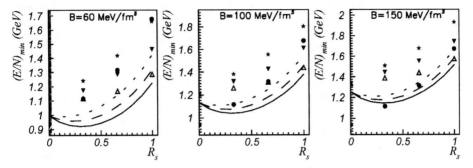

FIGURE 2. Minimal energy per baryon number in the MIT bag model, including the OGE potential with $\alpha_s = 0.5$, as a function of the strangeness fraction $R_s = \rho_s/\rho$. The continuous line corresponds to $m_s = 100$ MeV, the dashed line to $m_s = 200$ MeV and the dotted line to $m_s = 300$ MeV. Full circles represent the experimental masses, the other points refer to the masses evaluated in the model, with $m_s = 100$ MeV (open triangles), $m_s = 200$ MeV (full triangles), $m_s = 300$ MeV (stars), respectively.

In the above, ρ is the total baryon number density in the system ($\rho = A_B/V$), and the color degeneracy and baryon number $1/3$ of the quarks have been taken into account. From the above formulas we calculate the energy per baryon number to be:

$$\frac{E_{tot}}{A_B} = \frac{\varepsilon_{tot}}{\rho}. \tag{6}$$

In Fig. 1 the results of the minimal energy per baryon (6) corresponding to $B = 60, 100, 150$ MeV/fm^3 are shown as a function of R_s. For each value of B we explore three different values of the strange mass, $m_s = 100, 200, 300$ MeV, and we compare these results with the experimental nucleon and hyperon masses (full circles). We have also evaluated, according to formula (3.6) of Ref. [14], the baryonic masses which are obtained within the same model employed for bulk strange matter, using the same sets of bag parameter and strange quark mass. As it appears from the figure, the three lines corresponding to the different values of m_s are much lower than the experimental hyperon masses for $B = 60$ MeV/fm^3 and $B = 100$ MeV/fm^3, while this is not the case for $B = 150$ MeV/fm^3 and $m_s=300$ MeV; however, if we compare the energy of strange matter with the corresponding theoretical masses of the various hyperons, we find that strange matter is *always* lower in energy, and thus more stable. We can therefore conclude that the MIT bag model without perturbative gluon corrections allows the existence of strangelets.

MIT bag model with perturbative gluons

We consider now the effects of introducing in the calculation perturbative corrections due to the exchange of gluons . At first order in α_s, two contributions to the energy can be considered, the direct and the exchange one. Since the system is globally colorless

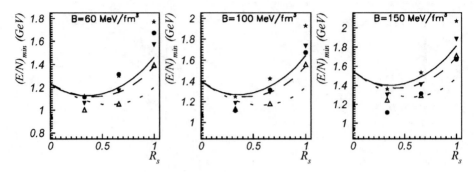

FIGURE 3. The same as in Fig. 2, but for $\alpha_s = 2.2$.

the direct term vanishes, while the exchange one gives the following contribution to the energy density of quarks of flavor f [10]:

$$\varepsilon_f^{OGE} = -\frac{\alpha_s}{\pi^3}m_f^4\left\{x_f^4 - \frac{3}{2}\left[\ln\left(\frac{x_f+\eta_f}{\eta_f}\right) - x_f\eta_f\right]^2 + \right.$$

$$\left. + \frac{3}{2}\ln^2\left(\frac{1}{\eta_f}\right) - 3\ln\left(\frac{\mu}{m_f\eta_f}\right)\left[\eta_f x_f - \ln\left(x_f+\eta_f\right)\right]\right\}. \quad (7)$$

Here:

$$x_f = \frac{k_{F_f}}{m_f}$$

$$\eta_f = \sqrt{1+x_f^2}.$$

and μ is a renormalization scale, for which we choose the value $\mu = 313$ MeV, according to Ref. [10]. For sake of illustration, we adopted two different values for α_s, a small perturbative value ($\alpha_s = 0.5$), which is in line with the choices and motivations of Fahri and Jaffe [10], and the canonical value which was employed by DeGrand *et al.* [14] ($\alpha_s = 2.2$), to reproduce the hyperon masses. The corresponding results are illustrated in Figs. 2 and 3. From Fig. 2 we can see that, even after the inclusion of perturbative gluons, strangelets are more stable than hyperons for almost all values of the model parameters. However, when we use the stronger coupling of Fig. 3 the stability of strange matter (and hence strangelets) becomes questionable, particularly for low values of the strange mass m_s. Only for $m_s = 300$ MeV the theoretical masses of hyperons always lie above the energy of bulk matter (not so the experimental masses).

From this analysis we can conclude (in agreement with previous findings) that, apart from rather extreme choices of the model parameters, metastable strangelets can exist in the MIT bag model.

STRANGELETS IN THE COLOR DIELECTRIC MODEL

The Color Dielectric Model provides absolute confinement of quarks through their interaction with a scalar field χ which represents a multi–gluon state and produces a density dependent constituent mass (see for example the review articles [15, 16, 17]) The typical Lagrangian of the CDM reads:

$$\mathscr{L} = \sum_{f=u,d,s} \bar{\psi}_f i\gamma^\mu \left(\partial_\mu - ig_s \frac{\lambda^a}{2} A_\mu^a \right) \psi_f - \frac{gf_\pi}{\chi} \sum_{f=u,d} \bar{\psi}_f \psi_f - m_s(\chi) \bar{\psi}_s \psi_s +$$

$$+ \frac{1}{2} (\partial_\mu \chi)^2 - U(\chi) - \frac{1}{4} \kappa(\chi) F_{\mu\nu}^a F^{a\mu\nu}, \tag{8}$$

where ψ_f are the quark fields, A_μ^a is the (effective) gluon field, $F_{\mu\nu}^a$ its strength tensor and χ is the color dielectric field; g_s is the strong (colour) coupling ($g_s^2/4\pi = \alpha_s$).

The u and d quark mass terms arise as a consequence of their interaction with the χ–field and read:

$$m_{u,d} = \frac{gf_\pi}{\chi}, \tag{9}$$

where g is a parameter of the model and f_π the pion decay constant, which is fixed to its experimental value, $f_\pi = 93$ MeV. For the strange quark mass we consider two different versions of the 3–flavors CDM, namely a *scaling model*, with

$$m_s = \frac{g'f_\pi}{\chi}, \tag{10}$$

and a *non–scaling model*, with a constant shift of the s–mass with respect to the u,d–one:

$$m_s = \frac{gf_\pi}{\chi} + \Delta m \equiv m_{u,d} + \Delta m. \tag{11}$$

In the above g' (or Δm) is another parameter of the model.

Concerning the color dielectric field, there exist in the literature several options, both for its coupling to the gluon tensor and for the potential $U(\chi)$. We adopt here both the single minimum (SM), quadratic potential:

$$U_{SM}(\chi) = \frac{1}{2}M^2\chi^2, \tag{12}$$

which introduces the third parameter of the model, M (the mass of the glueball), and the double minimum (DM), quartic potential:

$$U_{DM}(\chi) = \left(\frac{1}{2}\frac{M^2}{\chi_0^2} - \frac{3B}{\chi_0^4} \right)\chi^4 + \left(\frac{4B}{\chi_0^3} - \frac{M^2}{\chi_0} \right)\chi^3 + \frac{1}{2}M^2\chi^2. \tag{13}$$

The latter introduces an extra parameter, the bag pressure B, while the parameter χ_0 is used to make the ratio χ/χ_0 dimensionless. The color–dielectric function, $\kappa(\chi)$, is usually assumed to be a quadratic or quartic function of χ: we will use both options and hence we set:

$$\kappa(\chi) = \left(\frac{\chi}{\chi_0}\right)^{\beta}, \qquad \text{with} \qquad \beta = 2,4.\qquad(14)$$

The field equations are solved in the mean field approximation and neglecting the gluon fields: the latter are subsequently taken into account as a perturbation. The unperturbed (i.e. without gluon contribution) energy density reads:

$$\varepsilon_0 = \sum_{f=u,d,s} \frac{3}{8\pi^2} \left\{ m_f^4 \ln\left(\frac{m_f}{k_{F_f} + \sqrt{k_{F_f}^2 + m_f^2}}\right) \right.$$
$$\left. + k_{F_f}\sqrt{k_{F_f}^2 + m_f^2}\left(2k_{F_f}^2 + m_f^2\right) \right\} + U(\bar{\chi}),\qquad(15)$$

the quark masses being given by eqs. (9) and (10) [or (11)] with $\chi = \bar{\chi}$.

Beyond ε_0 we have perturbatively taken into account, to order α_s, the exchange of gluons, whose contribution to the energy density of an infinite, color singlet system is the analogous of eq. (7), but with the quark masses defined by Eqs. (9) and (10) [or (11)], and with an effective strong coupling constant (dressed by the colour dielectric function), which reads:

$$\tilde{\alpha}_s = \alpha_s \left(\frac{\chi_0}{\bar{\chi}}\right)^{\beta}\qquad(16)$$

as it can be deduced from the model Lagrangian. Eq. (7) only contains the exchange term of OGE, the direct one vanishing for infinite quark matter: at small baryonic densities the attractive electric contribution dominates the energy density; on the contrary the repulsive magnetic contribution becomes the dominant one at large densities.

Indeed the divergent behavior of the electric term for $\rho \to 0$ could prevent a perturbative treatment of OGE in this regime. We have overcome this difficulty by taking into account the Debye screening of the gluon propagator in the presence of a polarized medium. This can be achieved by replacing (16) with a new effective coupling:

$$\alpha_s^{eff}(q) = \tilde{\alpha}_s \frac{q^2}{q^2 + \frac{1}{2}\sum_{f=u,d,s} 16\tilde{\alpha}_s m_f k_{F_f}^2 \Pi(q/k_{F_f})},\qquad(17)$$

$\Pi(y)$ being the static limit of the polarization propagator [18]:

$$\Pi(y) = \frac{1}{2} - \frac{1}{2y}\left(1 - \frac{1}{4}y^2\right)\ln\left|\frac{1-\frac{1}{2}y}{1+\frac{1}{2}y}\right|.\qquad(18)$$

Actually this expression should be utilized in the momentum dependent $V_{OGE}(\vec{q})$ and then integrated to obtain the new expression for ε_f^{OGE}. For simplicity, since the q–integration is extended only up to k_{F_f} and the function $\Pi(y)$ varies at most by 9% in the range $0 \le y \le 1$, we have adopted $q = k_{F_f}$.

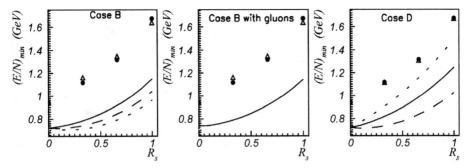

FIGURE 4. Minimal energy per baryon number in the CDM, as a function of $R_s = \rho_s/\rho$, for the cases B (with and without gluons) and D (solid lines). Full circles are the experimental hyperon masses, while triangular dots are the masses calculated in Ref. [19]. In the first panel the curves corresponding to $g' = 106.6$ MeV (dashed line) and $g' = 85.7$ MeV (dotted line) are also presented, while in the third panel the curves corresponding to $\Delta m = 312$ MeV (dotted line) and $\Delta m = 112$ MeV (dashed line) are shown. The remaining parameters of the cases B and D, respectively, are kept unaltered.

Stability of strangelets in the CDM: I

We consider here the work by Aoki *et al.* [19]: these authors solve self-consistently the mean field equations for quarks, color dielectric field and gluons, starting from a CDM Lagrangian with the Double Minimum potential (13) for the color dielectric field. This model is known to produce an unrealistic, too large binding energy in infinite quark matter [20, 21]; concerning the color dielectric function, Aoki *et al.* choose $\beta = 2$; they employ both the scaling and non–scaling version of the model, with two different sets for the model parameters whose values are dictated by two different and extreme choices for the "bag" parameter B: $B^{1/4} = 0$ MeV, with two degenerate vacua, and a large bag pressure, $B^{1/4} = 103.5$ MeV. The latter value of B is chosen to be as large as possible, but with the requirement that the two-phase picture must hold inside hadrons. In their calculation only the strange quark mass has to be considered as a truly free parameter, the remaining ones having been fixed in a previous work on the non–strange baryons [22].

We evaluate the minimum energy per baryon number using cases B and D (corresponding to $B \neq 0$) of the work of Aoki *et al.*, both without and with the perturbative exchange of a gluon.

As we can see from Fig. 4, this version of the CDM seems to favour strangelets as a (meta)–stable form of matter. This is due to the fact that when the DM potential is used to study hadrons, i.e. confined objects, a large contribution to the hadronic mass is given by the space fluctuations of the fields. When this version of the model is used to describe infinite quark matter, these contributions vanish due to the homogeneity of the system. For this reason, deconfined matter is favoured in this version of the model, which would even imply spontaneous decay of ordinary nuclei or two flavor nuclear matter into quark matter.

The effect of perturbative gluons in this model is very small, due to the rather strong Debye screening, which we have included. Whether or not we take into account gluon

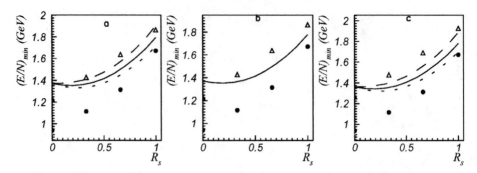

FIGURE 5. Minimal energy per baryon number as a function of $R_s = \rho_s/\rho$ for the Single Minimum version of the CDM. The various panels correspond to: (a) parameter set I without gluons, (b) parameter set I with gluons, (c) parameter set II without gluons. Full circles are the experimental baryon masses, while triangular dots are the masses calculated in Ref. [23]. In the first and third panels the calculations obtained with $g'/g = 1.89$ (long-dashed lines) and $g'/g = 1.37$ (short-dashed lines) are also shown.

corrections, strange matter always appears to be more stable than baryons.

Stability of strangelets in the CDM: II

In this subsection we follow the approach of J. McGovern [23], using the model Lagrangian reported in eq. (8) with $\beta = 4$ in the color dielectric function. McGovern employs only the scaling model and the Single Minimum potential, with different values of the parameters. In this case the behavior of $\alpha_s^{eff}(\bar{\chi})$ is even more divergent, for small densities, than in the case $\beta = 2$ previously considered: hence the use of Debye screening in the effective strong coupling constant is mandatory. In Ref. [23] two different sets for the model parameters are used: they allow to satisfactorily reproduce the splittings between hyperon masses, but the absolute values of the masses themselves are generally too large. In Fig. 5 we show our results, comparing our curves with both the experimental and the theoretical masses. As we can see, also in this case the inclusion of perturbative gluons is rather irrelevant. The curves corresponding to strange matter are well above the experimental masses, and below the theoretical ones. Yet, if we take into account surface effects, which would increase our curves of about $50 \div 100$ MeV, only for $R_s \simeq \frac{2}{3}$ strangelets are (marginally) allowed by the present calculation and a more refined one, taking into account surface energy contributions, is needed to clarify the situation. We notice that a larger strange quark mass ($g'/g = 1.89$) obviously excludes stable strangelets, while the smaller m_s value ($g'/g = 1.37$) does not substantially alter the above considerations.

CONCLUSIONS

The aim of this contribution was to compare the predictions about strangelet stability within the MIT bag model and the Color Dielectric Model, and to draw conclusions about their model dependence: we have compared the curves corresponding to the minimum energy per baryon number to the mass of hyperons having the same strangeness fraction and calculated within the same model and parameter values that we adopt in our calculations.

The analysis shows that the existence of (stable) strangelets is supported only by those models which entail a two–phase picture of hadrons, namely which maintain a false vacuum inside hadrons. This happens both in the MIT bag model, and in the Double Minimum version of the Color Dielectric Model. The Single Minimum version of the CDM does not allow the existence of strangelets, independently of the parameter sets used to perform the calculations.

The conclusions that we can draw indicate that the stability of strangelets depends rather crucially on the model employed; this fact can set serious challenges to the search for strangelets in heavy ion collisions.

REFERENCES

1. Schaffner, J., Greiner, C., and Stöcker, H., *Phys. Rev.*, **C46**, 322 (1992).
2. Schaffner, J. *et al.*, *Phys. Rev. Lett.*, **71**, 1328 (1993).
3. Schaffner, J. *et al.*, *Annals of Physics*, **235**, 35 (1994).
4. Chin, S. A., and Kerman, A. K., *Phys. Rev. Lett.*, **43**, 1292 (1979).
5. Liu, H., and Shaw, G. L., *Phys. Rev.*, **D30**, 1137 (1984).
6. Greiner, C., Koch, P., and Stöcker, H., *Phys. Rev. Lett.*, **58**, 1825 (1987).
7. Greiner, C., Rischke, D.-H., Koch, P., and Stöcker, H., *Phys. Rev.*, **D38**, 2797 (1988).
8. Greiner, C., and Stöcker, H., *Phys. Rev.*, **D44**, 3517 (1991).
9. Dover, C. B., Production of strange clusters in relativistic heavy ion collisions (1993), preprint BNL-48594, presented at HIPAGS 1993.
10. Fahri, E., and Jaffe, R. L., *Phys. Rev.*, **D30**, 1601 (1984).
11. Schaffner-Bielich, J., Greiner, C., Diener, A., and Stöcker, H., *Phys. Rev.*, **C55**, 3038 (1997).
12. Madsen, J., *Phys. Rev. Lett.*, **85**, 4687 (2000).
13. Alberico, W., Drago, A., and Ratti, C., *Nucl. Phys.*, **A706**, 143 (2002).
14. DeGrand, T., Jaffe, R. L., Johnson, K., and Kiskis, J., *Phys. Rev.*, **D12**, 2060 (1975).
15. Wilets, L., *Chiral Solitons*, ed. Liu, K.F. World Scientific, Singapore, 1987, p. 362.
16. Birse, M., *Prog. Part. Nucl. Phys.*, **25**, 1 (1990).
17. Pirner, H., *Prog. Part. Nucl. Phys.*, **29**, 33 (1992).
18. Fetter, A., and Walecka, J., *Quantum Theory of Many-particle systems*, McGraw–Hill, 1971.
19. Aoki, N., Nishikawa, K., and Hyuga, H., *Nucl. Phys.*, **A534**, 573 (1991).
20. Drago, A., Fiolhais, M., and Tambini, U., *Nucl. Phys.*, **A588**, 801 (1995).
21. Barone, V., and Drago, A., *J. Phys.*, **G21**, 1317 (1995).
22. Aoki, N., and Hyuga, H., *Nucl. Phys.*, **A505**, 525 (1989).
23. McGovern, J., *Nucl. Phys.*, **A533**, 553 (1991).

Antinuclei Production and Coalescence at RHIC

Christof Struck* for the STAR Collaboration

*University of Frankfurt, August-Euler-Str. 6, 60486 Frankfurt/Main, Germany

Abstract. The production of light antinuclei has been observed in Au+Au collisions at RHIC. STAR extracts an invariant yield for both $\bar{d}$ and $^3\overline{\text{He}}$ and, together with the invariant yield of antiprotons, calculates the coalescence parameters B_2 and B_3. The collision energy dependence of B_2 and B_3 shows no dramatic increase in freeze-out volume at RHIC compared to SPS. The interpretation of the results in the coalescence framework makes it possible to relate the coalescence parameters to the volumes of homogeneity extracted from HBT measurements.

INTRODUCTION

Ultrarelativistic heavy ion collisions are used to study nuclear matter in the laboratory at extreme density and temperatures (for a general overview see for example [1]). A small fraction of the emitted particles are light nuclei and antinuclei. The bulk consists of pions with an admixture of kaons and nucleons. The hot fireball formed in these collisions cools by expansion and the emitted nucleons cease interaction at a *freeze-out* temperature of the order 100 MeV. Due to the small binding energy compared to the fireball temperature, light nuclei produced during the hot and dense phase are likely to break up before they escape. The dominant production mechanism for (anti)nuclei in heavy ion collisions is thru final-state coalescence [2, 3, 4], where clusters of (anti)nucleons in close proximity and small relative momenta merge to form light (anti)nuclei during the final stages of kinetic freeze-out. We report the production of $\bar{d}$ and $^3\overline{\text{He}}$ in Au+Au collisions at a center-of-mass energy $\sqrt{s_{NN}} = 200$ GeV at RHIC. With a high multiplicity and a antiproton-to-proton ratio close to one are these collisions well suited for the production of antinuclei.

The invariant yield of nuclei or antinuclei with mass number A and momentum p_A can be related in the coalescence picture to the nucleon invariant yield at momentum $p_p = p_A/A$,

$$E_A \frac{d^3 N_A}{d^3 p_A} = B_A \left(E_p \frac{d^3 N_p}{d^3 p_p} \right)^A, \qquad (1)$$

where it is assumed, that (anti)neutrons and (anti)protons are produced with identical momentum spectra. The coalescence parameter B_A characterizes the probability that A nucleons with similar momentum form a bound state. Studies of light nuclei production in nucleus-nucleus collisions at BEVALAC and SIS energies of 0.2 to 2.0 AGeV, as well as high energy p+p and p+A collisions at FNAL, showed a constant coalescence parameter independent of system size and beam energy, where B_A can directly be related to the nuclear wave function of the produced (anti)nuclei [2]. However, in larger systems

CP644, *Exotic Clustering: 4th Catania Relativistic Ion Studies*, edited by S. Costa, A. Insolia, and C. Tuvè
© 2002 American Institute of Physics 0-7354-0099-7/02/$19.00

at AGS energies and above, B_A was found to be smaller by one order of magnitude, indicating that in high energy heavy ion collisions the spatial distribution of the nucleons needs to be taken into account in the coalescence picture [5]. Once the system size is larger than the intrinsic size of the produced (anti)nuclei, (anti)nucleons with small relative momenta are not always in close proximity and thus do not always form a bound state. Another important aspect are the space-momentum correlations induced by the strong collective flow observed in high energy heavy ion collisions [6], where particles in the same region tend to have similar momentum, thus increasing the coalescence rate.

The recent coalescence model of Scheibl and Heinz [7] includes collective flow in both transverse and longitudinal direction. They use the similarity between the physics of the coalescence mechanism and the Hanbury-Brown, Twiss (HBT) correlation method [8], to express the coalescence parameter in terms of the *volume of homogeneity* extracted from HBT measurements. Thus the coalescence parameter allows to probe the nucleon phase space distribution at the kinetic freeze-out stage and provides access to the source size complementary to the HBT measurements.

EXPERIMENT AND ANALYSIS

The antinuclei measurements presented here are from the STAR experiment. STAR is one of five experiments currently setup at RHIC/BNL. It's main tracking detector is a large Time Projection Chamber (TPC) with a pseudorapidity coverage from -1.5 to $+1.5$ and full azimuthal coverage. For the momentum determination the detector is housed by a solenoidal magnet, which can by operated at a field strength up to 0.5 T. Events are selected based on the detection of spectator neutrons in Zero Degree Calorimeters placed about 18 m away on both sides of the nominal interaction point. The centrality selection is done by the charged particle multiplicity measurement in the TPC. The data sample used in this analysis was taken during the run period in 2001 and consists of $\approx$3.0 M Au+Au events at $\sqrt{s_{NN}} = 200$ GeV and 10% centrality.

We identify (anti)nuclei with the ionisation energy loss dE/dx measured in the TPC gas. A charged particle with sufficient transverse momentum to leave the TPC volume generates a track length exceeding 1.4 m, which is sampled by up to 45 space-points. Important for the analysis of the relatively rare antinuclei are tight cuts to remove background in the particle identification. Only tracks with at least 30 space-point samples are accepted in the analysis. Space-points where more than one track contribute are excluded in the dE/dx calculation and tracks with more than 30% potentially merged ionisation clusters are discarded. To avoid the Landau tails in the energy loss measurement a truncated mean is calculated, where only the lower 70% of the dE/dx samples are taken into account. Figure 1 shows the truncated mean dE/dx as a function of the magnetic rigidity $|p/Z|$ for the negatively charged tracks considered in this analysis.

The Bethe-Bloch expectation for $\bar{t}$ and $^3\overline{He}$ is also plotted in Figure 1. A clear $\bar{d}$ band can be observed below a rigidity of 1 GeV/c and about 180 candidates clustered around the $^3\overline{He}$ expectation. Please note that the rigidity is plotted in Figure 1 and the momenta of the $^3\overline{He}$ candidates is twice as large. No clear $\bar{t}$ band is observed, but these are expected to have a similar momentum distribution as $^3\overline{He}$ and thus reside mainly in

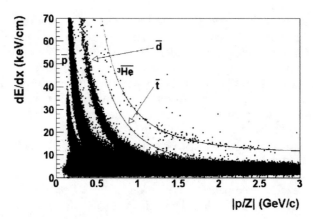

FIGURE 1. Ionization (dE/dx) as a function of rigidity ($|p/Z|$) for negatively charged tracks. Included are the Bethe-Bloch expectations for $\bar{t}$ and $^3\overline{\mathrm{He}}$.

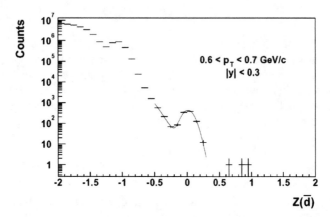

FIGURE 2. Z variable (see text) for antideuterons with $0.6 < p_t < 0.7$ GeV/c and $|y| < 0.3$. The exponential+Gaussian fit to extract signal and background is also show.

a region where the separation power of the dE/dx is not sufficient.

The $\bar{\mathrm{d}}$ yield is analyzed in the transverse momentum range $0.5 < p_t < 0.8$ GeV/c and rapidity range $|y| < 0.3$ of good particle identification and high reconstruction efficiency. The Z variable, defined as $Z = \ln([dE/dx]/I_{\bar{d}}(p))$ with the expected ionisation $I_{\bar{d}}(p)$ for a $\bar{\mathrm{d}}$ with momentum p, is used to extract the yield from the measured dE/dx distribution. Figure 2 shows the Z distribution for $\bar{\mathrm{d}}$ in the momentum range $0.6 < p_t < 0.7$ GeV/c. The yield is calculated from a combined exponential background and Gaussian signal fit, also show in the plot.

In contrast to the $\bar{\mathrm{d}}$, the dE/dx measurement provides a good particle separation for the $^3\overline{\mathrm{He}}$ over almost the full momentum range. A similar analysis based on Z

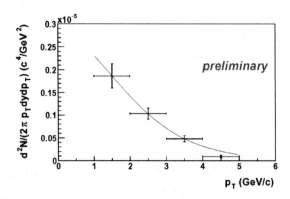

FIGURE 3. ${}^{3}\overline{\text{He}}$ invariant yield as a function of transverse momentum. The line shows the fit of the exponential function (see text).

shows the ${}^{3}\overline{\text{He}}$ signal nearly background free and all candidates were accepted in the analysis within two sigma of the distribution. Both the extracted $\bar{\text{d}}$ and the ${}^{3}\overline{\text{He}}$ yields were corrected for reconstruction efficiency, energy loss in the detector material and absorption.

A unique tool for the measurement of rare signals in heavy ion collisions is the high level software trigger in STAR [9]. Since the event rate in STAR is limited by the available data storage rate and not by the read-out rate of the detector systems, a high level trigger can be used to filter interesting events. The trigger decision is based on a full event reconstruction of the TPC data, performed by a Alpha-processor farm, and thus allows to trigger on physics observables. Amongst other things the trigger was used during the run 2001 to filter events with $Z = -2$, i.e. ${}^{3}\overline{\text{He}}$, candidates and scanned additional ≈ 0.7 M central Au+Au events. The trigger efficiency relative to the offline reconstruction is in the order of 80% [10]. The ${}^{3}\overline{\text{He}}$ signal was thereby enhanced by additional 30 candidates.

RESULTS AND DISCUSSION

We extract a $\bar{\text{d}}$ invariant yield in the transverse momentum range $0.5 < p_t < 0.8$ GeV/c of $[2.7 \pm 0.2(stat.)] \times 10^{-3}$, about a factor 50 more than the yield at SPS energies ($\sqrt{s_{NN}} = 17$ GeV). Unfortunately, due to the limited momentum range a slope parameter T (see below) cannot be extracted.

The ${}^{3}\overline{\text{He}}$ signal was observed with a mean transverse momentum of ≈ 2.5 GeV/c. We find an invariant yield at mean p_t of $[1.0 \pm 0.1(stat.)] \times 10^{-6}$. A factor of 10 more candidates have been found than at SPS and during the first year of RHIC [11]. With the large kinematic coverage in STAR this rather large sample can be used to extract the rapidity density dN/dy and the slope parameter T from the ${}^{3}\overline{\text{He}}$ spectrum shown in Figure 3 for the first time. Fitting an exponential function in the transverse mass

$m_t = \sqrt{p_t^2 + m_0^2}$ distribution,

$$E\frac{d^3N}{d^3p} = \frac{dN/dy}{2\pi T(m_0 + T)}e^{-\frac{m_t - m_0}{T}}, \tag{2}$$

in the momentum range of sufficient statistics ($1.0 < p_t < 4.0$ GeV) yields $dN/dy = [6.3 \pm 0.5(stat.)] \times 10^{-6}$ and $T = 965 \pm 140(stat.)$ MeV.

STAR has measured the invariant yields for antiprotons in the same centrality range [12]. This data allows to calculate the coalescence parameters B_2 and B_3 from Equation 1. To correct the antiproton yield for antihyperon feeddown a preliminary estimate was used based on the correction of the 130 GeV data [11]. The systematic error of the coalescence parameter calculation is in the order of 20% and dominated by this preliminary correction.

For the top 10% central Au+Au collisions we find $B_2 = [3.7 \pm 0.3(stat.) \pm 0.8(sys.)] \times 10^{-4}$ in the $\bar{d}$ kinematic range $0.5 < p_t < 0.8$ GeV/c and $|y| < 0.3$. At the mean p_t of the observed $^3\overline{He}$ sample $B_3 = [1.4 \pm 0.1(stat.) \pm 0.4(sys.)] \times 10^{-7}$ is found. Figure 4 shows the coalescence parameter as a function of collision energy. The qualitative trend for both B_2 and B_3 is very similar: The value is constant for small systems, i.e. p+p and p+A collisions, and decreases for A+A collisions as a function of collision energy.

Several models [23, 5, 24] were proposed to relate the coalescence parameter to a geometrical source size. In these models B_A scales with the volume as $B_A \propto 1/V^{A-1}$. If we use this simple expression and the measured coalescence parameter ratios to compare the volumes at RHIC and SPS, we get $V_{\bar{d}}(RHIC, 200GeV) = (1.3 \pm 0.2)V_{\bar{d}}(SPS)$ and $V_{3\overline{He}}(RHIC, 200GeV) = (2.4 \pm 0.4)V_{3\overline{He}}(SPS)$. Both measurements indicate only a small increase of the antinucleon freeze-out volume when going from SPS energy $\sqrt{s_{NN}} = 17$ GeV to $\sqrt{s_{NN}} = 200$ GeV at RHIC.

In the coalescence framework of Scheibl and Heinz [7] the coalescence parameter can be compared to the volume of homogeneity V_{hom} extracted from HBT measurements. The model is based on a hydrodynamical density matrix formulation of the coalescence process and includes collective flow in both beam and transverse direction. Assuming a box-like density profile, the coalescence parameters B_2 and B_3 in this model are given by

$$B_2 = \frac{3\pi^{3/2}\langle C_{\bar{d}}\rangle}{2m_t V_{hom}(m_t)}e^{2(m_t - m)(1/T_p - 1/T_{\bar{d}})}, \tag{3}$$

$$B_3 = \frac{2\pi^3\langle C_{3\overline{He}}\rangle}{\sqrt{3}m_t^2 V_{hom}^2(m_t)}e^{3(m_t - m)(1/T_p - 1/T_{3\overline{He}})}, \tag{4}$$

where the homogeneity volume V_{hom} is defined by the HBT radii $R_\perp$ and $R_\parallel$ as

$$V_{hom}(m_t) = R_\perp^2(m_t)R_\parallel(m_t). \tag{5}$$

Here, $\langle C_{\bar{d}}\rangle = 0.8$ and $\langle C_{3\overline{He}}\rangle = 0.7$ (taken from [7]) are quantum mechanical correction factors related to the internal structure of the $\bar{d}$ and $^3\overline{He}$, respectively. m_t denotes the

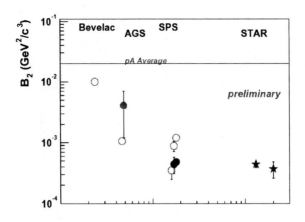

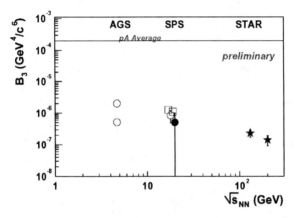

FIGURE 4. Coalescence parameter B_2 and B_3 as a function of collision energy for Au+Au or Pb+Pb collisions [13, 14, 15, 16, 17, 18, 19, 20, 21, 11, 22]. Hollow markers show nuclei and solid markers show antinuclei.

transverse mass of the (anti)proton and T_x the corresponding slope parameters. The effective freeze-out volume $V_{eff}(A, M_t)$ for (anti)nuclei with transverse mass M_t depends in this model explicit on the mass number A,

$$V_{eff}(A, M_t) = \left(\frac{2\pi}{A}\right)^{3/2} V_{hom}(m_t). \qquad (6)$$

Thermodynamical coalescence models [23] predict a constant freeze-out volume for all (anti)nucleon clusters and are inconsistent with the data [11].

To extract V_{hom} from B_2 using Equation 3, the exponential factor can be neglected since the yields are measured at very low transverse momentum. We find similar homogeneity volumes for both $\bar{d}$ and $^3\overline{He}$: $V_{hom,\bar{d}} = 135 \pm 20$ fm^3 and $V_{hom,^3\overline{He}} = 125 \pm 25$ fm^3.

TABLE 1. Antinuclei results for 130 GeV, taken from [11], and 200 GeV. The homogeneity volume calculation is based on [7] (see text).

$\sqrt{s_{NN}}$ (GeV)	A	$B_A \cdot 10^{-4}$ $(GeV^2 c^{-3})^{A-1}$	T_A^* (MeV)	$R_\perp^2 R_\parallel$ (fm^3)	V_{eff} (fm^3)
130	$\bar{d}$	4.5 ± 1.3		107 ± 7	600 ± 40
130	$^3\overline{He}$	0.0021 ± 0.0012	700 ± 250	99 ± 16	299 ± 50
200	$\bar{d}$	3.7 ± 1.1		135 ± 20	775 ± 100
200	$^3\overline{He}$	0.0014 ± 0.0005	965 ± 140	125 ± 25	375 ± 70

* not measured for $\bar{d}$

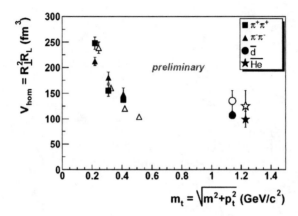

FIGURE 5. Homogeneity volume V_{hom}, extracted from HBT and the coalescence parameters, as a function of transverse mass. Filled symbols show the results for Au+Au collisions at $\sqrt{s_{NN}} = 130$ GeV and open symbols the preliminary data for 200 GeV, whereas the HBT results were taken from [25, 26].

Thus the effective volumes resulting from Equation 6 indicate a larger volume for the $\bar{d}$ source than for $^3\overline{He}$, with a ratio of $V_{eff,\bar{d}}/V_{eff,^3\overline{He}} = 2.1 \pm 0.2$. Table 1 summarizes the coalescence parameters and the extracted volumes. The results of the previous STAR measurement at $\sqrt{s_{NN}} = 130$ GeV [11] are given as well.

Figure 5 shows the volumes of homogeneity, extracted from two-pion HBT and the coalescence parameters B_2 and B_3, as a function of transverse mass for both 130 GeV and 200 GeV collision energy. The different methods agree rather well, however, it is not clear that the space-time geometry for pions and antinulceons should be the same and quantitative comparisons should be made carefully.

ACKNOWLEDGMENTS

We wish to thank the RHIC Operations Group and the RHIC Computing Facility at Brookhaven National Laboratory, and the National Energy Research Scientific Computing Center at Lawrence Berkeley National Laboratory for their support. This work was supported by the Division of Nuclear Physics and the Division of High Energy Physics of the Office of Science of the U.S. Department of Energy, the United States National Science Foundation, the Bundesministerium fuer Bildung und Forschung of Germany, the Institut National de la Physique Nucleaire et de la Physique des Particules of France, the United Kingdom Engineering and Physical Sciences Research Council, Fundacao de Amparo a Pesquisa do Estado de Sao Paulo, Brazil, the Russian Ministry of Science and Technology and the Ministry of Education of China and the National Natural Science Foundation of China.

REFERENCES

1. Harris, J. W., and Müller, B., *Annu. Rev. Nucl. Part. Sci*, **46**, 71 (1996).
2. Butler, S. T., and Pearson, C. A., *Phys. Rev.*, **129**, 836 (1963).
3. Schwarzschild, A., and Zupancic, C., *Phys. Rev.*, **129**, 854 (1963).
4. Gutbrod, H. H., et al., *Phys. Rev. Lett.*, **37**, 667 (1976).
5. Sato, H., and Yazaki, K., *Phys. Lett. B*, **98**, 153 (1981).
6. Schnedermann, E., and amd U.Heinz, J. S., *Phys. Rev. C*, **48**, 2462 (1993).
7. Scheibl, R., and Heinz, U., *Phys. Rev. C*, **59**, 1585 (1999).
8. Heinz, U., *Nucl. Phys.*, **A610**, 264c (1996).
9. Adler, C., et al. (2002), to be published in NIM A.
10. Struck, C., Ph.D. thesis, University of Frankfurt, Germany (2002), to be published.
11. Adler, C., et al., *Phys. Rev. Lett.*, **87**, 262301–1 (2001).
12. Barannikova, O., "Mid-rapidity π, k and $\bar{p}$ yields and spectra in Au+Au collisions at 130 and 200 GeV from STAR," in [27], to appear in Nucl. Phys. A.
13. Bearden, I. G., et al., *Phys. Rev. Lett.*, **85**, 2681 (2000).
14. Armstrong, T. A., et al., *Phys. Rev. Lett.*, **85**, 2685 (2000).
15. Armstrong, T. A., et al., *Phys. Rev. C*, **61**, 064908 (2000).
16. Wang, S., et al., *Phys. Rev. Lett.*, **74**, 2646 (1995).
17. Appelquist, G., et al., *Phys. Lett. B*, **376**, 245 (1996).
18. Weber, M., et al., *Nucl. Phys.*, **A661**, 177c (1999).
19. Ambrosini, G., et al., *Phys. Lett. B*, **417**, 202 (1998).
20. Hansen, A. G., et al., *Nucl. Phys.*, **A661**, 387c (1999).
21. Bennet, M. J., et al., *Phys. Rev. C*, **58**, 1155 (1998).
22. Afanasiev, S. V., et al., *Phys. Lett. B*, p. 22 (2000).
23. Mekjian, A. Z., *Phys. Rev. C*, **17**, 1051 (1978).
24. Llope, W. J., et al., *Phys. Rev. C*, **52**, 2004 (1995).
25. Lopez Noriega, M., "Identical Particle Interferometry at STAR," in [27], to appear in Nucl. Phys. A.
26. Adler, C., et al., *Phys. Rev. Lett.*, **87**, 082301 (2001).
27. Gutbrod, H. H., et al., editors, *Proceedings of the 16th Int. Conf. on Ultra-Relativistic Nucleus-Nucleus Collisions, Nantes, France, 18 - 24 July, 2002*, Elsevier Science, Amsterdam, 2002, to appear in Nucl. Phys. A.

Possible Hadronic Molecule Λ(1405) and Thermal Glueballs in SU(3) Lattice QCD

H. Suganuma*, N. Ishii†, H. Matsufuru**, Y. Nemoto‡ and T. T. Takahashi§

*Faculty of Science, Tokyo Institute of Technology, Tokyo 152-8552, Japan
†The Institute of Physical and Chemical Research (RIKEN), Wako 351-0198, Japan
**Yukawa Institute for Theoretical Physics, Kyoto University, Kyoto 606-8502, Japan
‡Brookhaven National Laboratory, RBRC, New York 11973-5000, USA
§RCNP, Osaka University, Mihogaoka 10-1, Osaka 567-0047, Japan

Abstract. We aim to construct quark hadron physics based on QCD. First, using lattice QCD, we study mass spectra of positive-parity and negative-parity baryons in the octet, the decuplet and the singlet representations of the SU(3) flavor. In particular, we consider the lightest negative-parity baryon, the Λ(1405), which can be an exotic hadron as the $N\bar{K}$ molecular state or the flavor-singlet three-quark state. We investigate the negative-parity flavor-singlet three-quark state in lattice QCD using the quenched approximation, where the dynamical quark-anitiquark pair creation is absent and no mixing occurs between the three-quark and the five-quark states. Our lattice QCD analysis suggests that the flavor-singlet three-quark state is so heavy that the Λ(1405) cannot be identified as the three-quark state, which supports the possibility of the molecular-state picture of the Λ(1405). Second, we study thermal properties of the scalar glueball in an anisotropic lattice QCD, and find about 300 MeV mass reduction near the QCD critical temperature from the pole-mass analysis. Finally, we study the three-quark potential, which is responsible to the baryon properties. The detailed lattice QCD analysis for the 3Q potential indicates the Y-type flux-tube formation linking the three quarks.

LATTICE QCD STUDY FOR Λ(1405)

Among a lot of hadrons, the Λ(1405) is a very special interesting hadron. In spite of the strange baryon, the Λ(1405) is the lightest negative-parity baryon. In fact, the Λ(1405) is much lighter than the low-lying non-strange negative-parity baryons, the $N(1520)$ with $J^P = \frac{3}{2}^-$ and the $N(1535)$ with $J^P = \frac{1}{2}^-$. Moreover, there are two interesting physical interpretations on the Λ(1405).

- In the quark-model framework, the Λ(1405) is described as the flavor-singlet three-quark system. (cf. The H-dibaryon is also a flavor-singlet candidate.)
- As another interpretation, the Λ(1405) is an interesting candidate of the hadronic molecule such as the $N\bar{K}$ bound state with a large binding energy about 30MeV. (cf. For deutrons, the binding energy is about 2.2MeV.)

In the valence picture, the $N\bar{K}$ hadronic molecule is described as qqq-$q\bar{q}$, and hence we call this state as the "5-quark (5Q) state" for the simple notation. Of course, in the real world, the 3Q state and the 5Q state would be mixed as

$$|\Lambda(1405)\rangle \simeq C_{3Q}|3Q\rangle + C_{5Q}|3Q - Q\bar{Q}\rangle \qquad (1)$$

CP644, Exotic Clustering: 4th Catania Relativistic Ion Studies, edited by S. Costa, A. Insolia, and C. Tuvè
© 2002 American Institute of Physics 0-7354-0099-7/02/$19.00

So, the more realistic question is as follows. Which is the dominant component of the $\Lambda(1405)$, the 3Q state or the 5Q state ? The answer to this question would be important for the argument of the hyper-nuclei and some related subjects in astrophysics. So, we investigate the $\Lambda(1405)$ using lattice QCD, which is the first-principle calculation of the strong interaction. However, also in lattice QCD, the 3Q and the 5Q states are mixed automatically through the q-$\bar{q}$ pair creation, and therefore it is almost impossible to distinguish the 3Q and the 5Q states in lattice QCD.

To overcome this difficulty, we use the advantage of the quenched approximation, which does not include the dynamical q-$\bar{q}$ pair creation. Then, in quenched QCD, we can investigate the 3Q and 5Q states, individually. Note here that quenched QCD reproduces various hadron masses, and is widely used in lattice QCD simulations. In fact, once the 3Q state is prepared as the initial state, the system continues to be the 3Q state during the time evolution in quenched QCD, with keeping essence of QCD. We report here the lattice QCD test for the $\Lambda(1405)$ in terms of the 3Q flavor-singlet state[1]. Also, we investigate all possible low-lying baryons with positive-parity and negative-parity in the flavor-octet, decuplet and singlet representations.

We adopt $SU(3)_c$ lattice QCD using the improved action with the clover fermion on the three different lattices with β=5.75, 5.95 and 6.10. We use anisotropic lattice with finer temporal lattice spacing as $a_s = 4a_t$.

Most of hadron masses of pseudo-scalar mesons, vector mesons, positive-parity baryons and negative-parity baryons are reproduced within about 10% deviation, as shown in Table 1. However, in lattice QCD, the flavor-singlet 3Q state seems rather heavy as

$$M(3\mathrm{Q}, \text{flavor singlet}, J^P = 1/2^-) \simeq 1.7 \mathrm{GeV}. \qquad (2)$$

In other words, the $\Lambda(1405)$ seems so light that it cannot be identified as the flavor-singlet 3Q state, which supports the $N\bar{K}$ molecule picture for the $\Lambda(1405)$.

For more definite conclusion, we need to investigate the 5Q state in lattice QCD at the quenched level, as well as the full lattice QCD calculation for the $\Lambda(1405)$.

	$\beta = 5.95$	$\beta = 6.10$	Exp.		$\beta = 5.95$	$\beta = 6.10$	Exp.
ρ	0.7965(53)	0.8005(65)	0.770	$N^{(-)}$	1.599(59)	1.618(57)	1.535
K^*	0.892 (fit)	0.892 (fit)	0.892	$\Sigma^{(-)}$	1.705(49)	1.717(49)	1.620
ϕ	0.9913(53)	0.9873(66)	1.020	$\Xi^{(-)}$	1.810(40)	1.816(42)	–
N	1.0781(81)	1.1055(72)	0.939	$\Lambda_{octet}^{(-)}$	1.703(48)	1.700(49)	1.670
Λ	1.1825(75)	1.2002(67)	1.116	$\Lambda_{singlet}^{(-)}$	1.646(49)	1.725(39)	1.405
Σ	1.1982(76)	1.2173(67)	1.193	$\Delta^{(-)}$	1.877(73)	1.833(80)	1.700
Ξ	1.3183(75)	1.3291(67)	1.318	$\Sigma^{*(-)}$	1.955(61)	1.913(69)	–
Δ	1.342(16)	1.3685(17)	1.232	$\Xi^{*(-)}$	2.032(50)	1.994(57)	–
Σ^*	1.440(14)	1.4586(15)	1.385	$\Omega^{(-)}$	2.109(39)	2.074(46)	–
Ξ^*	1.538(12)	1.5486(13)	1.530	$\Lambda_{singlet}^{(+)}$	2.292(46)	2.150(70)	–
Ω	1.635(11)	1.6387(12)	1.672				

TABLE 1. The lattice QCD result for various hadron masses in GeV.

SCALAR GLUEBALLS AND THEIR THERMAL PROPERTIES

The "Higgs particle" in QCD

"Can you imagine a molecule of photons ? [2]" Of course, there is no photonic molecule in QED, because of the absence of the self-interaction between photons. In QCD, however, due to its nonabelian nature, there appears the self-interaction among gluons, and gluonic bound states, glueballs, are predicted to exist like ordinary mesons. (In fact, two or more gluons can make color-singlet objects like Q-Q̄ mesons.) Anyway, the glueball, which was theoretically predicted in QCD, is an exotic bound-state of gauge fields.

The glueballs are color-singlet objects without valence quarks, so that they are flavor-singlet and baryonless as $I = S = B = 0$. Lattice QCD simulations predict the masses of glueballs as $M(J^{PC} = 0^{++}) = 1500 - 1700\text{MeV}$ and $M(J^{PC} = 2^{++}) = 2000 - 2200\text{MeV}$. The experimental candidate of the lowest scalar glueball is considered as $f_0(1500)$, which is hard to be explained with the quark model.

In a simple construction, the scalar glueball operator is expressed as $\Phi(x) = G^a_{\mu\nu}G^{\mu\nu}_a$, with $G^a_{\mu\nu}$ being the field strength. However, in the nonperturbative vacuum of QCD, this operator is known to be condensed like the Higgs scalar as $\frac{g^2}{32\pi}\langle G^a_{\mu\nu}G^{\mu\nu}_a \rangle \simeq (200\text{MeV})^4$, which is called as the gluon condensate. Like the Higgs particle, the physically observed field is expressed as the shifted operator, $\tilde{\Phi}(x) \equiv G^a_{\mu\nu}G^{\mu\nu}_a - \langle G^a_{\mu\nu}G^{\mu\nu}_a \rangle$. Thus, the scalar glueball is directly connected with the vacuum structure of nonperturbative QCD, and, as a consequence, the scalar glueball is expected to be rather sensitive to the QCD vacuum. For instance, we expect a large thermal effect on the scalar glueball [3], especially, near the critical temperature T_c of the QCD phase transition, because the nonperturbative aspect of QCD would be largely reduced near T_c.

For comparison, we mention about the thermal mesons and thermal baryons in lattice QCD. In spite of many model predictions on the thermal properties of mesons, the recent lattice QCD Monte Carlo simulations indicate no significant thermal effect on the various mesons masses as $M_{\text{meson}}(T \simeq T_c) \simeq M_{\text{meson}}(T = 0)$ from the direct measurement of the pole mass from the temporal correlation of the meson operators. As for the thermal baryon, there is no pole mass measurement in lattice QCD, yet.

Thermal Glueballs in lattice QCD

In general, the pole mass is measured from the temporal correlation of the operator as $G(t) \equiv \langle \tilde{\Phi}(t)\tilde{\Phi}^{\dagger}(0) \rangle$. If t is large enough, the pole mass m is obtained from the exponential decreasing as $G(t) \sim e^{-mt}$ in the Euclidean metric. At high temperature, however, the temporal distance is limited as $0 < t < 1/T$ in the imaginary-time formalism. Then, it is rather difficult to perform the accurate measurement of the pole mass at high temperature. (Here, do not confuse the pole mass with the screening mass, which is measured from the spatial correlation. The screening mass is not the physically observed particle mass.) For instance, in most lattice QCD simulations at finite temperature, the number of the lattice points is $4 - 6$ in the temporal direction. In such cases, the independent information on the temporal correlator $G(t)$ is only $2-3$, because of the reflection symmetry in the temporal direction. Then, it is almost impossible to measure the pole mass in these lattices.

To overcome this difficulty and to perform the accurate measurement of the thermal mass, we take the following prescriptions in lattice QCD.

- To get detailed information of the temporal correlation $G(t) = \langle \tilde{\Phi}(t)\tilde{\Phi}^{\dagger}(0) \rangle$, we adopt the anisotropic lattice, where the temporal lattice spacing is much finer than the spatial one as $a_s = 4a_t$.
- To reduce the statistical error, we use a large number of gauge configurations: 5,500-9,900 gauge configurations are used at each temperature. (cf. In most lattice studies, about 100 gauge configurations are used.)
- For the accurate measurement of the lowest glueball mass, we carefully construct the appropriate operator which largely overlaps with the lowest glueball state, using the smearing method.

Here, the smearing method is a standard technique to reduce the excited-state components in the operator, without any change of physics. In fact, we can single out the lowest-state operator on lattice with the smearing method.

We calculate the temporal correlation of the scalar glueball in lattice QCD with $\beta = 6.25$ and $20^3 \times N_t$ with N_t=33,34,35,36,37,38,40,43,45,50,72. From the pole-mass analysis of the lattice QCD data, we observe about 300 MeV mass reduction of the lowest scalar glueball near the critical temperature of the QCD phase transition[4], as shown in Fig.1. This result may suggest an observation of the large thermal effect on the scalar glueball in future RHIC experiments.

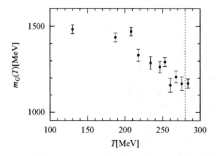

FIGURE 1. The lowest scalar-glueball mass $m_G(T)$ plotted against the temperature T. The vertical dotted line denotes the critical temperature $T_c \simeq 280\text{MeV}$ in quenched QCD.

THE THREE-QUARK POTENTIAL IN SU(3) LATTICE QCD

In the modern science, to be based on the experiment and to find out the principle are two important viewpoints. To extract the principle or the essence of phenomena, some simplification is to be done, if necessary. Of course, such simplification does not mean to treat only simple system. Nevertheless, many theoretical particle physicists misunderstand this point. The lattice QCD study of the three-quark baryonic potential was one of the typical blind spots in the elementary particle physics.

Using lattice QCD, the simple quark-antiquark (Q-$\bar{\text{Q}}$) potential has been studied in detail. In contrast, there was almost no lattice QCD calculation for the three-quark (3Q) potential, although it plays the essential role to determine the baryon properties. Note here that the three-body force among the quarks is one of the "primary force" reflecting the three colors in QCD. (cf. In most cases, three-body forces are regarded as residual interactions.)

In fact, QCD has two kinds of inter-quark potentials, the Q-$\bar{\text{Q}}$ potential and the 3Q potential, corresponding to mesons and baryons, respectively. The 3Q potential is also important for the study of the quark confinement, which is one of the most relevant features in QCD. For, one of the essential points of the quark confinement is the color-flux-tube formation and the appearance of the linear potential among quarks. The study of the 3Q potential tells us how to realize the quark confinement in baryons. Therefore, we study the 3Q potential in SU(3)$_c$ lattice QCD at the quenched level.

Now, let us consider the potential form in the Q-$\bar{\text{Q}}$ and 3Q systems with respect to QCD. In the short-distance region, perturbative QCD is applicable and the Coulomb-type potential appears through the one-gluon-exchange (OGE) process. In the long-distance distance at the quenched level, the flux-tube picture would be applicable from the argument of the strong-coupling expansion of QCD, and hence a linear confinement potential proportional to the total flux-tube length is expected to appear. Indeed, lattice QCD results for the Q-$\bar{\text{Q}}$ potential are well described by $V_{Q\bar{Q}}(r) = -\frac{A_{Q\bar{Q}}}{r} + \sigma_{Q\bar{Q}}r + C_{Q\bar{Q}}$ at the quenched level. In fact, $V_{Q\bar{Q}}$ is described by a sum of the short-distance OGE result and the long-distance flux-tube result.

Based on the short-distance OGE and the long-distance flux-tube picture[5], we theoretically deduce the functional form of the 3Q potential V_{3Q} as

$$V_{3Q} = -A_{3Q} \sum_{i<j} \frac{1}{|\mathbf{r}_i - \mathbf{r}_j|} + \sigma_{3Q}L_{\min} + C_{3Q}, \tag{3}$$

with $L_{\min}$ the minimal value of total length of flux tubes linking the three quarks: $L_{\min} = \text{AP} + \text{BP} + \text{CP}$ in Fig.2.

For more than 300 different patterns of the 3Q systems in total, we perform the accurate measurement of the 3Q potential V_{3Q} using SU(3)$_c$ lattice QCD with β=5.7, 5.8 and 6.0 at the quenched level, and compare the lattice QCD data with the theoretical form of Eq.(3) [6, 7]. As a technical progress for the accurate measurement of V_{3Q}, we adopt the gauge-covariant smearing technique for the ground-state enhancement by removing many excited-state components such as flux-tube vibrational modes.

As a remarkable fact, the three-quark potential V_{3Q} is well described by Eq.(3) within about 1 % deviation. (For detail, see Refs.[6, 7].) From the comparison with the Q-$\bar{\text{Q}}$ potential, we find a universal feature of the string tension as $\sigma_{3Q} \simeq \sigma_{Q\bar{Q}}$ and the one-gluon-exchange result for the Coulomb coefficient as $A_{3Q} \simeq \frac{1}{2}A_{Q\bar{Q}}$. In fact, the quark confinement in baryons is realized by the Y-type flux-tube formation, and the quark confining force is universal between mesons and baryons.

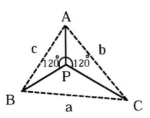

FIGURE 2. The flux-tube configuration of the 3Q system with the minimal value of the total flux-tube length. There appears a physical junction linking the three flux tubes at the Fermat point P.

SUMMARY AND CONCLUDING REMARKS

We have studied the three topics of the quark hadron physics with $SU(3)_c$ lattice QCD.
(1) We have studied negative-parity baryons with anisotropic lattice QCD. In particular, the $\Lambda(1405)$, the lightest negative-parity baryon, is an interesting candidate of the molecular state of $N\bar{K}$ (the five-quark system) or the flavor-singlet three-quark state. We have analyzed the flavor-singlet three-quark state corresponding to the $\Lambda(1405)$ in lattice QCD at the quenched level. The point is to use the quenched approximation, where the dynamical quark-anitiquark pair creation is absent, and no mixing occurs between the three-quark and the five-quark states. Our lattice QCD analysis suggests that the $\Lambda(1405)$ is too light to be identified as a flavor-singlet three-quark state, which supports the $N\bar{K}$ molecular picture of the $\Lambda(1405)$.
(2) We have studied the thermal properties of the glueballs in $SU(3)_c$ anisotropic lattice QCD. From the pole-mass analysis, we have observed about 300 MeV mass reduction of the lowest scalar glueball near the critical temperature of the QCD phase transition. This result may indicate an observation of the large thermal effect on the scalar glueball in future RHIC experiments.
(3) We have studied the static three-quark (3Q) potential for more than 300 different patterns of the 3Q systems in $SU(3)_c$ lattice QCD with β=5.7, 5.8 and 6.0. The lattice QCD data of the 3Q potential V_{3Q} are well reproduced within 1 % deviation by the sum of a constant, the two-body Coulomb term and the three-body linear confinement term proportional to the color-flux-tube length. From the comparison with the Q-$\bar{Q}$ potential, we have found a universal feature of the string tension as $\sigma_{3Q} \simeq \sigma_{Q\bar{Q}}$. Thus, the detailed analysis of the 3Q potential indicates the Y-type flux-tube, which links the three quarks.

REFERENCES

1. N. Nakajima, H. Matsufuru, Y. Nemoto and H. Suganuma, AIP Conf. Proc. **CP594**, 349 (2001).
2. K. Ishikawa, Scientific American **247**, 142 (1982).
3. H. Ichie, H. Suganuma and H. Toki, Phys. Rev. **D52**, 2994 (1995).
4. N. Ishii, H. Suganuma and H. Matsufuru, Phys. Rev. **D66**, 014507 (2002).
5. S. Capstick and N. Isgur, Phys. Rev. **D34**, 2809 (1986).
6. T.T. Takahashi, H. Matsufuru, Y. Nemoto and H. Suganuma, Phys. Rev. Lett. **86**, 18 (2001).
7. T. T. Takahashi, H. Suganuma, H. Matsufuru and Y. Nemoto, Phys. Rev. **D65**, 114509 (2002).

List of Participants

Attilio Agodi
Università di Catania
Dipartimento di Fisica e Astronomia
Via S. Sofia 64
95123 Catania (Italy)
Attilio.Agodi@ct.infn.it

Wanda Maria Alberico
Dipartimento di Fisica Teorica
Via P. Giuria 1
10125 Torino (Italy)
alberico@to.infn.it

Alejandro Algora
Instituto de Fisica Corpuscular
IFIC-Univ. Valencia
Edificio de Institutos de Paterna
Apartado Oficial 22085
46071 Valencia (Spain)
algora@ific.uv.es

Shigeyoshi Aoyama
Kitami Institute of Technology
Koentyo 165
Kitami 090-6507 (Japan)
aoyama@mail.kitami-it.ac.jp

Marcello Baldo
INFN, Sez. di Catania
Via S. Sofia 64
95123 Catania (Italy)
Marcell.Baldo@ct.infn.it

Wolfgang Bauer
Michigan State University
Dept. of Physics and Astronomy
S. Shaw Lane
East Lansing, MI 48824 (USA)
bauer@pa.msu.edu
www.pa.msu.edu/~bauer

David Blaschke
University Rostock
FB Physik
Universitaetsplatz 3
D-18051 Rostock (Germany)
david@thsun1.jinr.ru

Ignazio Bombaci
Dipartimento di Fisica, Università di Pisa
Via Buonarroti, 2
Pisa 56127 (Italy)
bombaci@pi.infn.it

Aldo Bonasera
LNS
V.S. Sofia 44
95123 Catania (Italy)
bonasera@lns.infn.it

Fiorella Burgio
INFN, Sez. di Catania
Via S. Sofia 64
95123 Catania (Italy)
Fiorella.Burgio@ct.infn.it

Francesco Catara
Dip. di Fisica e Astronomia
Univ. di Catania
Via S. Sofia 64
95123 Catania (Italy)
Francesco.Catara@ct.infn.it

Jean-Pierre Coffin
IReS de Strasbourg
Batiment 20, BP 28
67037-Strasbourg cedex 2 (France)
coffin@in2p3.fr

Salvatore Costa
Università di Catania
Dipartimento di Fisica e Astronomia
Via S. Sofia 64
95123 Catania (Italy)
Salvatore.Costa@ct.infn.it
sunct.ct.infn.it/~costa

Enrico De Filippo
INFN, Sez. di Catania
Via S. Sofia 64
95123 Catania (Italy)
defilippo@ct.infn.it
Enrico.Defilippo@ct.infn.it

Massimo Di Toro
LNS/INFN and University of Catania

Via S.Sofia 44
95123 Catania (Italy)
ditoro@lns.infn.it

Claudio Dorso
Depto. Fisica, Facultad de Ciencias
Universidad de Buenos Aires
Pab.1
Ciudad Universitaria
1428 Buenos Aires (Argentina)
codorso@df.uba.ar

Giuseppe Faraci
Dipartimento Fisica e Astronomia
Via S. Sofia 64
95129 Catania (Italy)
Giuseppe.Faraci@ct.infn.it

Dmitri Fedorov
IFA, Aarhus University
Ny Munkegade
8000 Aarhus C (Denmark)
fedorov@ifa.au.dk
www.ifa.au.dk/~fedorov

Lidia S. Ferreira
CFIF and Departamento de Fisica
Instituto Superior Tecnico
Av. Rovisco Pais
1049-001 Lisboa (Portugal)
flidia@ist.utl.pt

Roberto Giordano
Dip. di Fisica e Astronomia
Univ. di Catania
Via S. Sofia 64
95123 Catania (Italy)
Roberto.Giordano@ct.infn.it

Alberto Grasso
Universita' degli Studi di Catania
Dipartimento di Fisica
Via S. Sofia 64
95123 Catania (Italy)
Alberto.Grasso@ct.infn.it

Carsten Greiner
Institut für Theoretische Physik

Universität Giessen
Heinrich-Buff-Ring 16
D-35392 Giessen (Germany)
carsten.greiner@theo.physik.uni-giessen.de

Walter Greiner
University of Frankfurt
Institut fur Theoretische Physik
Postfach 11 19 32
Robert Mayer Strasse 8-10
60054 Frankfurt am Main (Germany)
greiner@th.physik.uni-frankfurt.de
www.th.physik.uni-frankfurt.de/~stoecker
/greiner.html

Konstantin Gridnev
St.Petersburg University
Institute of Physics
Uljanova 1, Peterhof
198504 St.Petersburg (Russia)
gridnev@nuclpc1.phys.spbu.ru

Dieter Gross
Hahn Meitner Institute
Bereich T/V
Glienicker Str.100
D-14109 Berlin (Germany)
gross@hmi.de

Alessandra Guglielmetti
Istituto di Fisica Generale Applicata
Universita' di Milano
Via Celoria 16
20133 Milano (Italy)
guglielmetti@mi.infn.it

Boris Hippolyte
Institut de Recherches Subatomiques
23, rue du Loess Bat20 BP28
67037 Strasbourg (France)
hippolyt@in2p3.fr

Josette Imme'
Dept. of Physics and Astronomy
University of Catania
Via S. Sofia 64
95123 Catania (Italy)
Josette.Imme@ct.infn.it

Antonio Insolia
Dept. of Physics and Astronomy
University of Catania
Via S. Sofia 64
95123 Catania (Italy)
Antonio.Insolia@ct.infn.it

Naoyuki Itagaki
Department of Physics
University of Tokyo
7-3-1 Hongo
113-0033 Tokyo (Japan)
itagaki@phys.s.u-tokyo.ac.jp
tkynt2.phys.s.u-tokyo.ac.jp/~itagaki

Aiichi Iwazaki
Nishogakusha University
Shonan-machi ohi 2590
277-8585 Chiba (Japan)
iwazaki@yukawa.kyoto-u.ac.jp

Sidney Kahana
Brookhaven National Laboratory
Physics Dept.
Upton, NY 11973 (USA)
kahana@bnl.gov

Yoshiko Kanada-En'yo
Institute of Particle and Nuclear Studies
High Energy Accel. Research Org.
1-1 Oho
305-0801 Tsukuba (Japan)
yoshiko.enyo@kek.jp
research.kek.jp/people/yenyo/

Feodor F. Karpeshin
V.A.Fock Institute of Physics
St.Petersburg State University
Ulianovskaya, 1, Petrodvirets
RU-198904 St. Petersburg (Russia)
karpesh@nuclpc1.phys.spbu.ru

Kiyoshi Kato
Division of Physics
Graduate School of Science
Hokkaido University
Kita 10, Nishi 8
060-0810 Sapporo (Japan)
kato@nucl.sci.hokudai.ac.jp
nucl.sci.hokudai.ac.jp/kato.html

Teng Lek Khoo
Argonne National Laboratory
9700 S. Cass Ave.
Argonne, IL 60517 (USA)
khoo@anl.gov

Reiner Klingenberg
University of Dortmund
Experimental Physics IV
D-44221 Dortmund (Germany)
Reiner.Klingenberg@cern.ch

Marco La Commara
Università di Napoli "Federico II"
Complesso Univ. di M.S. Angelo
Via Cinthia
I-80126 Napoli (Italy)
marco.lacommara@unina.it

Elena La Guidara
Università di Catania
Dipartimento di Fisica e Astronomia
Via S. Sofia 64
95123 Catania (Italy)
laguidara@lns.infn.it

Gaetano Lanzanò
INFN sez. di Catania
Via S. Sofia 64
95123 Catania (Italy)
Gaetano.Lanzano@ct.infn.it

Giuseppe Lo Re
Università di Catania
Dipartimento di Fisica e Astronomia
Via S. Sofia 64
95123 Catania (Italy)
giuseppe.lore@ct.infn.it

Umberto Lombardo
Università di Catania
Dipartimento di Fisica e Astronomia
Via S. Sofia 64
95123 Catania (Italy)
umberto.lombardo@ct.infn.it

Enrico Maglione
Dip. di Fisica G. Galilei
Via Marzolo 8
35131 Padova (Italy)
maglione@pd.infn.it

Concettina Maiolino
LNS
Via S. Sofia 64
95123 Catania (Italy)
maiolino@lns.infn.it

Ito Makoto
Div. of Phys., Grad. School of Sci.
Hokkaido Univ.
N10 W8, Kita-ku
060-0810 Sapporo (Japan)
itom@nucl.sci.hokudai.ac.jp

Toshiki Maruyama
Advanced Science Research Center
Japan Atomic Energy Research Institute
Tokai, Naka
319-1195 Ibaraki (Japan)
maru@hadron02.tokai.jaeri.go.jp
hadron31.tokai.jaeri.go.jp/~maru

Hiroshi Masui
Division of Phys., Dept. of Science
Hokkaido University
N10-W8, Kita-ku
060-0810 Sapporo (Japan)
masui@nucl.sci.hokudai.ac.jp

Luciano Moretto
Lawrence Berkeley National Laboratory
MS88
Berkeley, CA 94720 (USA)
lgmoretto@lbl.gov

Lysiane Mornas
Departamento de Fisica
Universidad de Oviedo
Avda Calco Sotelo s/n
E-33007 Oviedo (Asturias) (Spain)
lysiane@pinon.ccu.uniovi.es

Hidekatsu Nemura

Institute of Particle and Nuclear Studies
KEK
Oho 1-1
305-0801 Tsukuba (Japan)
hidekatsu.nemura@kek.jp

Orazio Nicotra
Univerisity of Messina
98100 Messina (Italy)
Orazio.Nicotra@ct.infn.it

Thomas Nilsson
CERN-ISOLDE
EP division
1204 Genève 23 (Switzerland)
Thomas.Nilsson@cern.ch
cern.ch/tnilsson

Grazyna Odyniec
Lawrence Berkeley National Laboratory
Nuclear Science Div. MS 70-319
1 Cyclotron Rd.
Berkeley, California 94720 (USA)
G_Odyniec@lbl.gov

Alexey Ogloblin
Kurchatov Institute
123182 Moscow (Russia)
aoglob@dni.polyn.kiae.su

Angelo Pagano
I.N.F.N. Sezione di Catania
Via S. Sofia 64
95123 Catania (Italy)
Angelo.Pagano@ct.infn.it

Giusimelissa Palazzo
Universitá di Catania
Dipartimento di Fisica e Astronomia
Via S.Sofia 64
I-95123 Catania (Italy)
gpal@sunct.ct.astro.it

Massimo Papa
I.N.F.N. Sez. Catania
Via S.Sofia 64
I-95123 Catania (Italy)
papa@lns.infn.it

Sara Pirrone
INFN Catania
Via S. Sofia 64
95123 Catania (Italy)
Sara.Pirrone@ct.infn.it

Dorin M. S. Poenariu
National Institute of Physics
and Nuclear Engineering
P. O. Box MG-6
76900 Bucharest-Magurele (Romania)
poenaru@ifin.nipne.ro

Giuseppe Politi
Dipartimento di Fisica e Astronomia
Via S. Sofia 64
95123 Catania (Italy)
Giuseppe.Politi@ct.infn.it

Renato Potenza
Università di Catania
Dipartimento di Fisica e Astronomia
Via S. Sofia 64
95123 Catania (Italy)
Renato.Potenza@ct.infn.it

Giovanni Raciti
Università di Catania
Dipartimento di Fisica & Astronomia
Via S. Sofia 64
95123 Catania (Italy)
raciti@lns.infn.it

Andrea Rapisarda
Università di Catania
Dipartimento di Fisica & Astronomia
Via S. Sofia 64
95123 Catania (Italia)
Andrea.Rapisarda@ct.infn.it
www.ct.infn.it/~rapis

Francesco Riggi
Università di Catania
Dipartimento di Fisica & Astronomia
Via S. Sofia 64
95123 Catania (Italy)
Franco.Riggi@ct.infn.it
www.axpfct.ct.infn.it/~riggi

Angelo Rinollo
Universita' di Catania and
INFN-Laboratori Nazionali del Sud
Via Santa Sofia, 44
95123 Catania (Italy)
rinollo@lns.infn.it

Guy Royer
Laboratoire Subatech
4 rue A. Kastler - La Chantrerie
44307 Nantes Cedex 03 (France)
royer@subatech.in2p3.fr

Gerd Röpke
University of Rostock
Deptm. Physics
Universitaetspl. 3
18051 Rostock (Germany)
gerd@darss.mpg.uni-rostock.de

Jüern Schmelzer
Fachbereich Physik
Universitaet Rostock
18051 Rostock (Germany)
juern-w.schmelzer@physik.uni-rostock.de

Hans Schulze
INFN, Sez. di Catania
Via S. Sofia 64
95123 (Italy)
Hans.Schulze@ct.infn.it

Christof Struck
University of Frankfurt
Institut fuer Kernphysik
August-Euler-Str. 6
60486 Frankfurt/Main (Germany)
struck@bnl.gov

Hideo Suganuma
Tokyo Institute of Technology
Ohokayama 2-12-1, Meguro
152-8551 Tokyo (Japan)
suganuma@th.phys.titech.ac.jp
www.afs.titech.ac.jp/ryudo/suganuma.html

Yury Tchuvil'sky
SINP, Moscow State University

Vorob'evy Gory
119992 Moscow (Russia)
tchuvl@nucl-th.sinp.msu.ru

Salvatrice Terranova
LNS
Via S. Sofia 44
95123 Catania (Italy)
terranova@lns.infn.it

Akihiro Tohsaki-Suzuki
Shinshu University
3-15-1 Tokida
386-8567 Ueda (Japan)
asuzuk1@giptc.shinshu-u.ac.jp

Svetlana Tretyakova
Joint Institute for Nuclear Research
St. Joliot Curie 6
141980 Dubna, Moscow region (Russia)
tsvetl@sungraph.jinr.ru

Cristina Tuvè
Department of Physics and Astronomy
University of Catania
Via S. Sofia 64
95123 Catania (Italy)
Cristina.Tuve@ct.infn.it

Isaac Vidana
Dipartimento di Fisica
Universita' di Pisa and INFN Sezione di
Pisa
Via Buonarroti 2
I-56127 Pisa (Italy)
vidana@df.unipi.it

Roman Wolski
FLNR, JINR
141980 Dubna (Russia)
wolski@nrsun.jinr.ru

Philip Woods
Edinburgh University
Kings, Buildings, Mayfield Road
Edinburgh EH9 3JZ (UK)
pjw@np.ph.ed.ac.uk

AUTHOR INDEX

A

Akaishi, Y., 56
Alamanos, N., 80
Alberico, W. M., 348
Algora, A., 96
Aoyama, S., 49

B

Baldo, M., 247
Balenzuela, P., 167
Bauer, W., 219
Bereziani, Z., 239
Bogdanov, D. D., 80
Bollenbach, T., 219
Bombaci, I., 239, 319
Bonasera, A., 40, 167, 206, 313
Bonilla, C., 34
Bugrov, V. P., 142
Burgio, G. F., 247

C

Chiba, Satoshi, 40
Cseh, J., 96

D

Delion, D. S., 12
Diaz Alonso, J., 233
Dorso, C. O., 167
Drago, A., 239

E

Elliott, J. B., 155

F

Faraci, G., 173
Fedorov, D. V., 20

Ferreira, L. S., 3
Fomichev, A. S., 80
Frontera, F., 239
Funaki, Y., 31
Furman, W. I., 142
Fynbo, H. O. U., 20

G

Gherghescu, R. A., 112
Greiner, C., 337
Greiner, W., 112, 196, 327
Gridnev, K. A., 196
Guglielmetti, A., 103

H

Hess, P. O., 96
Hippolyte, B., 296
Horiuchi, H., 31, 188

I

Ikeda, K., 49, 62, 68, 74, 86
Insolia, A., 12
Ishii, N., 366
Ison, M. J., 167
Itagaki, N., 62
Ito, M., 68
Iwazaki, A., 27

J

Jensen, A. S., 20

K

Kadmensky, S. G., 142
Kahana, D. E., 270
Kahana, S. H., 270
Kanada-En'yo, Y., 188
Kartavenko, V. G., 196

RETURN TO ➡ PHYSICS LIBRARY
351 LeConte Hall   642-3122

LOAN PERIOD 1	2	3
1-MONTH		
4	5	6

ALL BOOKS MAY BE RECALLED AFTER 7 DAYS
Overdue books are subject to replacement bills

DUE AS STAMPED BELOW

This book will be held in PHYSICS LIBRARY until MAR 2 4 2003		

FORM NO. DD 25   UNIVERSITY OF CALIFORNIA, BERKELEY
BERKELEY, CA 94720